PILOT'S HANDBOOK
of Aeronautical Knowledge

Revised 1980

Reprinted by

Aviation Supplies & Academics, Inc.

AC 61-23B

Foreword

The Pilot's Handbook of Aeronautical Knowledge contains essential information used in training and guiding pilots. This revised handbook suggests methods on how to use: (a) Flight Information Publications; (b) data in Aircraft Flight Manual and Pilot's Operating Handbook; and (c) basic instruments essential for airplane attitude control.

Except for Federal Aviation Regulations pertinent to civil aviation, those subject areas in which an applicant for private pilot certification may be tested are covered in this handbook. Not all topics which appear herein are discussed in depth, however. The handbook is intended to assist the applicant for pilot certification.

Advisory Circular 61–23A, dated 1971, is cancelled.

Comments regarding this publication should be directed to the Department of Transportation, Federal Aviation Administration, Flight Standards National Field Office, Examinations Standards Branch, AFO–590, P.O. Box 25082, Oklahoma City, Oklahoma 73125.

iii

Contents

Page

Chapter I—PRINCIPLES OF FLIGHT

Forces Acting on the Airplane in Flight 1
Lift 4
Gravity (Weight) 6
Thrust 6
Drag 7
Relationship Between Angle of Attack and Lift 9
Relationship of Thrust and Drag in Straight-and-Level Flight 10
Relationship Between Lift and Weight in Straight-and-Level Flight 10
Factors Affecting Lift and Drag 10
Effect of Wing Area on Lift and Drag 10
Effect of Airfoil Shape on Lift and Drag 10
Effect of Wing Design on Stall 11
Effect of Airspeed on Lift and Drag 11
Effect of Air Density on Lift and Drag 12
Turning Tendency (Torque Effect) 12
Reactive Force 12
Spiraling Slipstream 14
Gyroscopic Precession 14
"P" Factor or Asymmetric Propeller Loading 14
Corrections for Turning Tendency or Torque During Flight 14
Airplane Stability 15
Longitudinal Stability About the Lateral Axis 16
Longitudinal Control (Pitch) About the Lateral Axis 18
Lateral Stability About the Longitudinal Axis 18
Lateral Control (Roll) About the Longitudinal Axis 19
Lateral Stability or Instability in Turns 20
Directional Stability About the Vertical Axis (Yaw) 21
Directional Control About the Vertical Axis (Yaw) 21
Loads and Load Factors 22
Load Factors and Airplane Design 23
Effect of Turns on Load Factor 24
Effect of Load Factor on Stalling Speed 24
Effect of Speed on Load Factor 26

Page

Effect of Flight Maneuvers on Load Factor 26
Effect of Turbulence on Load Factor 27
Determining Load Factors in Flight 27
Forces Acting on the Airplane When at Airspeeds Slower than Cruise 28
Forces in a Climb 28
Forces in a Glide 29
Turns During Flight 30

Chapter II—AIRPLANES AND ENGINES

Airplane Structure 33
Flight Control Systems 33
Wing Flaps 33
Landing Gear 34
Electrical System 35
Engine Operation 37
How an Engine Operates 37
Cooling System 38
Ignition System 39
Fuel System 40
Fuel Tanks, Selectors, and Strainers 40
Fuel Primer 41
Fuel Pressure Gauge 41
Induction, Carburetion, and Injection Systems 41
Mixture Control 41
Carburetor Icing 42
Carburetor Air Temperature Gauge 43
Outside Air Temperature Gauge (OAT) 43
Fuel Injection System 43
Proper Fuel is Essential 43
Fuel Contamination 44
Refueling Procedures 44
Oil System 45
Propeller 46

	Page
Fixed-Pitch Propeller	47
Controllable-Pitch Propellers	48
Starting the Engine	48
Engines Equipped with a Starter	48
Engines not Equipped with a Starter	49
Idling the Engine During Flight	50
Exhaust Gas Temperature (EGT) Gauge	50
Superchargers or Turbochargers	50
Aircraft Documents, Maintenance, and Inspections	50
Aircraft Owner Responsibilities	50
Certificate of Aircraft Registration	51
Airworthiness Certificate	51
Aircraft Maintenance	52
Inspections	53
Preventive Maintenance	53
Repairs and Alterations	53
Special Flight Permits	54
Airworthiness Directives	54
Preflight Inspection	54
Ground Runup and Functional Check	56

Chapter III—FLIGHT INSTRUMENTS

	Page
The Pitot-Static System and Associated Instruments	59
The Altimeter	59
Types of Altitude	62
Vertical Speed Indicator	62
The Airspeed Indicator	63
Gyroscopic Flight Instruments	65
Sources of Power for Gyroscopic Operation	65
Gyroscopic Principles	66
Turn-and-Slip-Indicator	67
Turn Coordinator	68
The Heading Indicator	69
The Attitude Indicator	69
Magnetic Compass	70
Compass Errors	71
Using the Magnetic Compass	72

Chapter IV—AIRPLANE PERFORMANCE

	Page
Weight Control	73
Effects of Weight	73
Weight Changes	74
Balance, Stability, and Center of Gravity	74
Effects of Adverse Balance	74
Management of Weight and Balance Control	75
Terms and Definitions	75
Aircraft Weight Nomenclature	76
Control of Loading—General Aviation Airplanes	76
Basic Principles of Weight and Balance Computations	76
Useful Load Check	78
Weight and Balance Restrictions	79
Light Single-Engine Airplane Loading Problems	81
Airplane Performance	86
Use of Performance Charts	87

Chapter V—WEATHER

	Page
Weather Information for the Pilot	99
Services to the Pilot	99
Observations	99
Meteorological Centers and Forecast Offices	100
Service Outlets	101
Users	101
Nature of the Atmosphere	102
Oxygen and the Human Body	102
Significance of Atmospheric Pressure	102
Measurement of Atmospheric Pressure	103
Effect of Altitude on Atmospheric Pressure	103
Effect of Altitude on Flight	103
Effect of Differences in Air Density	105
Pressure Recorded in Millibars	105
Wind	106
The Cause of Atmospheric Circulation	106
Wind Patterns	107
Convection Currents	109
Effect of Obstructions on Wind	109

	Page
Low-Level Wind Shear	113
Wind and Pressure Representation on Surface Weather Maps	116
Moisture and Temperature	116
Relative Humidity	117
Temperature-Dewpoint Relationship	117
Methods by Which Air Reaches the Saturation Point	118
Effect of Temperature on Air Density	118
Effect of Temperature on Flight	118
Effect of High Humidity on Air Density	119
Effect of High Humidity on Flight	119
Dew and Frost	119
Fog	119
Clouds	119
Ceiling	123
Visibility	125
Precipitation	125
Air Masses and Fronts	125
Warm Front	125
Cold Front	126
Occluded Front	129
Aviation Weather Forecasts, Reports, and Weather Charts	134
Aviation Forecasts	135
Aviation Weather Reports	142
Surface Aviation Weather Reports	143
Pilot Weather Reports (PIREPS)	148
Radar Weather Reports (RAREPS)	148
Weather Charts	149

Chapter VI—BASIC CALCULATIONS USING NAVIGATIONAL COMPUTERS OR ELECTRONIC CALCULATORS

Determining En Route Time for a Flight	161
Determining Groundspeed During Flight	161
Determining Total Flight Time Available	162
Determining Total Fuel to be Used on a Flight	162
Determining True Airspeed	162
Converting Knots to Miles Per Hour	162
Solution of a Wind Triangle Problem	162

Chapter VII—NAVIGATION

	Page
Aeronautical Charts	165
Sectional Aeronautical Charts	166
Relief	166
Aeronautical Data	167
Airport and Air Navigation Lighting and Marking Aids	167
Meridians and Parallels	168
Measurement of Direction	169
Variation	169
Deviation	171
Basic Calculations	174
Effect of Wind	174
Calculating Time, Speed, Distance, and Fuel Consumption	176
Converting Minutes to Equivalent Hours	176
Converting Knots to Miles per Hour	176
Fuel Consumption	176
The Wind Triangle	177
Data for Return Trip	177
Radio Navigation	179
VHF Omnidirectional Range (VOR)	183
Using the VOR	183
Tracking with Omni	184
Tips on Using the VOR	185
Automatic Direction Finder (ADF)	186
Flight Planning	186
Assembling Necessary Materials	188
Weather Check	189
In-Flight Visibility and the VFR Pilot	189
Visibility vs. Time	189
How to Get a Briefing	190
Using the Aeronautical Chart	192
Use of the Airport/Facility Directory	193
Aircraft Flight Manual or Pilot's Operating Handbook	194
Using the Plotter, Computer, or Electronic Calculator, etc.	194
VFR Flight Plan	194

Chapter VIII—FLIGHT INFORMATION PUBLICATIONS

Flight Information Publication Policy	197

	Page
Aeronautical Information and the National Airspace System	198
Air Navigation Radio Aids	198
General	198
Nondirectional Radio Beacon (NDB)	198
VHF Omnidirectional Range (VOR)	199
VOR Receiver Check	199
Tactical Air Navigation (TACAN)	200
VHF Omnidirectional Range/Tactical Air Navigation (VORTAC)	200
Distance Measuring Equipment (DME)	201
Class of NAVAIDS	201
Maintenance of FAA NAVAIDS	202
Navaids with Voice	202
VHF/UHF Direction Finder	202
Radar	202
Air Traffic Control Radar Beacon System (ATCRBS)	203
Airport, Air Navigation Lighting and Marking Aids	204
Aeronautical (Light) Beacons	204
Rotating Beacon	204
Auxiliary Lights	204
Obstructions	205
Airway Beacons	205
Control of Lighting Systems	205
Visual Approach Slope Indicator (VASI)	206
Tri-Color Visual Approach Slope Indicator	207
Markings	208
Airspace	209
Uncontrolled Airspace	209
Controlled Airspace	210
Special Use Airspace	211

	Page
Other Airspace Areas	212
Air Traffic Control	213
Services Available to Pilots	213
Airport Advisory Practices at Nontower Airports	214
Automatic Terminal Information Service (ATIS)	216
Radar Traffic Information Service	217
Terminal Control Area Operation	221
Radar Service for VFR Aircraft in Difficulty	222
Transponder Operation	222
Radio Communication	225
Phraseology and Techniques	225
Airport Operations	229
Tower-Controlled Airports	230
Nontower Airports	230
Light Signals	234
Communications	234
Departure Delays	235
Taxiing	235
Special VFR Clearances	236
Preflight	237
Flight Plan—VFR	238
Emergency Procedures	239
Emergency Locator Transmitters	240
Search and Rescue	241
Wake Turbulence	242
Medical Facts for Pilots	246
Good Operating Practices	251
APPENDIX I—Obtaining FAA Publications	257

Illustrations

	Page
Parts of the airplane.	xiv
Figure 1-1. Forces acting on the airplane in flight.	1
Figure 1-2. Typical airfoil sections	2
Figure 1-3. Angle of attack and flightpath	3
Figure 1-4. Cross sectional view of an airfoil.	4
Figure 1-5. Nomenclature of airfoil section	4
Figure 1-6. Component forces.	4
Figure 1-7. Relationship between flightpath and relative wind.	5
Figure 1-8. Vectors	5
Figure 1-9. Wing planforms	5
Figure 1-10. Bernoulli's Principle applied to airfoils.	6
Figure 1-11. Wing deflecting the air downward	6
Figure 1-12. Relationship between relative wind, lift, and drag	7
Figure 1-13. Airplane suspended from the center of gravity.	7
Figure 1-14. Drag acts parallel to and in the same direction as the relative wind.	7
Figure 1-15. Form drag and skin friction drag.	8
Figure 1-16. Typical airplane drag curves	8
Figure 1-17. Flow of air around a wing at various angles of attack	9
Figure 1-18. Use of flaps increases lift and drag.	10
Figure 1-19. View of wingtip twist.	11
Figure 1-20. Slotted and plain wing	11
Figure 1-21. Stall strip.	12
Figure 1-22. Effect of altitude, temperature, and humidity on takeoff run and rate of climb.	13
Figure 1-23. Factors which cause left-turning tendency.	14
Figure 1-24. Static stability	15
Figure 1-25. Relationship of oscillation and stability.	16
Figure 1-26. Axes of rotation	17
Figure 1-27. Neutral stability	17
Figure 1-28. Negative stability	18
Figure 1-29. Positive stability.	18
Figure 1-30. Effect of elevators.	19
Figure 1-31. Effect of trim tabs.	20
Figure 1-32. Effect of dihedral.	20

	Page
Figure 1-33. Effect of sweepback	21
Figure 1-34. Keel effect.	21
Figure 1-35. Effect of ailerons	22
Figure 1-36. Effect of rudder	22
Figure 1-37. Forces acting on an airplane in a bank	24
Figure 1-38. Load factors in turns.	25
Figure 1-39. Load factor chart.	26
Figure 1-40. Stall speed chart.	26
Figure 1-41. Flight at airspeeds slower than cruise	28
Figure 1-42. Forces acting on an airplane in a climb.	29
Figure 1-43. Power available vs. power required.	29
Figure 1-44. Forces acting on an airplane in a glide.	30
Figure 1-45. Forces acting on an airplane in a turn.	31
Figure 2-1. Wing flaps.	34
Figure 2-2. Electrical system schematic.	36
Figure 2-3. Circuit breaker panel	36
Figure 2-4. Ammeter	37
Figure 2-5. Basic parts of a reciprocating engine	38
Figure 2-6. Four strokes of the piston	39
Figure 2-7. A float-type carburetor	41
Figure 2-8. Formation of ice in the fuel intake system	42
Figure 2-9. Normal combustion and explosive combustion	44
Figure 2-10. Factors affecting propellers	46
Figure 2-11. Changes in propeller blade angle from hub to tip.	46
Figure 2-12. Relationship of travel distance and speed of propeller blade.	47
Figure 2-13. Effective and geometric propeller pitch	47
Figure 2-14. Certificate of aircraft registration.	52
Figure 2-15. Standard airworthiness certificate	53
Figure 2-16. Preflight inspection chart	55
Figure 3-1. Pitot-static system with instruments.	59

Page

Figure 3–2.   Sensitive altimeter 60
Figure 3–3.   Vertical speed indicator 62
Figure 3–4.   Airspeed indicator 63
Figure 3–5.   Airspeed indicator showing color-coded marking system. 64
Figure 3–6.   Typical pump-driven vacuum system 66
Figure 3–7.   Precession of a gyroscope 66
Figure 3–8.   Turn and slip indicator 67
Figure 3–9.   Indications of the ball in various types of turns 68
Figure 3–10.   Turn coordinator 68
Figure 3–11.   Heading indicator 69
Figure 3–12.   Heading indicator 69
Figure 3–13.   Attitude indicator. 70
Figure 3–14.   Various indications on the attitude indicator. 70
Figure 3–15.   Earth's magnetic field. 71
Figure 3–16.   Magnetic compass 71

Figure 4–1.   Lateral or longitudinal unbalance 74
Figure 4–2.   Weight and balance illustrated 77
Figure 4–3.   Determining moments. 77
Figure 4–4.   Establishing a balance 78
Figure 4–5.   Airplane weight and balance 79
Figure 4–6.   Weight and balance data. 80
Figure 4–7.   Loading schedule placard. 80
Figure 4–8.   Airplane weight and balance diagram 82
Figure 4–9.   Loading graph. 82
Figure 4–10.   C.G. moment envelope 82
Figure 4–11.   Weight shifting diagram. 83
Figure 4–12.   Solution to proportion problem 84
Figure 4–13.   Pressure altitude and density altitude chart. 89
Figure 4–14.   Determining density altitude. 90
Figure 4–15.   Takeoff performance data chart. 91
Figure 4–16.   A cruise performance chart 92
Figure 4–17.   Cruise and range performance chart. 93
Figure 4–18.   Power setting table. 94
Figure 4–19.   Cruise performance chart 94
Figure 4–20.   Climb data chart. 95
Figure 4–21.   Climb data chart. 95

Page

Figure 4–22.   Maximum glide distance chart. 95
Figure 4–23.   Crosswind and headwind component chart 96
Figure 4–24.   Stall speed chart. 96
Figure 4–25.   Stall speed chart. 96
Figure 4–26.   A landing performance data chart 97
Figure 4–27.   Determining speed for best rate and best angle of climb. 97

Figure 5–1.   Data flow in the Aviation Weather Service 100
Figure 5–2.   The realm of flight 103
Figure 5–3.   Barometric pressure at a weather station. 104
Figure 5–4.   Effect of atmospheric density at sea level 105
Figure 5–5.   Takeoff distance increases with increase in field eleva-
               tion 106
Figure 5–6.   Effect of heat at the equator on atmospheric circulation. 107
Figure 5–7.   Principal air currents in the Northern Hemisphere 108
Figure 5–8.   Circulation of wind within a "low" 109
Figure 5–9.   Use of favorable winds in flight 109
Figure 5–10.   On-shore winds. 110
Figure 5–11.   Off-shore winds 111
Figure 5–12.   Avoiding turbulence caused by convection currents 112
Figure 5–13.   Varying surfaces affect the normal glidepath. 113
Figure 5–14.   Effect of descending currents on landings. 114
Figure 5–15.   Turbulence caused by obstructions. 115
Figure 5–16.   Effect of hills or mountains on air currents 116
Figure 5–17.   Wind as shown on weather map 116
Figure 5–18.   Flow of air around a "high" 117
Figure 5–19.   Temperature conversion chart. 118
Figure 5–20.   Cumulus clouds 120
Figure 5–21.   Stratus-type clouds 121
Figure 5–22.   Various types of bad weather clouds 122
Figure 5–23.   Cross section of a cumulonimbus cloud (thunderhead) 124
Figure 5–24.   A warm front 127
Figure 5–25.   A cold front 128
Figure 5–26.   Weather map wind shift line 129
Figure 5–27.   An occluded front. 130
Figure 5–28.   Stages in the development of an occlusion. 131
Figure 5–29.   Development of an occlusion 132

Page

Figure 5-30. Clouds and precipitation accompanying a typical occlu-
sion 133
Figure 5-31. Section of typical weather map. 134
Figure 5-32. Example of an area forecast. 135
Figure 5-33. Synopsis 135
Figure 5-34. Clouds and weather 136
Figure 5-35. Icing and freezing level. 137
Figure 5-36. Terminal forecasts 137
Figure 5-37. Portions of hourly sequence report. 138
Figure 5-38. Summary of sky cover designators 139
Figure 5-39. Ceiling designators 144
Figure 5-40. Weather symbols and meanings. 144
Figure 5-41. Obstructions to vision—symbols and meanings 145
Figure 5-42. Precipitation intensity and intensity trend 145
Figure 5-43. Radar weather report 148
Figure 5-44. Contractions reporting operational status of radar. 149
Figure 5-45. Symbols used on the weather map 149
Figure 5-46. Frontal symbols 150
Figure 5-47. Numerical classification of fronts. 151
Figure 5-48. Section of a surface weather map as transmitted on fac-
simile. 152
Figure 5-49. Weather depiction chart. 153
Figure 5-50. Notations used on weather depiction charts. 154
Figure 5-51. Radar summary chart. 155
Figure 5-52. Symbols used for echo intensity and trend. 156
Figure 5-53. Echo coverage symbols on the radar summary chart. 157
Figure 5-54. Weather symbols 157
Figure 5-55. U.S. low-level significant weather prog (SFC-400MB). ... 158
Figure 5-56. Some standard weather symbols. 159

Figure 7-1. Index of sectional and VFR terminal area charts 160

Figure 7-2. Altitude, form, and slope of terrain indicated on chart ... 166
Figure 7-3. Meridians and parallels. 167
Figure 7-4. Time zones 168
Figure 7-5. The compass rose 169
Figure 7-6. Course determination by reference to meridians 169
Figure 7-7. Variation, the angle between a magnetic and geographic
meridian 170

Page

Figure 7-8. A typical isogonic chart. 170
Figure 7-9. Areas of variation. 171
Figure 7-10. Relationship between true heading, magnetic heading,
and variation 172
Figure 7-11. Cause of deviation 172
Figure 7-12. Compass deviation card 172
Figure 7-13. Relationship between true, magnetic, and compass
headings 173
Figure 7-14. Relationship between groundspeed and airspeed. 173
Figure 7-15. Effect of wind on flightpath. 174
Figure 7-16. Effects of wind drift on maintaining desired course 174
Figure 7-17. Establishing a wind correction angle 175
Figure 7-18. Principle of the wind triangle. 176
Figure 7-19. The wind triangle as it is drawn 176
Figure 7-20. Steps in drawing the wind triangle 177
Figure 7-21. Finding true heading by direct measurement 177
Figure 7-22. Finding true heading by the wind correction angle 178
Figure 7-23. Pilot's planning sheet and visual flight log 179
Figure 7-24. Computations for a round-trip flight. 179
Figure 7-25. Steps in constructing the wind triangle. 180
Figure 7-26. VHF transmission follows a line-of-sight course 181
Figure 7-27. Omnihead 182
Figure 7-28. Tracking a radial in a crosswind. 184
Figure 7-29. ADF with fixed azimuth 184
Figure 7-30. ADF terms 185
Figure 7-31. ADF tracking. 187
Figure 7-32. Rule of thumb 187
Figure 7-33. Cockpit cut-off angle. 188
Figure 7-34. Flight plan form 191
Figure 7-35. Reverse side of flight plan form 192

Figure 8-1. VASI. 195
Figure 8-2. Basic runway 196

Figure 8-3. Nonprecision instrument runway 207
Figure 8-4. Precision instrument runway 208
Figure 8-5. Threshold/displaced threshold markings. 208
Figure 8-6. Overrun/stopway and blast pad area. 208
Figure 8-7. Closed runway or taxiway 208

208
209

		Page
Figure 8-8.	STOL runway.	209
Figure 8-9.	Minimum visibility and distance from clouds—VFR	209
Figure 8-10.	Altitudes and flight levels	210
Figure 8-11.	General dimensions of control zones, airport traffic areas, and vertical extent of airspace segments.	211
Figure 8-12.	Radar traffic information	218
Figure 8-13.	Radar traffic information	218
Figure 8-14.	Phonetic alphabet and Morse code.	229
Figure 8-15.	Traffic pattern	230

		Page
Figure 8-16.	Recommended traffic patterns at nontower airports	232
Figure 8-17.	Light signals	234
Figure 8-18.	Vortex generation.	242
Figure 8-19.	Vortex roll	243
Figure 8-20.	Vortices.	243
Figure 8-21.	Vortex sink	244
Figure 8-22.	Vortex—no wind	244
Figure 8-23.	Vortex movement in wind	244

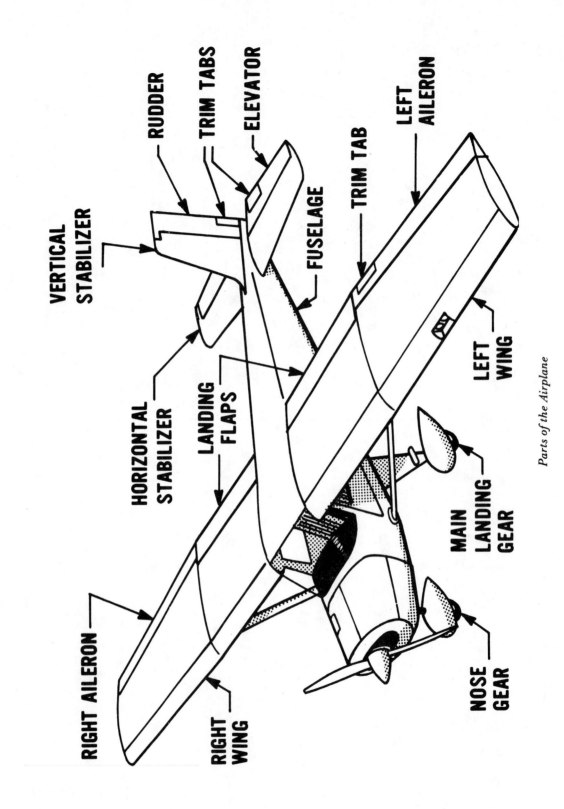

Parts of the Airplane

CHAPTER 1—PRINCIPLES OF FLIGHT

There are certain laws of nature or physics that apply to any object that is lifted from the earth and moved through the air. To analyze and predict airplane performance under various operating conditions, it is important that pilots gain as much knowledge as possible concerning the laws and principles that apply to flight.

The principles of flight discussed in this chapter are intended primarily for beginning pilots, and are not intended as a detailed and complete explanation of the complexities of aerodynamics. However, this information should encourage interested individuals to further their study.

Forces Acting on the Airplane in Flight

When in flight, there are certain favorable forces and other unfavorable forces acting on the airplane. It is the primary task of a pilot to control these forces so as to direct the airplane's speed and flightpath in a safe and efficient manner. To do this the pilot must understand these forces and their effects.

Among the aerodynamic forces acting on an airplane during flight, four are considered to be basic because they act upon the airplane during all maneuvers. These basic forces are *lift*, the upward acting force; *weight* (or gravity), the downward acting force; *thrust*, the forward acting force; and *drag*, the rearward acting, or retarding, force (Fig. 1-1).

While in steady state flight the attitude, direction, and speed of the airplane will remain constant until one or more of the basic forces changes in magnitude. In unaccelerated flight (steady flight) the opposing forces are in equilibrium. Lift and thrust are considered as positive forces (+), while weight and drag are considered as negative forces (−), and the sum of the opposing forces is zero. In other words, lift equals weight and thrust equals drag.

When pressure is applied to the airplane controls, one or more of the basic forces change in magnitude and become greater than the opposing force, causing the airplane to accelerate or move in the direction of the applied force. For example, if power is applied (increasing thrust) and altitude is maintained, the airplane will accelerate. As speed increases, drag increases, until a point is reached where drag again equals thrust, and the airplane will continue in steady flight at a higher speed. As another example, if power is applied while in level flight, and a climb attitude is established, the force of lift would increase during the time back elevator pressure is applied, but after a

steady state climb is established, the force of lift would be *approximately* equal to the force of weight. The airplane does not climb because lift is greater than in level flight, but because thrust is greater than drag, and because a component of thrust is developed which acts upward, perpendicular to the flightpath.

Airplane designers make an effort to increase the performance of the airplane by increasing the efficiency of the desirable forces of lift and thrust while reducing, as much as possible, the undesirable forces of weight and drag. Nonetheless, compromise must be made to satisfy the function and desired performance of the airplane.

Before discussing the four forces further, it will be helpful to define some of the terms used extensively in this section.

a. *Acceleration*—the force involved in overcoming inertia, and which is defined as a change of velocity per unit of time. It means changing speed and/or changing direction, including starting from rest (positive acceleration) and stopping (deceleration or negative acceleration).

b. *Airfoil*—any surface, such as an airplane wing, designed to obtain reaction such as lift from the air through which it moves. Typical airfoil sections are shown in Fig. 1-2.

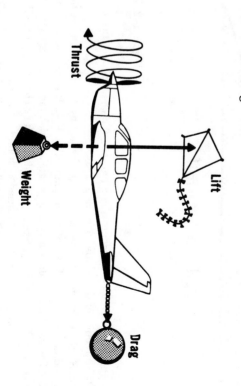

Figure 1-1. *Forces acting on the airplane in flight.*

Thrust

Lift

Weight

Drag

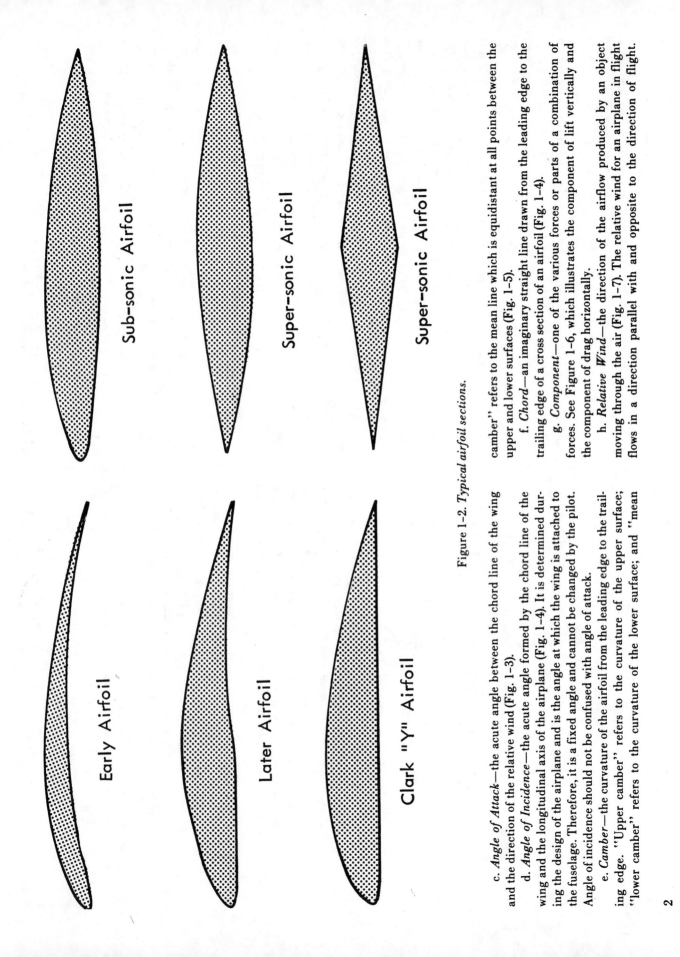

Figure 1–2. *Typical airfoil sections.*

c. *Angle of Attack*—the acute angle between the chord line of the wing and the direction of the relative wind (Fig. 1–3).

d. *Angle of Incidence*—the acute angle formed by the chord line of the wing and the longitudinal axis of the airplane (Fig. 1–4). It is determined during the design of the airplane and is the angle at which the wing is attached to the fuselage. Therefore, it is a fixed angle and cannot be changed by the pilot. Angle of incidence should not be confused with angle of attack.

e. *Camber*—the curvature of the airfoil from the leading edge to the trailing edge. "Upper camber" refers to the curvature of the upper surface; "lower camber" refers to the curvature of the lower surface; and "mean camber" refers to the mean line which is equidistant at all points between the upper and lower surfaces (Fig. 1–5).

f. *Chord*—an imaginary straight line drawn from the leading edge to the trailing edge of a cross section of an airfoil (Fig. 1–4).

g. *Component*—one of the various forces or parts of a combination of forces. See Figure 1–6, which illustrates the component of lift vertically and the component of drag horizontally.

h. *Relative Wind*—the direction of the airflow produced by an object moving through the air (Fig. 1–7). The relative wind for an airplane in flight flows in a direction parallel with and opposite to the direction of flight.

2

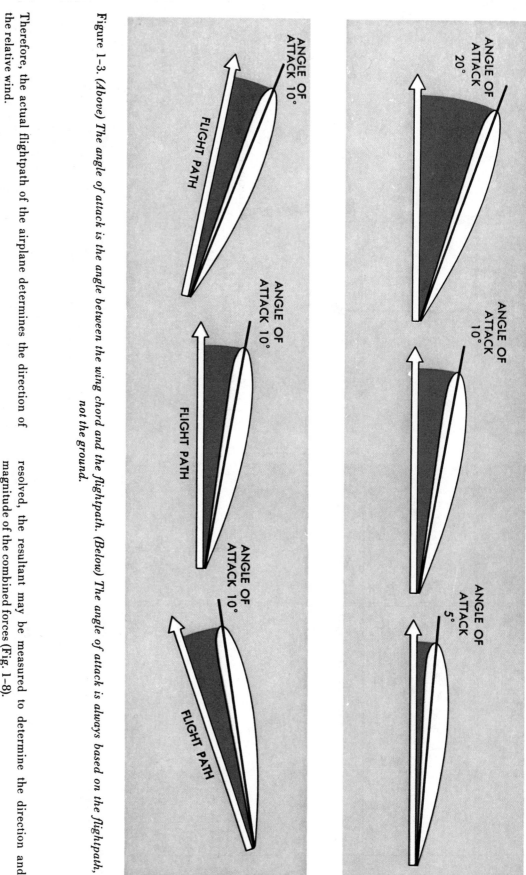

Figure 1-3. *(Above)* The angle of attack is the angle between the wing chord and the flightpath. *(Below)* The angle of attack is always based on the flightpath, not the ground.

Therefore, the actual flightpath of the airplane determines the direction of the relative wind.

  i. *Speed*—the distance traveled in a certain time.

  j. *Vectors*—the graphic representation of a force drawn in a straight line which indicates direction by an arrow and magnitude by its length. When an object is being acted upon by two or more forces, the combined effect of these forces may be represented by a resultant vector. After the vectors have been

resolved, the resultant may be measured to determine the direction and magnitude of the combined forces (Fig. 1-8).

  k. *Velocity*—the speed or rate of movement in a certain direction.

  l. *Wing Area*—the plan surface of the wing, which includes control surfaces and may include wing area covered by the fuselage (main body of the airplane), and engine nacelles.

  m. *Wing Planform*—the shape or form of a wing as viewed from above. It

3

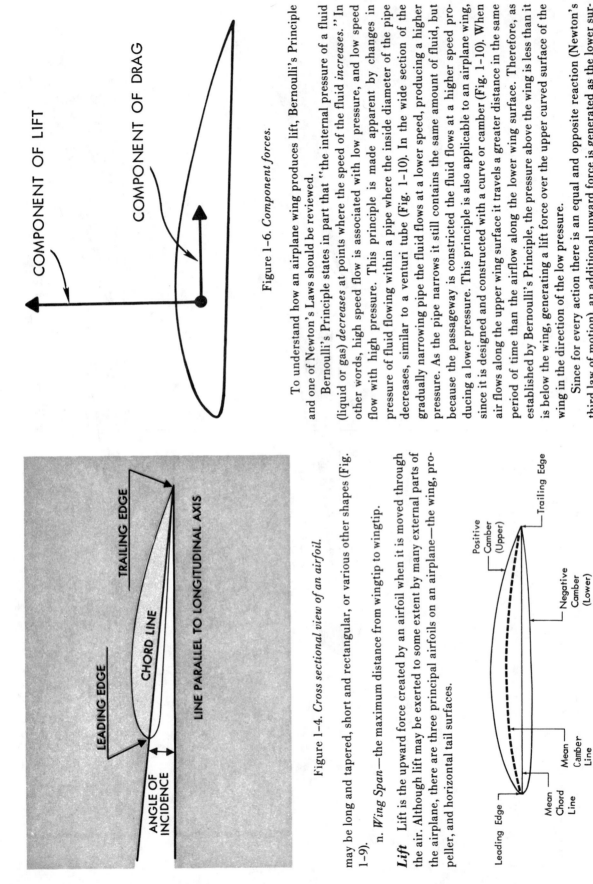

COMPONENT OF LIFT

COMPONENT OF DRAG

Figure 1-6. *Component forces.*

To understand how an airplane wing produces lift, Bernoulli's Principle and one of Newton's Laws should be reviewed.

Bernoulli's Principle states in part that "the internal pressure of a fluid (liquid or gas) *decreases* at points where the speed of the fluid *increases*." In other words, high speed flow is associated with low pressure, and low speed flow with high pressure. This principle is made apparent by changes in pressure of fluid flowing within a pipe where the inside diameter of the pipe decreases, similar to a venturi tube (Fig. 1-10). In the wide section of the pipe the fluid flows at a lower speed, producing a higher pressure. As the pipe narrows it still contains the same amount of fluid, but because the passageway is constricted the fluid flows at a higher speed producing a lower pressure. This principle is also applicable to an airplane wing, since it is designed and constructed with a curve or camber (Fig. 1-10). When air flows along the upper wing surface it travels a greater distance in the same period of time than the airflow along the lower wing surface. Therefore, as established by Bernoulli's Principle, the pressure above the wing is less than it is below the wing, generating a lift force over the upper curved surface of the wing in the direction of the low pressure.

Since for every action there is an equal and opposite reaction (Newton's third law of motion), an additional upward force is generated as the lower surface of the wing deflects the air downward (Fig. 1-11). Thus both the develop-ment of low pressure above the wing and reaction to the force and direction

---

may be long and tapered, short and rectangular, or various other shapes (Fig. 1-9).

n. *Wing Span*—the maximum distance from wingtip to wingtip.

Lift Lift is the upward force created by an airfoil when it is moved through the air. Although lift may be exerted to some extent by many external parts of the airplane, there are three principal airfoils on an airplane—the wing, pro-peller, and horizontal tail surfaces.

TRAILING EDGE

LEADING EDGE

CHORD LINE

LINE PARALLEL TO LONGITUDINAL AXIS

ANGLE OF INCIDENCE

Figure 1-4. *Cross sectional view of an airfoil.*

Positive Camber (Upper)

Trailing Edge

Negative Camber (Lower)

Mean Camber Line

Mean Chord Line

Leading Edge

Figure 1-5. *Nomenclature of airfoil section.*

4

of air as it is deflected from the wing's lower surface contribute to the total lift generated.

The amount of lift generated by the wing depends upon several factors: (1) speed of the wing through the air, (2) angle of attack, (3) planform of the wing, (4) wing area, and (5) the *density* of the air.

Lift acts upward and perpendicular to the relative wind and to the wing span (Fig. 1-12). Although lift is generated over the entire wing, an imaginary point is established which represents the resultant of all lift forces. This single point is the center of lift, sometimes referred to as the center of pressure (CP).

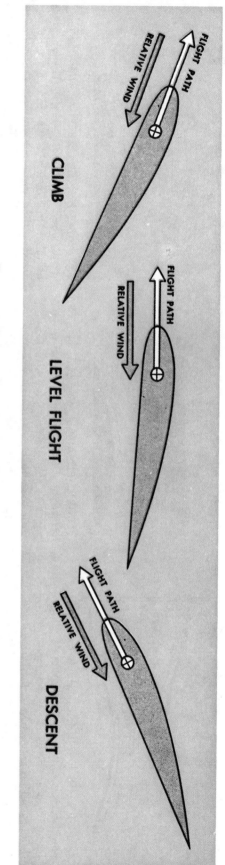

Figure 1-7. *Relationship between flightpath and relative wind.*

CLIMB

LEVEL FLIGHT

DESCENT

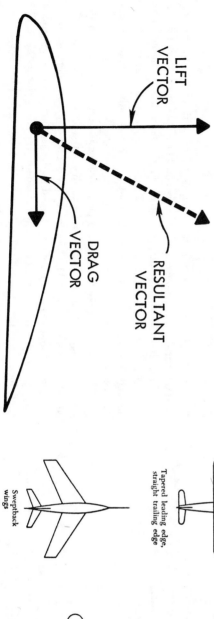

LIFT VECTOR

DRAG VECTOR

RESULTANT VECTOR

Figure 1-8. *Vectors.*

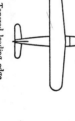

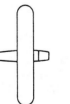

Sweptback wings

Tapered leading edge, straight trailing edge

Tapered leading and trailing edges

Straight leading and trailing edges

Delta wing

Straight leading edge, tapered trailing edge

Figure 1-9. *Wing planforms.*

5

BERNOULLI'S PRINCIPLE, which explains how lift is created by an airplane's wing, is depicted in these three diagrams. A fluid traveling through a constriction in a pipe (above) speeds up, and at the same time the pressure it exerts on the pipe decreases.

THE CONSTRICTED AIRFLOW shown here, formed by two opposed airplane wings, is analogous to the pinched-pipe situation at left: air moving between the wings accelerates, and this increase in speed results in lower pressure between the curved surfaces.

THE SAME PRINCIPLE applies when the air is distrubed by a single wing. The accelerating airflow over the top surface exerts less pressure than the airflow across the bottom. It is this continuing difference in pressure that creates and sustains lift.

Figure 1–10. *Bernoulli's Principle applied to airfoils.*

The location of the center of pressure relative to the center of gravity (weight) is very important from the standpoint of airplane stability. Stability will be covered in more detail later.

Gravity (Weight)   Gravity is the downward force which tends to draw all bodies vertically toward the center of the earth. The airplane's center of gravity (CG) is the point on the airplane at which all weight is considered to be concentrated. It is the point of balance. For example, if an airplane were suspended from a rope attached to the center of gravity, the airplane would balance (Fig. 1–13).

The center of gravity is located along the longitudinal centerline of the airplane (imaginary line from the nose to the tail) and somewhere near the center of lift of the wing. The location of the center of gravity depends upon the location and weight of the load placed in the airplane. This is controlled through weight and balance calculations made by the pilot prior to flight. The exact location of the center of gravity is important during flight, because of its effect on airplane stability and performance.

Thrust   The propeller, acting as an airfoil, produces the thrust, or forward force that drives the airplane through the air. It receives its power directly

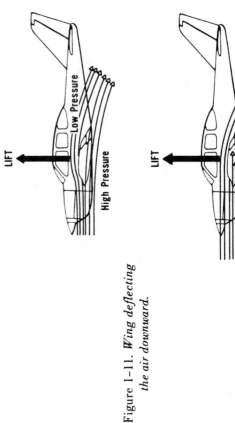

Figure 1–11. *Wing deflecting the air downward.*

6

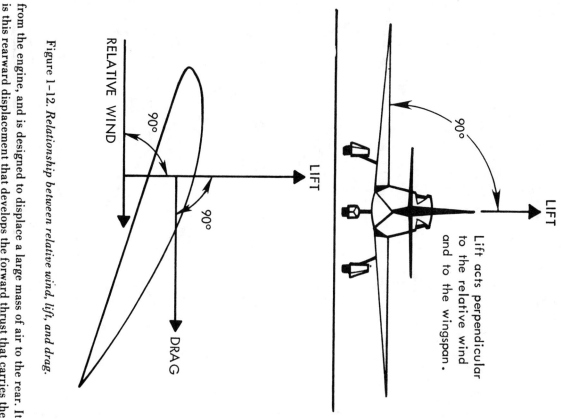

Figure 1-13. *Airplane suspended from the center of gravity.*

Lift acts perpendicular to the relative wind and to the wingspan.

Total weight of Aircraft

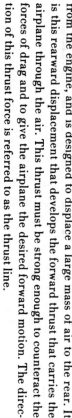

Figure 1-12. *Relationship between relative wind, lift, and drag.*

from the engine, and is designed to displace a large mass of air to the rear. It is this rearward displacement that develops the forward thrust that carries the airplane through the air. This thrust must be strong enough to counteract the forces of drag and to give the airplane the desired forward motion. The direction of this thrust force is referred to as the thrust line.

Drag   Drag is the rearward acting force which resists the forward movement of the airplane through the air. Drag acts parallel to and in the same direction as the relative wind (Fig. 1-14).

Every part of the airplane which is exposed to the air while the airplane is in motion produces some resistance and contributes to the total drag. Total drag may be classified into two main types: induced drag and parasite drag.

Induced drag is the undesirable but unavoidable by-product of lift, and increases in direct proportion to increases in angle of attack. The greater the angle of attack up to the critical angle, the greater the amount of lift developed, and the greater the induced drag. The airflow around the wing is deflected downward, producing a rearward component to the lift vector which

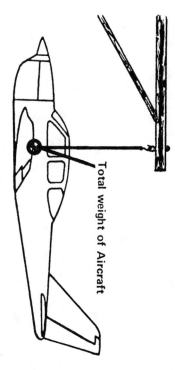

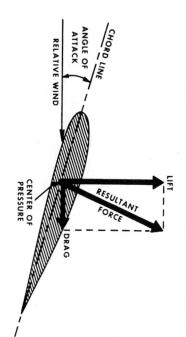

Figure 1-14. *Drag acts parallel to and in the same direction as the relative wind.*

is induced drag. The amount of air deflected downward increases greatly at higher angles of attack; therefore, the higher the angle of attack or the slower the airplane is flown, the greater the induced drag.

Parasite drag is the resistance of the air produced by any part of the airplane that does not produce lift.

Several factors affect parasite drag. When each factor is considered independently it must be assumed that other factors remain constant. These factors are (1) the more streamlined an object is, the less the parasite drag; (2) the more dense the air moving past the airplane, the greater the parasite drag; (3) the larger the size of the object in the airstream, the greater the parasite drag; and (4) as speed increases, the amount of parasite drag increases. If the speed is doubled, four times as much drag is produced.

Parasite drag can be further classified into form drag, skin friction, and interference drag. Form drag is caused by the frontal area of the airplane components being exposed to the airstream. A similar reaction is illustrated by Figure 1-15, where the side of flat plate is exposed to the airstream. This drag is caused by the form of the plate, and is the reason streamlining is

necessary to increase airplane efficiency and speed. Figure 1-15 also illustrates that when the face of the plate is parallel to the airstream the largest part of the drag is skin friction.

Skin friction drag is caused by air passing over the airplane's surfaces, and increases considerably if the airplane surfaces are rough and dirty.

Interference drag is caused by interference of the airflow between adjacent parts of the airplane such as the intersection of wings and tail sections with the fuselage. Fairings are used to streamline these intersections and decrease interference drag.

It is the airplane total drag that determines the amount of thrust required at a given airspeed. Figure 1-16 illustrates the variation in parasite, induced, and total drag with speed for a typical airplane in steady level flight.

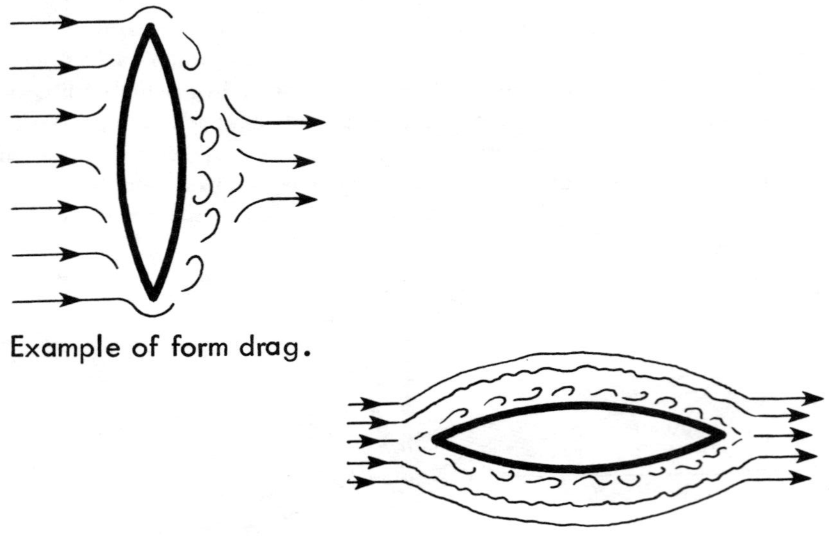

Example of form drag.

Example of skin friction drag.

Figure 1-15. *Form drag and skin friction drag.*

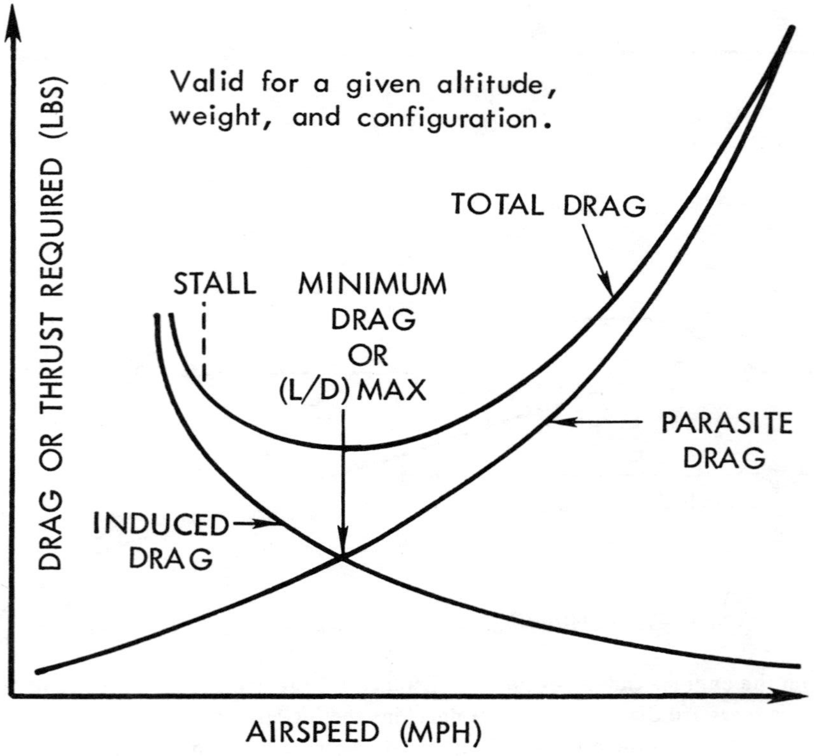

Figure 1-16. *Typical airplane drag curves.*

8

Thrust must equal drag in steady flight; therefore the curve for the total drag also represents the thrust required. Also note in Figure 1–16, that the airspeed at which minimum drag occurs is the same airspeed at which the maximum lift/drag ratio (L/D) takes place. At this point least power is required for both maximum lift and minimum total drag. This is important for determining maximum endurance and range for the airplane.

The force of drag can be controlled to a certain extent by the pilot. Loading the airplane properly, retracting the landing gear and flaps when not used, and keeping the surface of the airplane clean, all help to reduce the total drag.

Relationship Between Angle of Attack and Lift As stated previously, the angle of attack is the acute angle between the relative wind and the chord line of the wing. At small angles of attack most of the wing lift is a result of the difference in pressure between the upper and lower surfaces of the wing (Bernoulli's Principle). Additional lift is generated by the equal and opposite reaction of the airstream being deflected downward from the wing (Newton's Law).

As the angle of attack is increased, the airstream is forced to travel faster because of the greater distance over the upper surface of the wing; creating a greater pressure differential between the upper and lower surfaces. At the same time the airstream is deflected downward at a greater angle, causing an increased opposite reaction. Both the increased pressure differential and increased opposite reaction increase lift and also drag. Therefore as angle of attack is increased, lift is increased up to the critical angle of attack (Figure 1–17).

When the angle of attack is increased to approximately 18° to 20° (critical angle of attack) on most airfoils, the airstream can no longer follow the upper curvature of the wing because of the excessive change in direction. As the critical angle of attack is approached, the airstream begins separating from the rear of the upper wing surface. As the angle of attack is further increased, the airstream is forced to flow straight back, away from the top surface of the wing and from the area of highest camber. This causes a swirling or burbling of the air as it attempts to follow the upper surface of the wing.

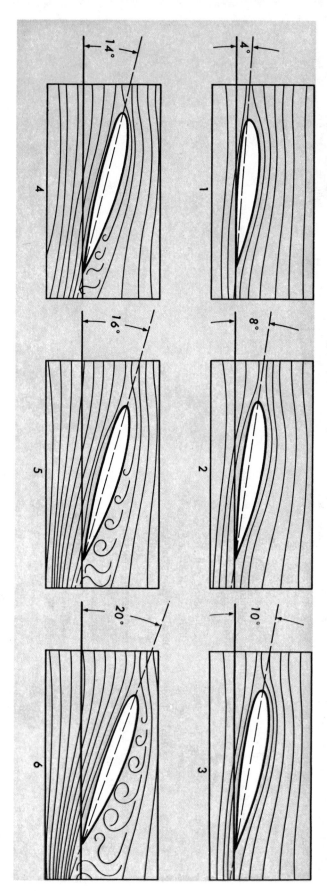

Figure 1–17. Flow of air around a wing at various angles of attack.

Factors Affecting Lift and Drag

A number of the factors that influence lift and drag include: wing area, shape of the airfoil, angle of attack, speed of the air passing over the wing (airspeed), and density of the air moving over the wing. A change in any of these factors affects the relationship between lift and drag. When lift is increased, drag is increased, or when lift is decreased, drag is decreased.

Effect of Wing Area on Lift and Drag   The lift and drag acting on a wing are proportional to the wing area. This means that if the wing area is doubled, other variables remaining the same, the lift and drag created by the wing will be doubled.

Effect of Airfoil Shape on Lift and Drag   Generally, the more curvature there is to the upper surface of an airfoil, the more lift is produced (up to a point). High-lift wings have a large convex curvature on the upper surface and a concave lower surface. Most airplanes have wing flaps which, when lowered, cause an ordinary wing to approximate this condition by increasing the curvature of the upper surface and creating a concave lower surface, thus increasing lift on the wing (Figure 1-18). A lowered aileron also accomplishes this by

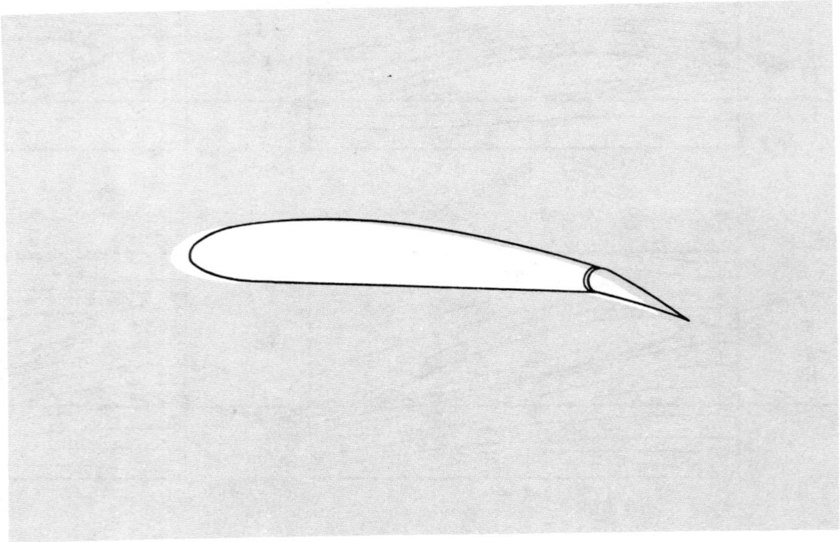

Figure 1-18. *Use of flaps increases lift and drag.*

When the critical angle of attack is reached, the turbulent airflow, which appeared near the trailing edge of the wing at lower angles of attack, quickly spreads forward over the entire upper wing surface (Figure 1-17). This results in a sudden increase in pressure on the upper wing surface and a considerable loss of lift. Due to the loss of lift and increase in form drag, the remaining lift is insufficient to support the airplane, and the wing stalls.

To recover from a stall, the angle of attack must be decreased so that the airstream can once again flow smoothly over the wing surface. Remember that the angle of attack is the angle between the chord line and the relative wind, not the chord line and the horizon. Therefore, an airplane can be stalled in any attitude of flight with respect to the horizon, if the angle of attack is increased up to and beyond the critical angle of attack.

Relationship of Thrust and Drag in Straight-and-Level Flight   During straight-and-level flight, thrust and drag are equal in magnitude if a constant airspeed is being maintained. When the thrust of the propeller is increased, thrust momentarily exceeds drag and the airspeed will increase, provided straight-and-level flight is maintained. As stated previously, with an increase in airspeed drag increases very rapidly. At some new and higher airspeed, thrust and drag forces again become equalized and speed again becomes constant.

If all the available power is used, thrust will reach its maximum, airspeed will increase until drag equals thrust, and once again the airspeed will become constant. This will be the top speed for that airplane in that configuration and attitude.

When thrust becomes less than drag, the airplane will decelerate to a slower airspeed, provided straight-and-level flight is maintained, and thrust and drag again become equal. Of course if the airspeed becomes too slow, or more precisely, if the angle of attack is too great, the airplane will stall.

Relationship Between Lift and Weight in Straight-and-Level Flight   A component of lift, the upward force on the wing, always acts perpendicular to the direction of the relative wind. In straight-and-level flight (constant altitude) lift counterbalances the airplane weight. When lift and weight are in equilibrium, the airplane neither gains nor loses altitude. If lift becomes less than weight, the airplane will enter a descent; if lift becomes greater than weight, the airplane will enter a climb. Once a steady-state climb or descent is established, the relationship of the four forces will no longer be the same as in straight-and-level flight. However, for all practical purposes lift still equals weight for small angles of climb or descent.

increasing the curvature of a portion of the wing and thereby increasing the angle of attack, which in turn increases lift and also drag. A raised aileron reduces lift on the wing by decreasing the curvature of a portion of the wing and decreasing the angle of attack. The elevators can change the curvature and angle of attack of the horizontal tail surfaces, changing the amount and direction of lift. The rudder accomplishes the same thing for the vertical tail surfaces.

Many people believe that the only hazard of in-flight icing is the weight of the ice which forms on the wings. It is true that ice formation will increase weight, but equally important is that ice formation will alter the shape of the airfoil and adversely affect all aspects of airplane performance and control.

As the ice forms on the airfoil, especially the leading edge, the flow of air over the wing is disrupted. This disruption of the smooth airflow causes the wing to lose part or all of its lifting efficiency. Also, drag is increased substantially.

Even a slight coating of frost on the wings can prevent an airplane from becoming airborne because the smooth flow of air over the wing surface is disrupted and the lift capability of the wing is destroyed. Even more hazardous is becoming airborne with frost on the wing because again performance and control could be adversely affected. This is why it is extremely important that all frost, snow, and ice be removed from the airplane before takeoff.

Effect of Wing Design on Stall

The type of wing design for a particular airplane depends almost entirely on the purpose for which that airplane is to be used.

If speed is the prime consideration, a tapered wing is more desirable than a rectangular wing, but a tapered wing with no twist has undesirable stall characteristics. Assuming equal wing area, the tapered wing produces less drag than the rectangular wing. The elliptical wing is more efficient (greater lift for the amount of drag), but does not have as good stall characteristics as the rectangular wing.

To achieve good stall characteristics, the root of the wing should stall first, with the stall pattern progressing outward to the tip. This type of stall pattern decreases undesirable rolling tendencies and increases lateral control when approaching a stall. It is undesirable that the wingtip stalls first, particularly if the tip of one wing stalls before the tip of the other wing, which usually happens.

A desirable stall pattern can be accomplished by: (1) designing the wing with a twist so that the tip has a lower angle of incidence and therefore a lower angle of attack when the root of the wing approaches the critical angle of attack (Fig. 1-19); (2) designing slots near the leading edge of the wingtip to allow air to flow smoothly over that part of the wing at higher angles of attack, therefore stalling the root of the wing first (Fig. 1-20); and (3) attaching stall or spoiler strips on the leading edge near the wing root. This strip breaks up the airflow at higher angles of attack and produces the desired effect of the root area of the wing stalling first (Fig. 1-21).

Effect of Airspeed on Lift and Drag

An increase in the velocity of the air passing over the wing (airspeed) increases lift and drag. Lift is increased

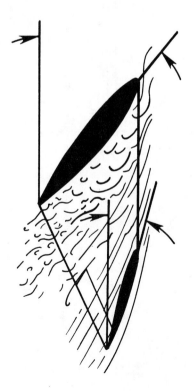

Figure 1-19. *View of wingtip twist. Ailerons are still effective even though wing root is in the stalled condition.*

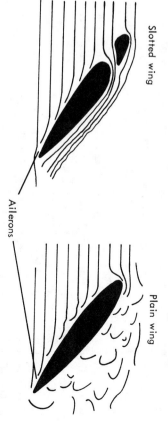

Slotted wing

Ailerons

Plain wing

Figure 1-20. *Slotted and plain wings at equal angles of attack.*

11

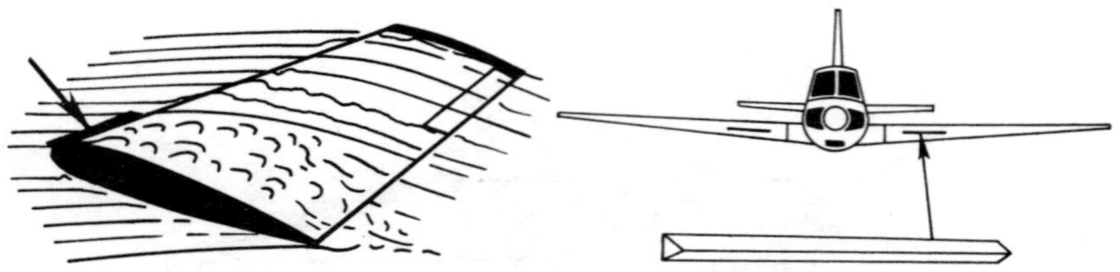

vapor than on a cool day. The more moisture in the air, the less dense the air.

Less dense air also produces other performance losses beside the loss of lift. Engine horsepower falls off and propeller efficiency decreases because of power loss and propeller blades—being airfoils—are less effective when air is less dense. Since the propeller is not pulling with the force and efficiency it would were the air dense, it takes longer to obtain the necessary forward speed to produce the required lift for takeoff—thus the airplane requires a longer takeoff run. The rate of climb will also be less for the same reasons.

From the above discussion it is obvious that *a pilot should beware of high, hot, and humid conditions*—high altitudes, hot temperatures, and high moisture content (high relative humidity). A combination of these three conditions could be disastrous, especially when combined with a short runway, a heavily loaded airplane, or other takeoff-limiting conditions.

Figure 1-21. *The stall strip ensures that the root section stalls first.*

because (1) the increased impact of the relative wind on the wing's lower surface creates a greater amount of air being deflected downward; (2) the increased speed of the relative wind over the upper surface creates a lower pressure on top of the wing (Bernoulli's Principle); and (3) a greater pressure differential between the upper and lower wing surface is created. Drag is also increased, since any change that increases lift also increases drag.

Tests show that lift and drag vary as the square of the velocity. The velocity of the air passing over the wing in flight is determined by the airspeed of the airplane. This means that if an airplane doubles its speed, it quadruples the lift and drag (assuming that the angle of attack remains the same).

Effect of Air Density on Lift and Drag   Lift and drag vary directly with the density of the air—as air density increases, lift and drag increase; as air density decreases, lift and drag decrease. Air density is affected by pressure, temperature, and humidity. At an altitude of 18,000 feet the density of the air is half the air density at sea level. Therefore, if an airplane is to maintain the same lift at high altitudes, the amount of air flowing over the wing must be the same as at lower altitudes. To do this the speed of the air over the wings (airspeed) must be increased. This is why an airplane requires a longer takeoff distance to become airborne at higher altitudes than with similar conditions at lower altitudes (Fig. 1-22).

Because air expands when heated, warm air is less dense than cool air. When other conditions remain the same, an airplane will require a longer takeoff run on a hot day than on a cool day (Fig. 1-22).

Because water vapor weighs less than an equal amount of dry air, moist air (high relative humidity) is less dense than dry air (low relative humidity). Therefore, when other conditions remain the same, the airplane will require a longer takeoff run on a humid day than on a dry day (Fig. 1-22). This is especially true on a hot, humid day because the air can hold much more water

Turning Tendency (Torque Effect)

By definition, "torque" is a force, or combination of forces, that produces or tends to produce a twisting or rotating motion of an airplane.

An airplane propeller spinning clockwise, as seen from the rear, produces forces that tend to twist or rotate the airplane in the opposite direction, thus turning the airplane to the left. Airplanes are designed in such a manner that the torque effect is not noticeable to the pilot when the airplane is in straight-and-level flight with a cruise power setting.

The effect of torque increases in direct proportion to engine power, airspeed, and airplane attitude. If the power setting is high, the airspeed slow, and the angle of attack high, the effect of torque is greater. During takeoffs and climbs, when the effect of torque is most pronounced, the pilot must apply sufficient right rudder pressure to counteract the left-turning tendency and maintain a straight takeoff path.

Several forces are involved in the insistent tendency of an airplane of standard configuration to turn to the left. All of these forces are created by the rotating propeller. How they are actually created varies greatly from one explanation to the next. Individual explanation of these forces is perhaps the best approach to understanding the reason for the left turning tendency.

The four forces are: reactive force, spiraling slipstream, gyroscopic precession, and "P" factor.

Reactive Force   This is based on Newton's Law of action and reaction. Applying this law to an airplane with a propeller rotating in a clockwise direc-

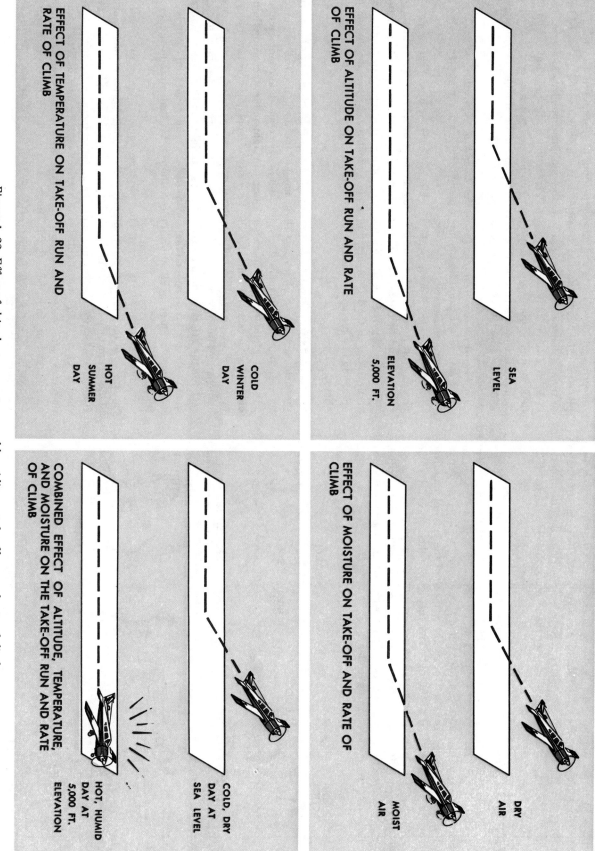

Figure 1–22. *Effect of altitude, temperature, and humidity on takeoff run and rate of climb.*

13

tion, as seen from the rear, a force is produced which tends to roll the entire airplane about its longitudinal axis in a counterclockwise direction. To better understand this concept, consider the air through which the propeller rotates as a restraining force. This restraining force acts opposite to the direction the propeller rotates, creating a tendency for the airplane to roll to the left (Fig. 1-23).

Spiraling Slipstream This theory is based on the reaction of the air to a rotating propeller blade. As the airplane propeller rotates through the air in a clockwise direction, as viewed from the rear, the propeller blade forces the air rearward in a spiraling clockwise direction. A portion of this spiraling slipstream strikes the left side of the vertical stabilizer forcing the airplane's tail to the right and the nose to the left, causing the airplane to rotate around the vertical axis (Fig. 1-23). The portion of the spiraling slipstream traveling under the fuselage is not obstructed; therefore, creating a different resistance between the obstructed and the unobstructed flow which causes the left turning tendency.

Gyroscopic Precession This theory is based on one of the gyroscopic properties which apply to any object spinning in space, even a rotating airplane propeller. As the nose of the airplane is raised or lowered, or moved left or right, a deflective force is applied to the spinning propeller which results in a reactive force known as precession. Precession is the resultant action or deflection of a spinning wheel (propeller in this case) when a force is applied to its rim. This resultant force occurs 90° ahead in the direction of rotation, and in the direction of the applied force (Fig. 1-23).

"P" Factor or Asymmetric Propeller Loading The effects of "P" factor or asymmetric propeller loading usually occur when the airplane is flown at a high angle of attack.

The downward moving blade, which is on the right side of the propeller arc, as seen from the rear, has a higher angle of attack, greater action and reaction, and therefore higher thrust than the upward-moving blade on the left (Fig. 1-23). This results in a tendency for the airplane to yaw around the vertical axis to the left. Again this is most pronounced when the engine is operating at a high power setting and the airplane is flown at a high angle of attack.

Corrections for Turning Tendency or Torque During Flight Since the airplane is flown in cruising flight most of the time, airplane manufacturers design the airplane with certain built-in corrections that counteract the left-turning tendency or torque effect during straight-and-level cruising flight only. This correction eliminates the necessity of applying constant rudder pressure. Because the effect of torque varies to such an extent during climbs and changes in angle of attack, it is impractical for airplane designers to correct for the effect of torque except during straight-and-level flight. Consequently the pilot is provided other means such as rudder and trim controls to counteract the turning effect during conditions other than straight-and-level flight.

Many manufacturers "cant" the airplane engine slightly so that the thrust line of the propeller points slightly to the right. This counteracts much of the left turning tendency of the airplane during various conditions of flight. Other manufacturers, when designing the airplane, increase the angle of incidence of the left wing slightly, which increases the angle of attack and therefore increases the lift on this wing. The increased lift counteracts left-turning tendency in cruising flight. The increase in lift will, however, increase drag on the left wing and, to compensate for this, the vertical stabilizer is off-set slightly to the left.

FORCE

FORCE

ACTION

PRECESSION

ASCENDING LEFT BLADE   DESCENDING RIGHT BLADE   NOSE HIGH ATTITUDE

P-FACTOR

REACTION

ACTION

TORQUE REACTION

SLIPSTREAM

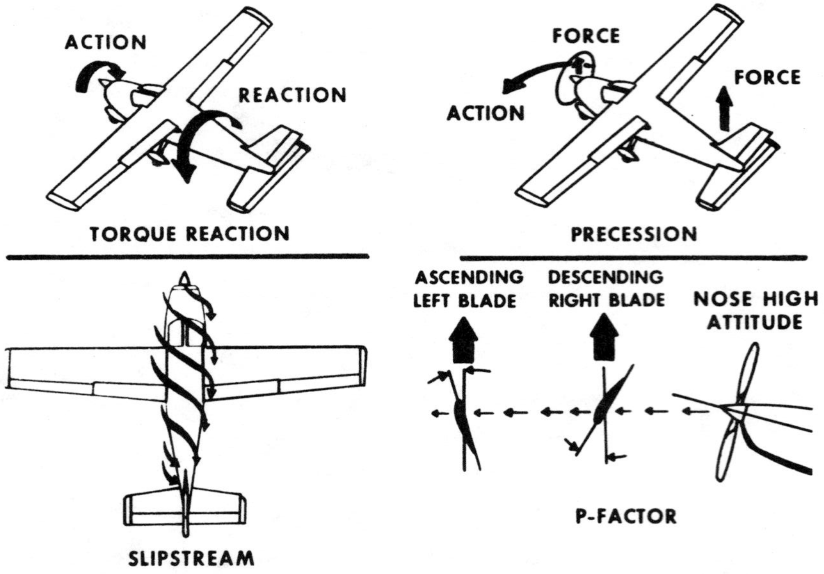

Figure 1-23. *Factors which cause left-turning tendency.*

14

Torque corrections in flight conditions other than cruising flight must be accomplished by the pilot. This is done by applying sufficient rudder to overcome the left-turning tendency. For example, in a straight climb, right rudder pressure is necessary to keep the airplane climbing straight.

When thinking of "torque" such things as reactive force, spiraling slipstream, gyroscopic precession, and asymmetric propeller loading ("P" factor) must be included, as well as any other power-induced forces that tend to turn the airplane.

Airplane Stability

Stability is the inherent ability of a body, after its equilibrium is disturbed, to develop forces or moments that tend to return the body to its original position. In other words, a stable airplane will tend to return to the original condition of flight if disturbed by a force such as turbulent air. This means that a stable airplane is easy to fly; however, this does not mean that a pilot can depend entirely on stability to return the airplane to the original condition. Even in the most stable airplanes, there are conditions that will require the use of airplane controls to return the airplane to the desired attitude. However, a pilot will find that a well designed airplane requires less effort to control the airplane because of the inherent stability.

Stability is classified into three types: (1) positive, (2) neutral, and (3) negative.

Positive stability can be illustrated by a ball inside of a bowl (Fig. 1-24). If the ball is displaced from its normal resting place at the bottom of the bowl, it will eventually return to its original position at the bottom of the bowl.

Neutral stability can be illustrated by a ball on a flat plane (Fig. 1-24). If the ball is displaced, it will come to rest at some new, neutral position and show no tendency to return to its original position.

Negative stability is in fact instability and can be illustrated by a ball on the top of an inverted bowl (Fig. 1-24). Even the slightest displacement of the ball will activate greater forces which will cause the ball to continue to move in the direction of the applied force. It should be obvious that airplanes should display positive stability, or perhaps neutral stability, but never negative stability.

Stability may be further classified as static and/or dynamic. *Static* stability means that if the airplane's equilibrium is disturbed, forces will be activated which will initially tend to return the airplane to its original position. However, these restoring forces may be so great that they will force the

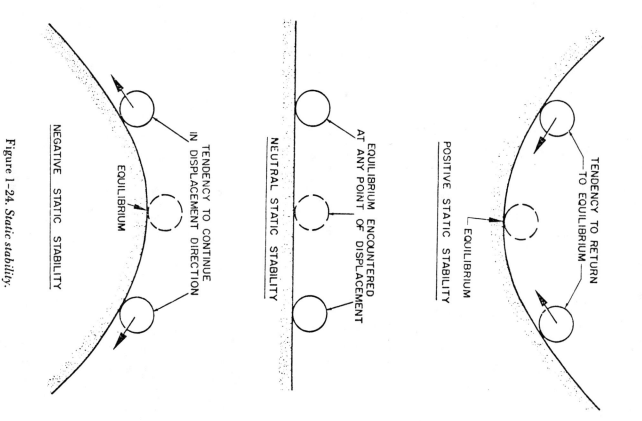

POSITIVE STATIC STABILITY

TENDENCY TO RETURN TO EQUILIBRIUM

EQUILIBRIUM

NEUTRAL STATIC STABILITY

EQUILIBRIUM ENCOUNTERED AT ANY POINT OF DISPLACEMENT

NEGATIVE STATIC STABILITY

TENDENCY TO CONTINUE IN DISPLACEMENT DIRECTION

EQUILIBRIUM

Figure 1-24. *Static stability.*

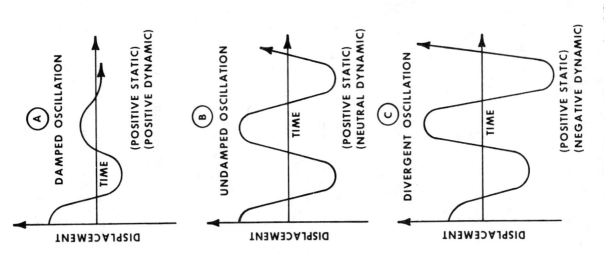

Figure 1-25. *Relationship of oscillation and stability.*

airplane beyond the original position and continue in that direction (Fig. 1-25).

On the other hand, *dynamic* stability is a property which dampens the oscillations set up by a statically stable airplane, enabling the oscillations to become smaller and smaller in magnitude until the airplane eventually settles down to its original condition of flight (Fig. 1-25).

Therefore an airplane should possess *positive* stability which is both *static* and *dynamic* in nature.

Before further discussion on stability, the axes of rotation will be reviewed because that is where stability has its effect.

The airplane has three axes of rotation around which movement takes place. These are (1) lateral axis—an imaginary line from wingtip to wingtip, (2) longitudinal axis—an imaginary line from the nose to the tail, and (3) vertical axis—an imaginary line extending vertically through the intersection of the lateral and longitudinal axes. The airplane can rotate around all three axes simultaneously or it can rotate around just one axis (Fig. 1-26). Think of these axes as imaginary axles around which the airplane turns, much as a wheel would turn around axles positioned in these same three planes. The three axes intersect at the center of gravity and each one is perpendicular to the other two.

Rotation about the lateral axis is called pitch, and is controlled by the elevators. This rotation is referred to as longitudinal control or longitudinal stability.

Rotation about the longitudinal axis is called roll, and is controlled by the ailerons. This rotation is referred to as lateral control or lateral stability.

Rotation about the vertical axis is called yaw and is controlled by the rudder. This rotation is referred to as directional control or directional stability.

Stability of the airplane then, is the combination of forces that act around these three axes to keep the pitch attitude of the airplane in a normal level flight attitude with respect to the horizon, the wings level, and the nose of the airplane directionally straight along the desired path of flight.

Longitudinal Stability About the Lateral Axis Longitudinal stability is important to the pilot because it determines to a great extent the pitch characteristics of the airplane, particularly as this relates to the stall characteristics. It would be unsafe and uncomfortable for the pilot if an airplane continually displayed a tendency to either stall or dive when the pilot's attention was diverted for some reason. If properly designed, the airplane will not display these unstable tendencies when the airplane is loaded according to the manufacturer's recommendations.

16

The location of the center of gravity with respect to the center of lift determines to a great extent the longitudinal stability of the airplane.

Fig. 1–27 illustrates neutral longitudinal stability. Note that the center of lift is directly over the center of gravity or weight. An airplane with neutral stability will produce no inherent pitch moments around the center of gravity.

Fig. 1–28 illustrates the center of lift in front of the center of gravity. This airplane would display negative stability and an undesirable pitchup moment during flight. If disturbed, the up and down pitching moment will tend to increase in magnitude. This condition can occur, especially if the airplane is

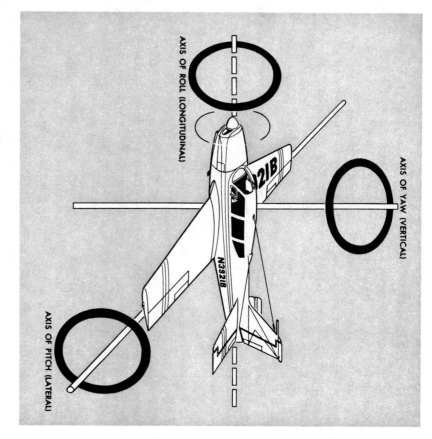

Figure 1–26. *Axes of rotation.*

loaded so that the center of gravity is rearward of the airplane's aft loading limits.

Fig. 1–29 shows an airplane with the center of lift behind the center of gravity. Again, this produces negative stability. Some force must balance the down force of the weight. This is accomplished by designing the airplane in such a manner that the air flowing downward behind the trailing edge of the wing strikes the upper surface of the horizontal stabilizer (except on T-tails). This creates a downward tail force to counteract the tendency to pitch down and provides positive stability.

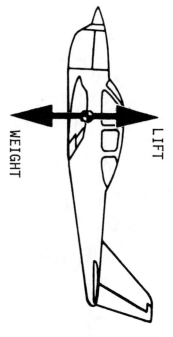

Figure 1–27. *Neutral stability.*

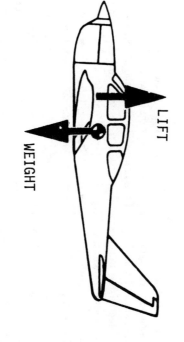

Figure 1–28. *Negative stability.*

17

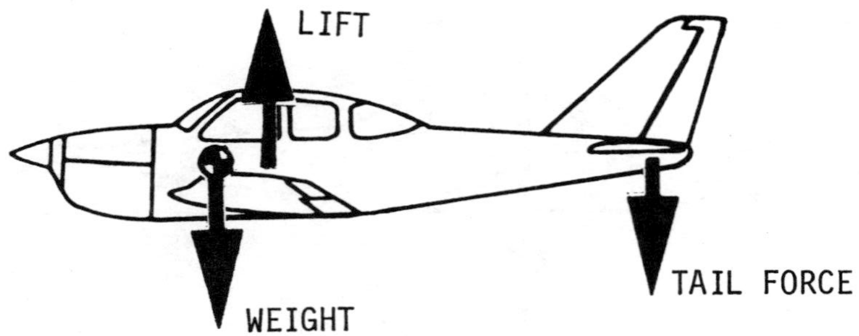

LIFT

WEIGHT

TAIL FORCE

Figure 1–29. *Positive stability.*

To further explain, if the nose is pitched down and the control released, the airspeed will increase. This in turn, will increase the downwash on the tail's horizontal stabilizer forcing the nose up (except on T-tails). Conversely, if the nose is pitched up and the control released, the airspeed will diminish, thus decreasing the downwash on the horizontal stabilizer. This permits the nose to pitch downward. There is one speed only for each degree of angle of attack and eventually, after several pitch oscillations, the airplane tends to stabilize at the airspeed (angle of attack) for which it is trimmed.

The above concept is of prime importance to the pilot. A common misconception about longitudinal stability is that an airplane is stable in respect to the horizon. This would be an undesirable characteristic of an airplane. Keep in mind that longitudinal stability is with respect only to airspeed (angle of attack).

The foregoing explanation of longitudinal stability needs some qualification because during certain flight maneuvers the airplane is not entirely "speed seeking," but "angle of attack seeking." This can be demonstrated by placing the airplane in a power-off glide and trimming the airplane for a specific speed. Then if the throttle is opened suddenly, the airplane will nose up and finally assume an attitude that results in a speed considerably less than that of the power-off glide. This is because of additional forces developed by the propeller blast over the horizontal stabilizer (except T-tails), and the fact that the airplane is stable only with relation to airflow, or the relative wind. In other words, the stable airplane is not concerned with its own attitude relative to the earth or horizon, but with the relative wind. It will always tend to maintain an alignment with the relative wind.

Longitudinal Control (Pitch) About the Lateral Axis In the previous discussion, the *one* speed or angle of attack concept was used to explain how longitudinal stability was attained. It is important for the pilot to know that the airplane is stable at various speeds or angles of attack, not just one. The controls which allow the pilot to depart from the one speed or angle of attack concept *or* the controls used to give the pilot longitudinal control around the lateral axis, are the elevators (Fig. 1–30) and the elevator trim tab (Fig. 1–31).

The function of the elevator control is to provide a means by which the wing's angle of attack may be changed.

On most airplanes the elevators are movable control surfaces hinged to the horizontal stabilizer, and attached to the control column in the cockpit by mechanical linkage. This allows the pilot to change the angle of attack of the entire horizontal stabilizer. The horizontal stabilizer normally has a negative angle of attack to provide a downward force rather than a lifting force. If the pilot applies back elevator pressure the elevator is raised, increasing the horizontal stabilizer's negative angle of attack and consequently increasing the downward tail force. This forces the tail down, increasing the angle of attack of the wings. Conversely, if forward pressure is applied to the elevator control, the elevators are lowered, decreasing the horizontal stabilizer's negative angle of attack and consequently decreasing the downward force on the tail. This decreases the angle of attack of wings (Fig. 1–30).

The elevator trim tab is a small auxiliary control surface hinged at the trailing edge of the elevators. The elevator trim tab acts on the elevators, which in turn acts upon the entire airplane. This trim tab is a part of the elevator but may be moved upward or downward independently of the elevator itself. It is controlled from the cockpit by a control which is separate from the elevator control. The elevator trim tab allows the pilot to adjust the angle of attack for a constant setting and therefore eliminates the need to exert continuous pressure on the elevator control to maintain a constant angle of attack. An upward deflection of the trim tab will force the elevator downward with the same result as moving the elevator downward with the elevator control, and conversely a downward deflection of the trim tab will force the elevator upward. The direction the trim tab is deflected will always cause the entire elevator to be deflected in the opposite direction (Fig. 1–31).

Lateral Stability About the Longitudinal Axis Lateral stability is the stability displayed around the longitudinal axis of the airplane. An airplane that tends to return to a wings-level attitude after being displaced from a level attitude by some force such as turbulent air, is considered to be laterally stable.

Three factors affect lateral stability; (1) dihedral, (2) sweepback, and (3) keel effect.

Dihedral is the angle at which the wings are slanted upward from the root to the tip (Fig. 1-32). The stabilizing effect of dihedral occurs when the airplane sideslips slightly as one wing is forced down in turbulent air. This sideslip results in a difference in the angle of attack between the higher and lower wing with the greatest angle of attack on the lower wing. The increased angle of attack produces increased lift on the lower wing with a tendency to return the airplane to wings-level flight. Note the direction of the relative wind during a slip by the arrows in Fig. 1-32.

Sweepback is the angle at which the wings are slanted rearward from the root to the tip (Fig. 1-33). The effect of sweepback in producing lateral stability is similar to that of dihedral, but not as pronounced. If one wing lowers in a slip, the angle of attack on the low wing increases, producing greater lift. This results in a tendency for the lower wing to rise, and return the airplane to level flight. Sweepback augments dihedral to achieve lateral stability. Another

reason for sweepback is to place the center of lift farther rearward, which affects longitudinal stability more than it does lateral stability.

Keel effect depends upon the action of the relative wind on the side area of the airplane fuselage. In a slight slip the fuselage provides a broad area upon which the relative wind will strike, forcing the fuselage to parallel the relative wind. This aids in producing lateral stability (Fig. 1-34).

Lateral Control (Roll) About the Longitudinal Axis Lateral control is obtained through the use of ailerons, and on some airplanes the aileron trim tabs. The ailerons are movable surfaces hinged to the outer trailing edge of the wings, and attached to the cockpit control column by mechanical linkage. Moving the control wheel or stick to the right raises the aileron on the right wing and lowers the aileron on the left wing. Moving the control wheel or stick to the left reverses this and raises the aileron on the left wing and lowers the aileron on the right wing. When an aileron is lowered, the angle of attack on that wing will increase, which increases the lift. This permits rolling the airplane laterally around the longitudinal axis (Fig. 1-35).

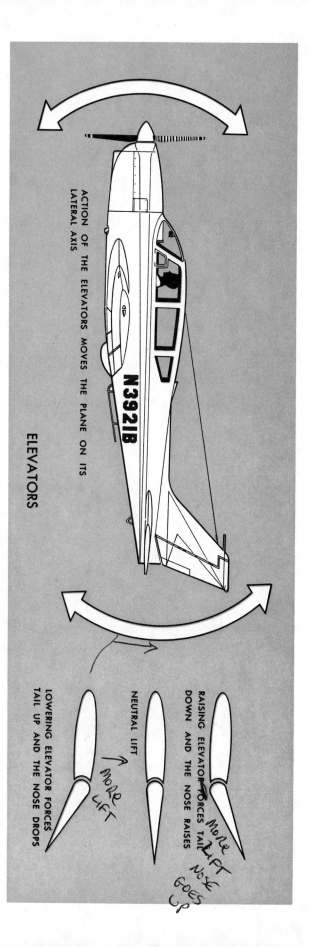

Figure 1-30. *Effect of elevators.*

ACTION OF THE ELEVATORS MOVES THE PLANE ON ITS LATERAL AXIS

ELEVATORS

RAISING ELEVATOR FORCES TAIL DOWN AND THE NOSE RAISES

NEUTRAL LIFT

LOWERING ELEVATOR FORCES TAIL UP AND THE NOSE DROPS

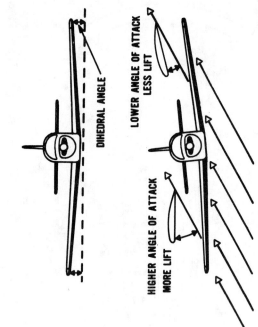

DIHEDRAL ANGLE

HIGHER ANGLE OF ATTACK
MORE LIFT

LOWER ANGLE OF ATTACK
LESS LIFT

Figure 1–32. *Effect of dihedral.*

Many airplanes are equipped with an aileron trim tab which is a small movable part of the aileron hinged to the trailing edge of the main aileron. These trim tabs can be moved independently of the ailerons. Aileron trim tabs function similar to the elevator trim tabs. Moving the trim tabs produces an effect on the aileron which in turn affects the entire airplane. If the trim tab is deflected upward the aileron is deflected downward, increasing the angle of attack on that wing, resulting in greater lift on that wing. The reverse is true if the trim tab is deflected downward.

Lateral Stability or Instability in Turns Because of lateral stability, most airplanes will tend to recover from shallow banks automatically. However, as the bank is increased, the wing on the outside of the turn travels faster than the wing on the inside of the turn. The increased speed increases the lift on the outside wing, causing a destabilizing rolling moment or an overbanking tendency. The angle of bank will continue to increase into a steeper and steeper bank unless the pilot applies a slight amount of control pressure to counteract this tendency. The overbanking tendency becomes increasingly significant when the angle of bank reaches more than 30°.

During a medium banked turn (a bank angle between the shallow bank and steep bank), an airplane tends to hold its bank constant and requires less

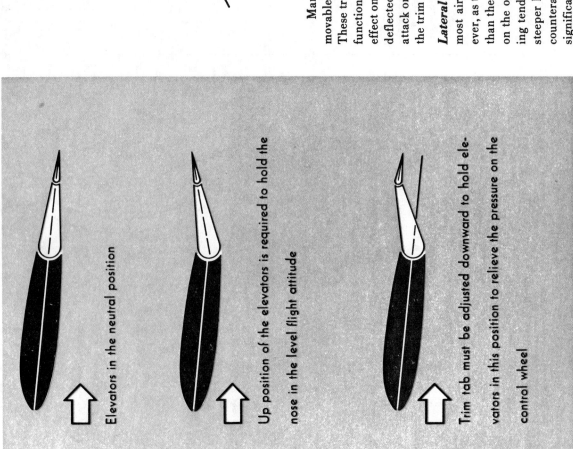

Elevators in the neutral position

Up position of the elevators is required to hold the nose in the level flight attitude

Trim tab must be adjusted downward to hold elevators in this position to relieve the pressure on the control wheel

Figure 1–31. *Effect of trim tabs.*

control input on the part of the pilot. This is because the stabilizing moments of lateral stability and the destabilizing moments of overbanking very nearly cancel each other out. A pilot can discover these various areas of bank through experimentation.

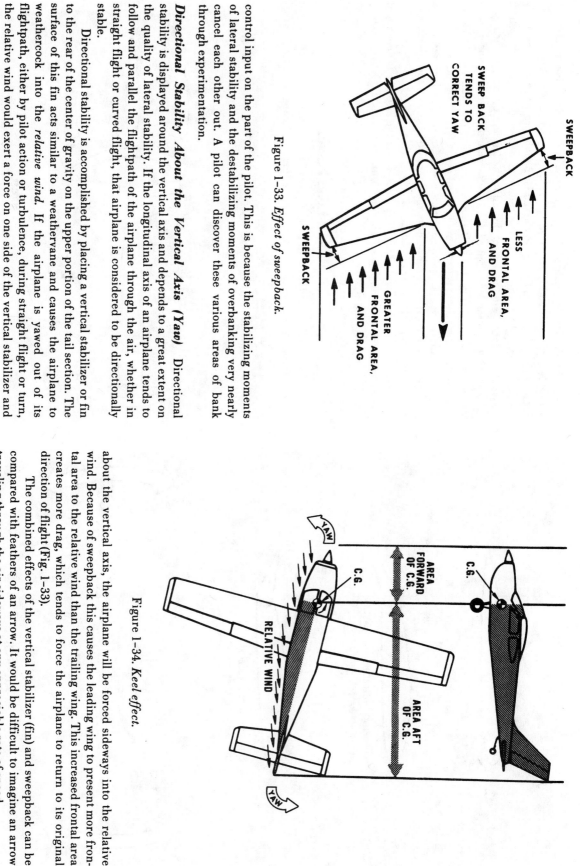

Figure 1-33. *Effect of sweepback.*

Directional Stability About the Vertical Axis (Yaw) Directional stability is displayed around the vertical axis and depends to a great extent on the quality of lateral stability. If the longitudinal axis of an airplane tends to follow and parallel the flightpath of the airplane through the air, whether in straight flight or curved flight, that airplane is considered to be directionally stable.

Directional stability is accomplished by placing a vertical stabilizer or fin to the rear of the center of gravity on the upper portion of the tail section. The surface of this fin acts similar to a weathervane and causes the airplane to weathercock into the *relative wind*. If the airplane is yawed out of its flightpath, either by pilot action or turbulence, during straight flight or turn, the relative wind would exert a force on one side of the vertical stabilizer and return the airplane to its original direction of flight.

Wing sweepback aids in directional stability. If the airplane is rotated about the vertical axis, the airplane will be forced sideways into the relative wind. Because of sweepback this causes the leading wing to present more frontal area to the relative wind than the trailing wing. This increased frontal area creates more drag, which tends to force the airplane to return to its original direction of flight (Fig. 1-33).

The combined effects of the vertical stabilizer (fin) and sweepback can be compared with feathers of an arrow. It would be difficult to imagine an arrow traveling through the air sideways at any appreciable rate of speed.

Directional Control About the Vertical Axis (Yaw) Directional con-

Figure 1-34. *Keel effect.*

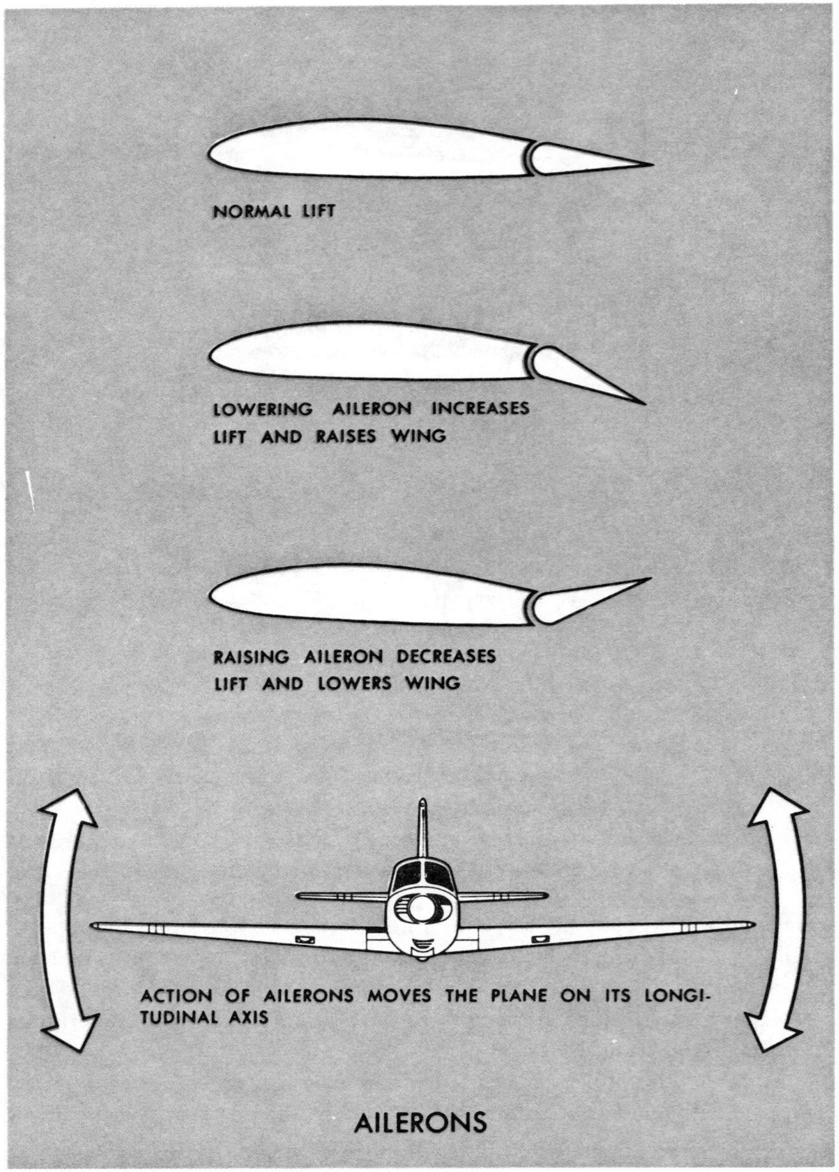

trol of the airplane is obtained through the use of the rudder. The rudder is a movable surface hinged to the trailing edge of the vertical stabilizer (fin) and attached by mechanical linkage to the rudder pedals located in the cockpit. By pressing the right rudder pedal, the rudder is deflected to the right, which causes the relative wind to deflect the tail to the left and the nose to the right. If left rudder pressure is applied the reverse action occurs and the nose is deflected to the left (Fig. 1–36). It should be understood that the purpose of the rudder during flight is to control yaw and not to turn the airplane.

Some airplanes are equipped with a rudder trim tab, which reacts in a similar manner on the rudder as does the aileron trim tab on the aileron and the elevator trim tab on the elevator.

The amount of control which the pilot has over the airplane is dependent upon the speed of the airflow striking the control surfaces. Effective airplane stability also depends upon speed of the airplane through the air. The greater the airspeed the greater the effect of stability as a restoring force.

Loads and Load Factors

An airplane is designed and certificated for a certain maximum weight during flight. This weight is referred to as the maximum certificated gross

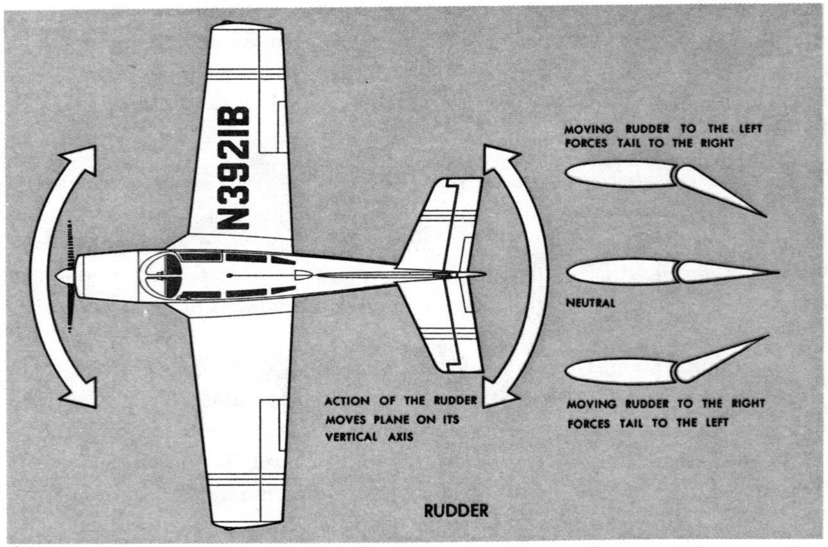

Figure 1–36. *Effect of rudder.*

Figure 1–35. *Effect of ailerons.*

weight. It is important that the airplane be loaded within the specified weight limits before flight, because certain flight maneuvers will impose an extra load on the airplane structure which may, particularly if the airplane is overloaded, impose stresses which will exceed the design capabilities of the airplane. Overstressing the airplane can also occur if the pilot engages in maneuvers creating high loads, regardless of how the airplane is loaded. These maneuvers not only increase the load that the airplane structure must support, but also increase the airplane's stalling speed.

The following will explain how extra load is imposed upon the airplane during flight.

During flight the wings of an airplane will support the maximum allowable gross weight of the airplane. So long as the airplane is moving at a steady rate of speed and in a straight line the load imposed upon the wings will remain constant.

A change in speed during straight flight will not produce any appreciable change in load, but when a change is made in the airplane's flightpath, an additional load is imposed upon the airplane structure. This is particularly true if a change in direction is made at high speeds with rapid forceful control movements.

According to certain laws of physics a mass (airplane in this case) will continue to move in a straight line unless some force intervenes, causing the mass (airplane) to assume a curved path. During the time the airplane is in a curved flightpath, it still attempts, because of inertia, to force itself to follow straight flight. This tendency to follow straight flight, rather than curved flight, generates a force known as centrifugal force which acts toward the outside of the curve.

Any time the airplane is flying in a curved flightpath with a positive load, the load the wings must support will be equal to the weight of the airplane plus the load imposed by centrifugal force. A positive load occurs when back pressure is applied to the elevator, causing centrifugal force to act in the same direction as the force of weight. A negative load occurs when forward pressure is applied to the elevator control, causing centrifugal force to act in a direction opposite to that of the force of weight.

Curved flight producing a positive load is a result of increasing the angle of attack and consequently the lift. Increased lift always increases the positive load imposed upon the wings. However, the load is increased only at the time the angle of attack is being increased. Once the angle of attack is established, the load imposed on the wings in flight are stated in terms of *load factor*.

Load factor is the ratio of the total load supported by the airplane's wing to the actual weight of the airplane and its contents; i.e., the actual load supported by the wings divided by the total weight of the airplane. For example, if an airplane has a gross weight of 2,000 lbs. and during flight is subjected to aerodynamic forces which increase the total load the wing must support to 4,000 lbs., the load factor would be 2.0. ($4{,}000 \div 2{,}000 = 2$.) In this example the airplane wing is producing "lift" that is equal to twice the gross weight of the airplane.

Another way of expressing load factor is the ratio of a given load to the pull of gravity; i.e., to refer to a load factor of three, as "three G's," where "G" refers to the pull of gravity. In this case the weight of the airplane is equal to "one G," and if a load of three times the actual weight of the airplane were imposed upon the wing due to curved flight, the load factor would be equal to "three G's."

Load Factors and Airplane Design To be certificated by the Federal Aviation Administration, the structural strength (load factor) of airplanes must conform with prescribed standards set forth by Federal Aviation Regulations.

All airplanes are designed to meet certain strength requirements depending upon the intended use of the airplanes. Classification of airplanes as to strength and operational use is known as the category system.

The category of each airplane can be readily identified by a placard or document (Airworthiness Certificate) in the cockpit which states the operational category or categories in which that airplane is certificated.

The category, maneuvers that are permitted, and the maximum safe load factors (limit load factors) specified for these airplanes are as follows:

Category	Permissible Maneuvers	Limit Load Factor*
Normal	1—Any maneuver incident to normal flying. 2—Stalls (except whip stalls). 3—Lazy eights, chandelles, and steep turns in which the angle of bank does not exceed 60°.	3.8

* To the limit loads given, a safety factor of 50% is added.

Continued

constant altitude coordinated turn the load factor (resultant load) is the result of two forces: (1) pull of gravity, and (2) centrifugal force (Fig. 1-37).

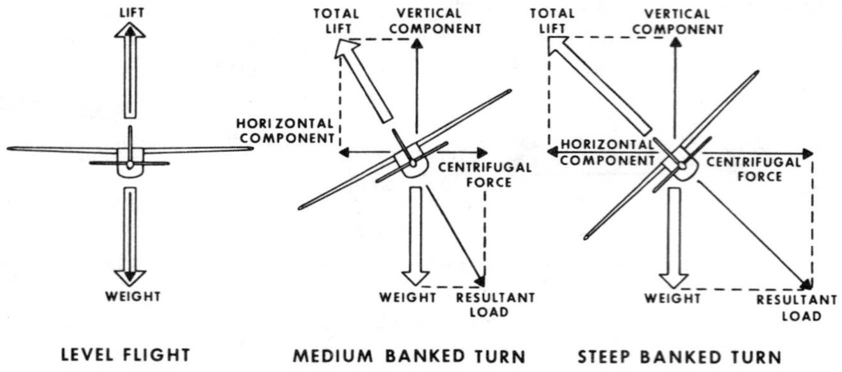

It is not within the scope of this handbook to discuss the mathematics of the turn. However, in any airplane at any airspeed, if a constant altitude is maintained during the turn, the load factor for a given degree of bank is the same, which is the resultant of gravity and centrifugal force. For any given angle of bank the rate of turn varies with the airspeed. In other words, if the angle of bank is held constant and the airspeed is increased, the rate of turn will decrease; or if the airspeed is decreased, the rate of turn will increase. Because of this, there is no change in centrifugal force for any given bank. Therefore, the load factor remains the same.

Fig. 1-38 and Fig. 1-39 reveal an important fact about load factor in turns. The load factor increases at a rapid rate after the angle of bank reaches 50°. The wing must produce lift equal to this load factor if altitude is to be maintained.

It should also be noted how rapidly load factor increases as the angle of bank approaches 90°. The 90° banked, constant altitude turn is not mathematically possible. An airplane can be banked to 90°, but a continued coordinated turn is impossible at this bank angle without losing altitude.

At an angle of bank of slightly more than 80° the load factor exceeds 6, which is the limit load factor of an acrobatic airplane.

The approximate maximum bank for conventional light airplanes is 60° which produces a load factor of 2. This bank reaches the limit of a normal category airplane. An additional 10° of bank will increase the load factor by approximately 1 G (Fig. 1-39), bringing it dangerously close to the point at which structural damage or complete failure may occur in these airplanes.

Effect of Load Factor on Stalling Speed Any airplane, within the limits of its structure and the strength of the pilot can be stalled at any airspeed. At a given airspeed the load factor increases as angle of attack increases, and the wing stalls because the angle of attack has been increased to a certain angle. Therefore, there is a direct relationship between the load factor imposed upon the wing and its stalling characteristics.

When a sufficiently high angle of attack is reached, the smooth flow of air over an airfoil breaks up and tears away, producing the abrupt change of characteristics and loss of lift which is defined as a stall.

A rule for determining the speed at which a wing will stall is that the stalling speed increases in proportion to the square root of the load factor. To further explain, the load factor produced in a 75° banked turn is 4 (Fig. 1-39). Applying the rule the square root of 4 is 2. This means that an airplane with a

Category	Permissible Maneuvers	Limit Load Factor*
Utility	1—All operations in the normal category. 2—Spins (if approved for that airplane). 3—Lazy eights, chandelles, and steep turns in which the angle of bank is more than 60°.	4.4
Acrobatic	No restrictions except those shown to be necessary as a result of required flight tests.	6.0

* To the limit loads given, a safety factor of 50% is added.

It should be noted that there is an increase in limit load factor with an increasing severity of maneuvers permitted. Small airplanes may be certificated in more than one category if the requirements for each category are met.

This system provides a means for the pilot to determine what operations can be performed in a given airplane without exceeding the load limit. Pilots are cautioned to operate the airplane within the load limit for which the airplane is designed so as to enhance safety and still benefit from the intended utilization of the airplane.

Effect of Turns on Load Factor A turn is made by banking the airplane so that lift from the wings pulls the airplane from its straight flightpath. In a

LEVEL FLIGHT   MEDIUM BANKED TURN   STEEP BANKED TURN

Figure 1-37. *Forces acting on an airplane in a bank.*

Figure 1–38. *The load supported by the wings increases as the angle of bank increases. The increase is shown by the relative lengths of the white arrows. Figures below the arrows indicate the increase in load factor. For example, the load factor during a 60-degree bank is 2.00, and the load supported by the wings is twice the weight of the airplane in level flight.*

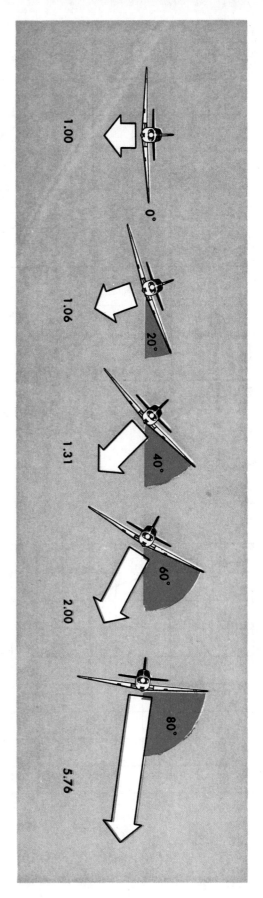

normal unaccelerated stalling speed of 50 knots can be stalled at twice that speed or 100 knots, by inducing a load factor of 4. If the airplane were capable of withstanding a load factor of 9, this airplane could be stalled at a speed of 150 knots.

Since the load factor squares as the stalling speed doubles, tremendous loads may be imposed on structures by stalling an airplane at relatively high airspeeds. An airplane which has a normal unaccelerated stalling speed of 50 knots will be subjected to a load factor of 4 G's when forced into an accelerated stall at 100 knots. As seen from this example, it is easy to impose a load beyond the design strength of the conventional airplane.

Reference to the chart in Fig. 1–40 will show that banking an airplane just over 75° in a steep turn increases the stalling speed by 100%. If the normal unaccelerated stalling speed is 45 knots, the pilot must keep the airspeed above 90 knots in a 75° bank to prevent sudden entry into a violent power stall. This same effect will take place in a quick pullup from a dive or a maneuver producing load factors above 1G. Accidents have resulted from sudden, unexpected loss of control, particularly in a steep turn near the ground.

The maximum speed at which an airplane can be safely stalled is the design maneuvering speed. The design maneuvering speed is a valuable reference point for the pilot. When operating below this speed a damaging positive flight load should not be produced because the airplane should stall before the load becomes excessive. Any combination of flight control usage, including full deflection of the controls, or gust loads created by turbulence should not create an excessive air load if the airplane is operated below maneuvering speed. (Pilots should be cautioned that certain adverse wind shear or gusts may cause excessive loads even at speeds below maneuvering speed.)

Design maneuvering speed can be found in the Pilot's Operating Handbook or on a placard within the cockpit. It can also be determined by multiplying the normal unaccelerated stall speed by the square root of the limit load factor. A rule of thumb that can be used to determine the maneuvering speed is approximately 1.7 times the normal stalling speed. Thus, an airplane which normally stalls at 35 knots should never be stalled when the airspeed is above 60 knots (35 knots × 1.7 = 59.5 knots).

A knowledge of this must be applied from two points of view by the competent pilot: the danger of inadvertently stalling the airplane by increasing

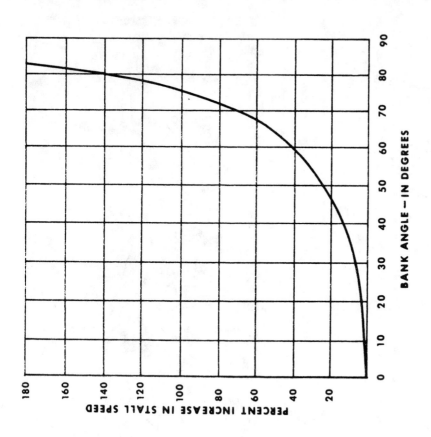

Figure 1–40. *Stall speed chart.*

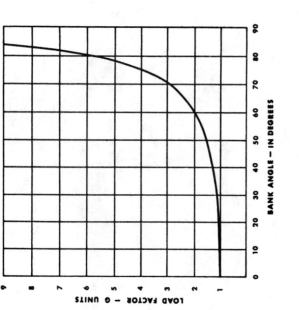

Figure 1–39. *Load factor chart.*

the load factor such as in a steep turn or spiral; and that intentionally stalling an airplane above its design maneuvering speed imposes a tremendous load factor on the structure.

Effect of Speed on Load Factor The amount of excess load that can be imposed on the wing depends on how fast the airplane is flying. At slow speeds, the maximum available lifting force of the wing is only slightly greater than the amount necessary to support the weight of the airplane. Consequently, the load factor should not become excessive even if the controls are moved abruptly or the airplane encounters severe gusts, as previously stated. The reason for this is that the airplane will stall before the load can become excessive. However, at high speeds, the lifting capacity of the wing is so great that a sudden movement of the elevator controls or a strong gust may increase the load factor beyond safe limits. Because of this relationship between speed and safety, certain "maximum" speeds have been established. Each airplane is restricted in the speed at which it can safely execute maneuvers, withstand abrupt application of the controls, or fly in rough air. This speed is referred to as the design maneuvering speed, which was discussed previously.

Summarizing, at speeds below design maneuvering speed, the airplane

should stall before the load factor can become excessive. At speeds above maneuvering speed, the limit load factor for which an airplane is stressed can be exceeded by abrupt or excessive application of the controls or by strong turbulence.

Effect of Flight Maneuvers on Load Factor Load factors apply to all flight maneuvers. In straight-and-level, unaccelerated flight a load factor of 1G is always present, but certain maneuvers are known to involve relatively high load factors.

Turns. As previously discussed, increased load factors are a characteristic of all banked turns. Load factors become significant both to flight per-

26

formance and to the load on wing structure as the bank increases beyond approximately 45°.

Stalls. The normal stall entered from straight-and-level flight, or an unaccelerated straight climb, should not produce added load factors beyond the 1 G of straight-and-level flight. As the stall occurs, however, this load factor may be reduced toward zero, the factor at which nothing seems to have weight, and the pilot has the feeling of "floating free in space." In the event recovery is made by abruptly moving the elevator control forward, a negative load is created which raises the pilot from the seat. This is a negative wing load and usually is so small that there is little effect on the airplane structure. The pilot should be cautioned, however, to avoid sudden and forceful control movements because of the possibility of exceeding the structural load limits.

During the pullup following stall recovery, however, significant load factors are often encountered. These may be increased by excessively steep diving, high airspeed, and abrupt pullups to level flight. One usually leads to the other, thus increasing the resultant load factor. The abrupt pullup at a high diving speed may easily produce critical loads on structures, and may produce recurrent or secondary stalls by building up the load factor to the point that the speed of the airplane reaches the stalling airspeed during the pullup.

Advanced Maneuvers. Spins, chandelles, lazy eights, and snap maneuvers will not be covered in this handbook. However, before attempting these maneuvers, pilots should be familiar with the airplane being flown, and know whether or not these maneuvers can be safely performed.

Effect of Turbulence on Load Factor Turbulence in the form of vertical air currents can, under certain conditions, cause severe load stress on an airplane wing.

When an airplane is flying at a high speed with a low angle of attack, and suddenly encounters a vertical current of air moving upward, the relative wind changes to an upward direction as it meets the airfoil. This increases the angle of attack of the wing.

If the air current is well defined and travels at a significant rate of speed upward (15 to 30 feet per second), a sharp vertical gust is produced which will have the same effect on the wing as applying sudden sharp back pressure on the elevator control.

All certificated airplanes are designed to withstand loads imposed by turbulence of considerable intensity. Nevertheless, gust load factors increase with increasing airspeed. Therefore it is wise, in extremely rough air, as in thunderstorm or frontal conditions, to reduce the speed to the design maneuvering speed. As a general rule, when severe turbulence is encountered, the

airplane should be flown at the *maneuvering speed* shown in the FAA-approved Airplane Flight Manual, Pilot's Operating Handbook, or placard in the airplane. This is the speed least likely to result in structural damage to the airplane, even if full control travel is used, and yet allows a sufficient margin of safety above stalling speed in turbulent air.

Placarded "never exceed speeds" are determined for smooth air only. High dive speeds or abrupt maneuvering in gusty air at airspeeds above the *maneuvering speed* may place damaging stress on the whole structure of an airplane.

Stress on the structure means stress on any vital part of the airplane. The most common failures due to load factors involve rib structure within the leading and trailing edges of wings.

The cumulative effect of such loads over a long period of time may tend to loosen and weaken vital parts so that actual failure may occur later when the airplane is being operated in a normal manner.

Determining Load Factors in Flight The leverage in the control systems of different airplanes varies; some types are balanced control surfaces while others are not. (A balanced control surface is an aileron, rudder, or elevator designed in such a manner as to put each side of its hinged axis in balance with the other side.) Therefore the pressure exerted by the pilot on the controls cannot be used as a means to determine the load factor produced in different airplanes. Load factors are best judged by feel through experience. They can be measured by an instrument called an accelerometer, but since this instrument is not commonly used in general aviation-type airplanes, developing the ability to judge load factors from the feel of their effect on the body is important. One indication the pilot will have of increased load factor is the feeling of increased body weight. In a 60° bank the body weight would double. A knowledge of the principles outlined above is essential to estimate load factors.

In view of the foregoing discussion on load factors, a few suggestions can be made to avoid overstressing the structure of the airplane:

1. Operate the airplane in conformance with the Pilot's Operating Handbook.
2. Avoid abrupt control usage at high speeds.
3. Reduce speed if turbulence of any great intensity is encountered in flight, or abrupt maneuvers are to be performed.
4. Reduce weight of airplane before flight if intensive turbulence or abrupt maneuvering is anticipated.
5. Avoid turns using an angle of bank in excess of 60°.

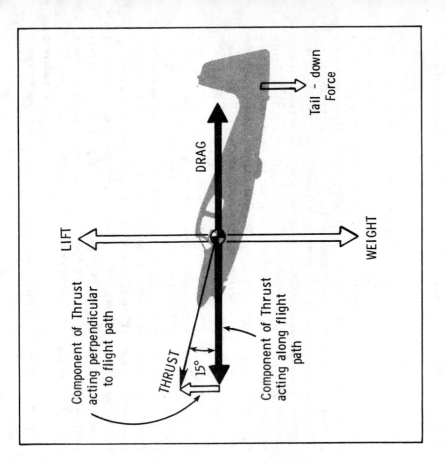

Figure 1-41. *The forces on the airplane in straight-and-level flight at airspeeds slower than cruise.*

Forces Acting on the Airplane When at Airspeeds Slower than Cruise

At a constant cruise airspeed, maintaining straight-and-level flight, the force of thrust and drag acts opposite to each other and parallel to the flightpath. These opposing forces are equal in magnitude. Also, the force of lift is equal in magnitude to the force of weight.

While maintaining straight-and-level flight at constant airspeeds slower than cruise, the opposing forces must still be equal in magnitude, but some of these forces are separated into components. In this flight condition the actual thrust no longer acts parallel and opposite to the flightpath and drag. Actual thrust is inclined upward as illustrated in Fig. 1-41. Note that now thrust has two components; one acting perpendicular to the flightpath in the direction of lift, while the other acts along the flightpath. Because the actual thrust is inclined, its magnitude must be greater than drag if its component of thrust along the flightpath is to equal drag. Also note that a component of thrust acts 90° to the flightpath, and thus acts in the same direction as wing lift. Fig. 1-41 also illustrates that the forces acting upward (wing lift and the component of thrust) equals the forces acting downward (weight and tail down force).

Wing loading (wing lift) is actually less at slow speeds, the actual thrust is greater than drag and wing lift is less than at cruise speed.

To summarize, in straight-and-level flight at slow speeds, the actual thrust is greater than drag and wing lift is less than at cruise speed.

Forces in a Climb

The forces acting on an airplane during a climb are illustrated in Fig. 1-42. When the airplane is in equilibrium, the weight can be resolved into two components: one opposing the lift, and the other acting in the same direction as the drag along the line of the relative wind. The requirements for equilibrium are: the thrust must equal the sum of the drag and the opposing component of the weight; and the lift must equal its opposing component of the weight. The steeper the angle of climb the shorter becomes the length of the component of lift, and simultaneously the component of drag becomes longer. Therefore, the lift requirement decreases steadily as the angle of climb steepens until, in a true vertical climb, if this were possible, the wings would supply no lift and the thrust would be the only force opposing both the drag and the weight, which would be acting downward in opposition.

At a constant power setting, a given rate of climb can be obtained either by climbing steeply at a low airspeed or by climbing on a shallow path at high airspeed. At one extreme, if the airspeed is too low the induced drag rises to a figure at which all thrust available is required to overcome the drag and none is available for climbing. At the other extreme, if the speed is the maximum

obtainable in level flight, again all the power is being used to overcome the drag and there is no rate of climb. Between these two extremes lies a speed, or a small band of speeds, which will achieve the best rate of climb. The best rate of climb is achieved not at the steepest angle, but at some combination of moderate angle and optimum airspeed at which the greatest amount of excess power is available to climb the airplane after the drag has been balanced.

Fig. 1-43 shows that the speed for minimum drag or the lowest point on the power-required curve, although low, is not the lowest possible that can be

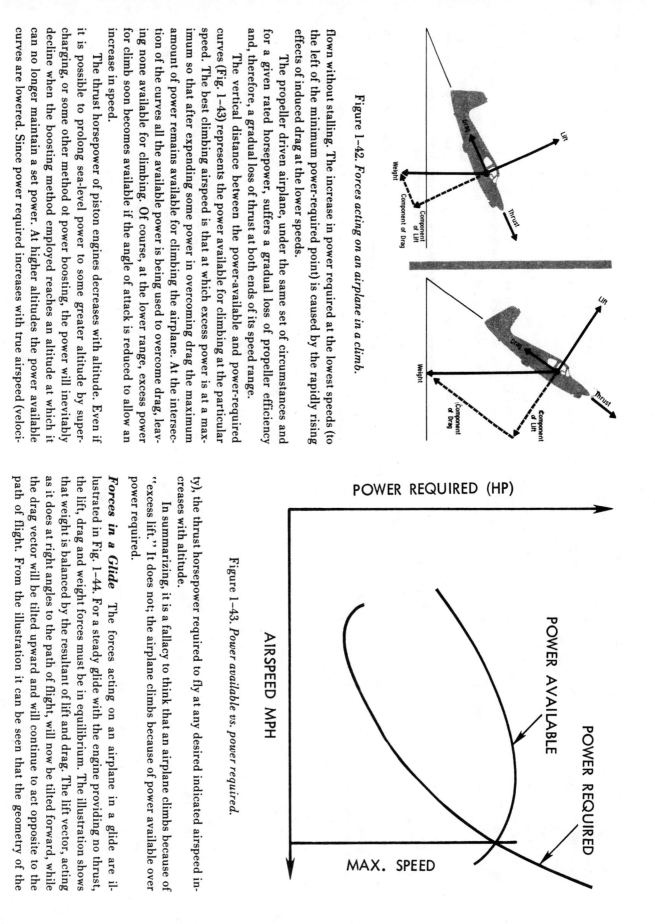

Figure 1–42. *Forces acting on an airplane in a climb.*

Figure 1–43. *Power available vs. power required.*

flown without stalling. The increase in power required at the lowest speeds (to the left of the minimum power-required point) is caused by the rapidly rising effects of induced drag at the lower speeds.

The propeller driven airplane, under the same set of circumstances and for a given rated horsepower, suffers a gradual loss of propeller efficiency and, therefore, a gradual loss of thrust at both ends of its speed range.

The vertical distance between the power-available and power-required curves (Fig. 1–43) represents the power available for climbing at the particular speed. The best climbing airspeed is that at which excess power is at a maximum so that after expending some power in overcoming drag the maximum amount of power remains available for climbing the airplane. At the intersection of the curves all the available power is being used to overcome drag, leaving none available for climbing. Of course, at the lower range, excess power for climb soon becomes available if the angle of attack is reduced to allow an increase in speed.

The thrust horsepower of piston engines decreases with altitude. Even if it is possible to prolong sea-level power to some greater altitude by supercharging, or some other method of power boosting, the power will inevitably decline when the boosting method employed reaches an altitude at which it can no longer maintain a set power. At higher altitudes the power available curves are lowered. Since power required increases with true airspeed (veloci-

ty), the thrust horsepower required to fly at any desired indicated airspeed increases with altitude.

In summarizing, it is a fallacy to think that an airplane climbs because of "excess lift." It does not; the airplane climbs because of power available over power required.

Forces in a Glide   The forces acting on an airplane in a glide are illustrated in Fig. 1–44. For a steady glide with the engine providing no thrust, the lift, drag and weight forces must be in equilibrium. The illustration shows that weight is balanced by the resultant of lift and drag. The lift vector, acting as it does at right angles to the path of flight, will now be tilted forward, while the drag vector will be tilted upward and will continue to act opposite to the path of flight. From the illustration it can be seen that the geometry of the

29

vectors is such that the angle between the lift vector and the resultant is the same as that between the glide path and the horizontal. This angle (X) between the glide path and the horizontal is called the glide angle. Further examination of this diagram will show that as drag is reduced and speed increased, the smaller will be the glide angle; therefore, the steepness of the glide path depends on the ratio of lift to drag. When gliding at the angle of attack for best lift-drag ratio, least drag is experienced, and the flattest glide will result. The lift-drag ratio (L/D) is a measure of the gliding efficiency or aerodynamic cleanness of the airplane. If the L/D is 11/1, it means that lift is 11 times greater than drag.

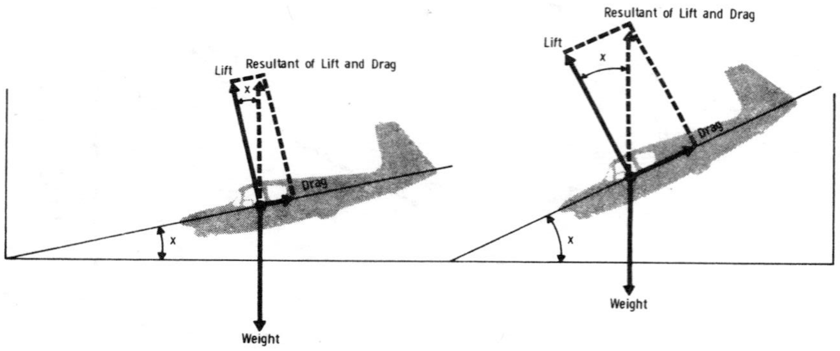

Figure 1–44. *Forces acting on an airplane in a glide.*

If the gliding airplane is flying at an airspeed just above the stall, it is operating at maximum angle of attack and therefore, maximum lift. This, however, does not produce the best glide angle for maximum glide distance because the induced drag at this point is high. By reducing the angle of attack, the airspeed increases and, although lift is less at the lower angle of attack, the airplane travels farther per increment of altitude lost because of greatly reduced drag. The increased range can be accomplished *up to a point,* by decreasing angle of attack and induced drag. At some point the best glide angle will be achieved. If airspeed continues to increase, the parasite drag begins to rise sharply and the airplane will again start losing more altitude per increment of distance traveled. The extreme of this is when the nose is pointed straight down.

It can be shown that best glide distance is obtained when L/D is at max-imum. This optimum condition is determined for each type of airplane and the speed at which it occurs is used as the recommended best range glide speed for the airplane. It will vary somewhat for different airplane weights, so the airspeed for a representative operational condition is generally selected.

If several instances of the optimum glide path were plotted by an observer on the ground under varying conditions of wind, they would be found to be inconsistent. However, the actual gliding angle of the airplane with respect to the moving air mass remains unchanged. Starting from a given altitude, a glide into the wind at optimum glide airspeed covers less distance over the ground than a glide downwind. Since in both cases the rate of descent is the same, the measured angle as seen by a ground observer is governed only by the groundspeed, being steeper at the lower groundspeed when gliding into the wind. The effect of wind, therefore, is to decrease range when gliding with a headwind component, and to increase it when gliding downwind. The endurance of the glide is unaffected by wind.

Variations in gross weight do not affect the gliding angle provided the optimum indicated airspeed for each gross weight is used. The fully loaded airplane will sink faster but at a greater forward speed, and, although it would reach the ground much quicker, it would have traveled exactly the same distance as the lighter airplane, and its glide angle would have been the same.

An inspection of Fig. 1–44 will show that an increase in the weight factor is equivalent to adding thrust to the weight component along the glide path. This means more speed and, therefore, more lift and drag which lengthen the resultant vector until the geometric balance of the diagram is restored. This is done without affecting the gliding angle. The higher speed corresponding to the increased weight is provided automatically by the larger component of weight acting along the glide path, and this component grows or diminishes in proportion to the weight. Since the gliding angle is unaffected, range also is unchanged.

Although range is not affected by changes in weight, endurance decreases with addition of weight and increases with reduction of weight. If two airplanes having the same lift-drag ratio, but different weights, start a glide from the same altitude, the heavier airplane, gliding at a higher airspeed, will cover the distance between the starting point and touchdown in a shorter time. Both, however, will cover the same distance. Therefore, the endurance of the heavier airplane is less.

Turns During Flight

Many pilots do not reach a complete understanding of what makes an airplane turn. Such an understanding is certainly worthwhile, since many accidents occur as a direct result of losing control of the airplane while in turning flight.

In review, the airplane is capable of movement around the three axes. It can be pitched around the lateral axis, rolled around the longitudinal axis, and yawed around the vertical axis. Yawing around the vertical axis causes most misunderstanding about how and why an airplane turns. First, it should be kept in mind that the *rudder does not turn* the airplane in flight.

Although most pilots know that an airplane is banked to make a turn, few know the reason why. The answer is quite simple. The airplane must be banked because the same force (lift) that sustains the airplane in flight is used to make the airplane turn. The airplane is banked and back elevator pressure is applied. This changes the direction of lift and increases the angle of attack on the wings, which increases the lift. The increased lift pulls the airplane around the turn. The amount of back elevator pressure applied, and therefore the amount of lift, varies directly with the angle of bank used. As the angle of bank is steepened the amount of back elevator pressure must be increased to hold altitude.

In level flight the force of lift acts opposite to and exactly equal in magnitude to the force of gravity. Gravity tends to pull all bodies to the center of the earth, therefore this force *always* acts in a vertical plane with respect to the earth. On the other hand, total lift always acts perpendicular to the relative wind, which for the purpose of this discussion is considered to be the same as acting perpendicular to the lateral axis of the wind.

With the wings level, lift acts directly opposite to gravity. However, as the airplane is banked, gravity still acts in a vertical plane, but lift will now act in an inclined plane.

As illustrated in Fig. 1–45, the force of lift can be resolved into two components—vertical and horizontal. During the turn entry the vertical component of lift still opposes gravity, and the horizontal component must overcome centrifugal force; consequently, the total lift must be sufficient to counteract both of these forces. The airplane is then pulled around the turn, not sideways, because the tail section acts as a weathervane which continually keeps the airplane streamlined with the curved flightpath.

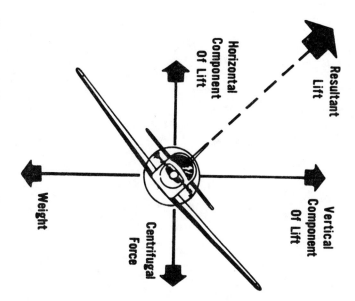

Figure 1–45. *Forces acting on an airplane in a turn.*

Also note in Fig. 1–45 that as the turn develops, centrifugal force will act opposite to the horizontal component of lift and the vertical component of lift will act opposite to gravity. The total resultant lift acts opposite to the total resultant load. So long as these opposing forces are equal to each other in magnitude the airplane will maintain a constant rate of turn. If the pilot moves the controls in such a manner as to change the magnitude of any of the forces, the airplane will accelerate or decelerate in the direction of the applied force. This will result in changing the rate at which the airplane turns.

CHAPTER II—AIRPLANES AND ENGINES

One of the most important activities in promoting safety in aviation is the airworthiness certification of airplanes. Each airplane certificated by the Federal Aviation Administration has been manufactured under rigid specifications of design, materials, workmanship, construction, and performance. This certification process provides adequate assurance that the airplane will not fail from a structural standpoint if the airplane is properly maintained and flown within the limitations clearly specified. However, this may not be true if the airplane is abused, improperly maintained, or flown without regard to its limitations.

The goal of airplane designers and manufacturers is to obtain maximum efficiency, combined with adequate strength. Excessive strength requires additional weight which lowers the efficiency of the airplane by reducing its speed and the amount of useful load it can carry.

This chapter covers airplane structure, including flight control systems, wing flaps, landing gear, engine operation, engine accessories, and associated engine instruments. Also included is material related to aircraft documents, aircraft maintenance, and inspection procedures.

Airplane Structure

As stated in Chapter I, the required structural strength is based on the intended use of the airplane. An airplane which is to be used for normal flying does not need the strength of an airplane which is intended to be used for acrobatic flight or other special purposes, some of which involve severe inflight stresses.

Numerous wing designs have been developed in an effort to determine the best type for a specific purpose. Basically, all wings are similar to those used by the Wright brothers and other pioneers. Modifications have been made, however, to increase lifting capacity, reduce friction, increase structural strength, and generally improve flight characteristics. Wing designs are subjected to thorough analysis before being approved for use on certificated airplanes. Strength tests determine the effect of strains and stresses which might be encountered in flight.

Airplane strength is measured basically by the total load which the wings are capable of carrying without permanent damage to the wing structure. The load imposed upon the wings depends upon the type of flight in which the

airplane is engaged. The wing must support not only the weight of the airplane, but the additional loads caused during certain flight maneuvers such as turns and pullouts from dives. Turbulent air also creates additional loads and these loads increase as the severity of the turbulence increases.

To permit utmost efficiency of construction without sacrificing safety, the FAA has established several categories of airplanes with minimum strength requirements for each. Limitations of each airplane are available to the pilot through markings on instruments, placards on instrument panels, operating limitations attached to airworthiness certificates, Aircraft Flight Manual, or Pilot's Operating Handbook.

Flight Control Systems The flight control systems in most general aviation airplanes consist of the cockpit controls, cables, pulleys, and linkages connected to the movable control surfaces outside the airplane.

There are three primary and two secondary flight control systems. The primary flight control systems consist of the elevator, aileron, and rudder, which are essential in controlling the aircraft. The secondary control systems consist of the trim tabs and wing flaps. The trim tabs enable the pilot to trim out control pressures, and the flaps enable the pilot to change the lifting characteristics of the wing and also to decrease the speed at which the wing stalls. All of the flight control systems, except the wing flaps, were discussed in Chapter I. The flaps will be discussed at this point.

Wing Flaps Wing flaps are a movable part of the wing, normally hinged to the inboard trailing edge of each wing. Flaps are extended or retracted by the pilot. Extending the flaps increases the wing camber, wing area (some types), and the angle of attack of the wing. This increases wing lift and also increases induced drag. The increased lift enables the pilot to make steeper approaches to a landing without an increase in airspeed. Their use at recommended settings also provides increased lift under certain takeoff conditions. When the flaps are no longer needed, they can be retracted.

Pilots are cautioned to operate the flaps within the airspeed limitations set forth for the particular airplane being flown. If the speed limitations are exceeded, the increased drag forces created by extending the flaps could result in structural damage to the airplane.

Fig. 2–1 shows the three types of flaps in general use: (A) the plain or simple flap is a portion of the trailing edge of the wing on a hinged pivot which

steeper angle of descent without an increase in airspeed. Extended flaps also permit a slower speed to be used on an approach and landing, thus reducing the distance of the landing roll.

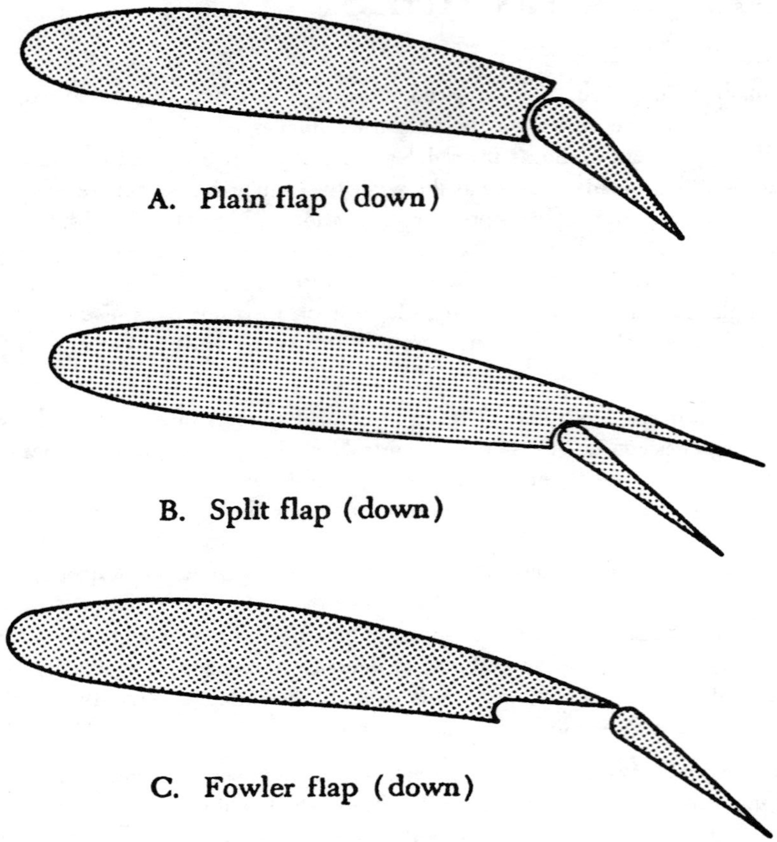

Landing Gear   The landing gear system supports the airplane during the takeoff run, landing, taxiing, and when parked. These ground operations require that the landing gear be capable of steering, braking, and absorbing shock.

A steerable nose gear or tailwheel permits the airplane to be controlled by the pilot throughout all operations while on the ground. Individual brakes installed on each main wheel permit the pilot to use either brake individually as an aid to steering or, by applying both brakes simultaneously, the pilot can decelerate or stop the airplane. Hydraulic shock struts or leaf springs are installed in the various types of landing gear systems to absorb the impact of landings, or the shock of taxiing over rough ground.

There are two basic types of landing gear used on light airplanes. These are the conventional landing gear and the tricycle landing gear.

The conventional landing gear, which was used on most airplanes manufactured years ago, is still used on some airplanes designed for operations on rough fields. This landing gear system consists of two main wheels and a tailwheel. Shock absorption is usually provided on the main landing gear by inflated tires and shock absorbers while it is provided on the tailwheel by a spring assembly to which the tailwheel is bolted. The tailwheel is usually steerable by the rudder pedals through at least 15 degrees on each side of a center point beyond which it becomes full swiveling.

The tricycle landing gear is used on most airplanes produced today. This gear has advantages over the conventional gear because it provides easier ground handling characteristics. The main landing gear is constructed similar to the main landing gear on the conventional system, but is located further rearward on the airplane. The nose gear is usually steerable by the rudder pedals through at least 10 degrees on each side of the center, beyond which it becomes free swiveling through a limited range. This permits sharper turns during taxiing. Shock absorption is provided on the nose gear by a shock strut.

Some light airplanes are equipped with retractable landing gear. Retracting the gear reduces the drag, and increases the airspeed without additional power. The landing gear normally retracts into the wing or fuselage through an opening which is covered by doors after the gear is retracted. This provides for the unrestricted flow of air across the opening which houses the gear. The retraction or extension of the landing gear is accomplished either electrically

A. Plain flap (down)

B. Split flap (down)

C. Fowler flap (down)

Figure 2-1. *Wing flaps.*

allows the flap to be moved downward, thereby changing the chord line, angle of attack, and the camber of the wing; (B) the split flap is a hinged portion of the bottom surface of the wing only, which when extended increases the angle of attack by changing the chord line; (C) the Fowler flap, which when extended not only tilts downward but also slides rearward on tracks. This increases angle of attack, wing camber, and wing area, thereby providing added lift without significantly increasing drag.

With all three types of flaps the practical effect of the flap is to permit a

or hydraulically by landing gear controls from within the cockpit to indicate whether the wheels are extended and locked, or retracted. In nearly all retractable landing gear installations, a system is provided for emergency gear extension in the event landing gear mechanism fails to lower the gear.

Electrical System

Electrical energy is required to operate navigation and communication radios, lights, and other airplane equipment.

Many airplanes in the past were not equipped with an electrical system. They were equipped with a magneto system which supplied electrical energy to the engine ignition system only. Modern airplanes still use an independent magneto system, but in addition are equipped with an electrical system. The magneto system does not depend upon the airplane electrical system for operation. In other words, the airplane electrical system can be turned off in flight and the engine will continue to operate efficiently, utilizing the electrical energy provided by the magnetos.

Most light airplanes are equipped with a 12-volt direct-current electrical system. Larger airplanes are equipped with a 24-volt system to provide an electrical reserve capacity for more complex systems, including additional electrical energy for starting.

A basic airplane electrical system consists of the following components:

1. Alternator or generator.
2. Battery.
3. Master switch or battery switch.
4. Bus bar, fuses, and circuit breakers.
5. Voltage regulator.
6. Ammeter.
7. Starting motor.
8. Associated electrical wiring.
9. Accessories.

Engine-driven generators or alternators supply electric current to the electrical system and also maintain a sufficient electrical charge in the battery which is used primarily for starting.

There are several basic differences between generators and alternators. Most generators will not produce a sufficient amount of electrical current at low engine r.p.m.'s to operate the entire electrical system. Therefore, during operations at low engine r.p.m.'s the electrical needs must be drawn from the battery, which in a short time may be depleted.

An alternator, however, produces a sufficient amount of electrical current at slower engine speeds by first producing alternating current which is con-verted to direct current. Another advantage is that the electrical output of an alternator is more constant throughout the ranges of engine speeds. Alternators are also lighter in weight, less expensive to maintain, and less prone to become overloaded during conditions of heavy electrical loads.

Electrical energy stored in a battery provides a source of electricity for use in the event the alternator or generator fails.

Some airplanes are equipped with receptacles to which external auxiliary power units (APU) can be connected to provide electrical energy for starting. These are very useful, especially during cold weather starting. Care must be exercised in starting engines using auxiliary power units when the battery is dead. If this is done, electrical energy will be forced into the dead battery, causing the battery to overheat and possibly explode, resulting in damage to the airplane.

A master switch is installed on airplanes to provide a means for the pilot to turn the electrical system "on" and "off." Turning the master switch "on" provides electrical energy to all the electrical equipment circuits with the exception of the ignition system. Although additional electrical equipment may be found in some airplanes, the following lists the equipment most commonly found which uses the electrical system for its source of energy:

1. Position lights.
2. Landing lights.
3. Taxi lights.
4. Anticollision lights.
5. Interior cabin lights.
6. Instrument lights.
7. Radio equipment.
8. Turn indicator.
9. Fuel gauges.
10. Stall warning system.
11. Pitot heat.
12. Cigarette lighter.

Some airplanes are equipped with a battery switch which controls the electrical power to the airplane in a manner similar to the master switch. In addition, an alternator switch is installed which permits the pilot to exclude the alternator from the electrical system in the event of alternator failure. With the alternator switch "off," the entire electrical load is placed on the battery. Therefore, all nonessential electrical equipment should be turned off to conserve the energy stored in the battery.

of power. This simplifies the wiring system and provides a common point from which voltage can be distributed throughout the system (Fig. 2-2).

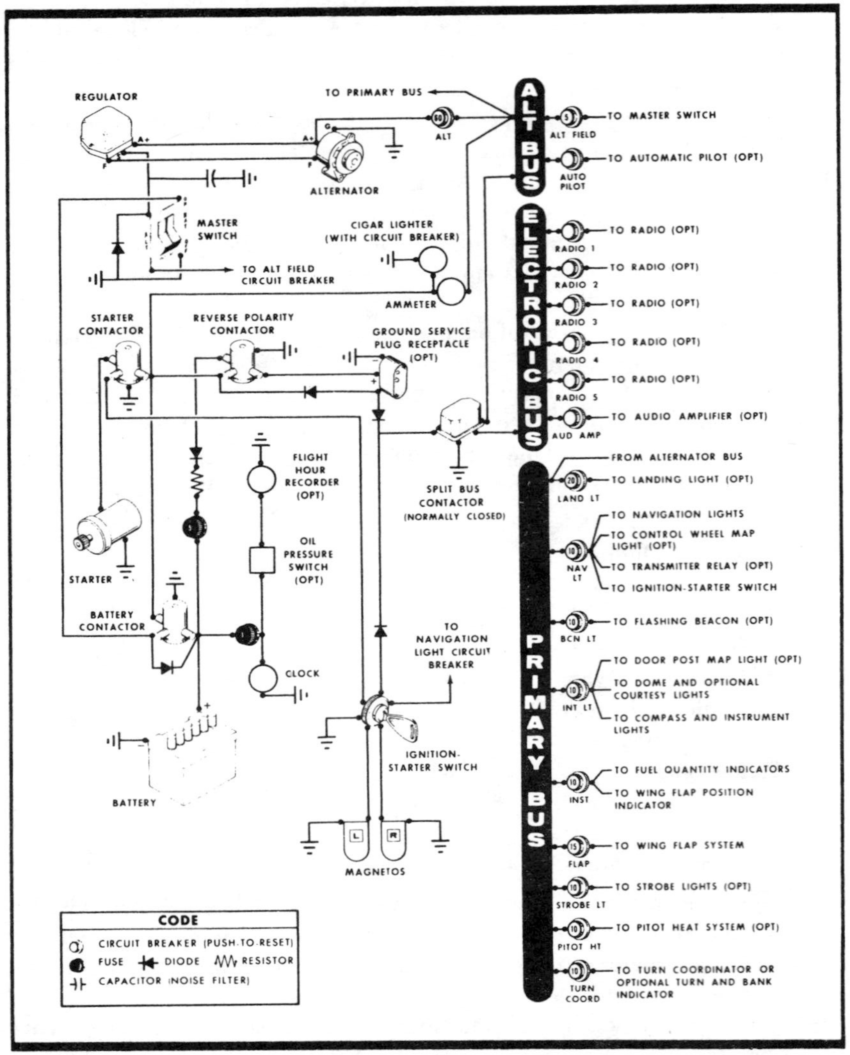

Fuses or circuit breakers are used in the electrical system to protect the circuits and equipment from electrical overload. Spare fuses of the proper amperage limit should be carried in the airplane to replace defective or blown fuses. Circuit breakers have the same function as a fuse but can be manually reset, rather than replaced, if an overload condition occurs in the electrical system. Placards at the fuse or circuit breaker location identify the circuit by name and if fuses are used, show the amperage limit of the fuse (Fig. 2-3).

An ammeter is an instrument used to monitor the performance of the airplane electrical system. Not all airplanes are equipped with an ammeter. Some are equipped with a light which, when lighted, indicates a discharge in the system as a generator/alternator malfunction.

An ammeter shows if the generator/alternator is producing an adequate supply of electrical power to the system by measuring the amperes of electricity. This instrument also indicates whether the battery is receiving an electrical charge. The face of most ammeters is designed with a zero point in the upper center of the dial and a plus value to the right of center; a negative value is to the left (Fig. 2-4). A vertical needle swings to the right or left, depending upon the performance of the electrical system. If the needle indicates a plus value, it means that the battery is being charged. After power is drawn from the bat-

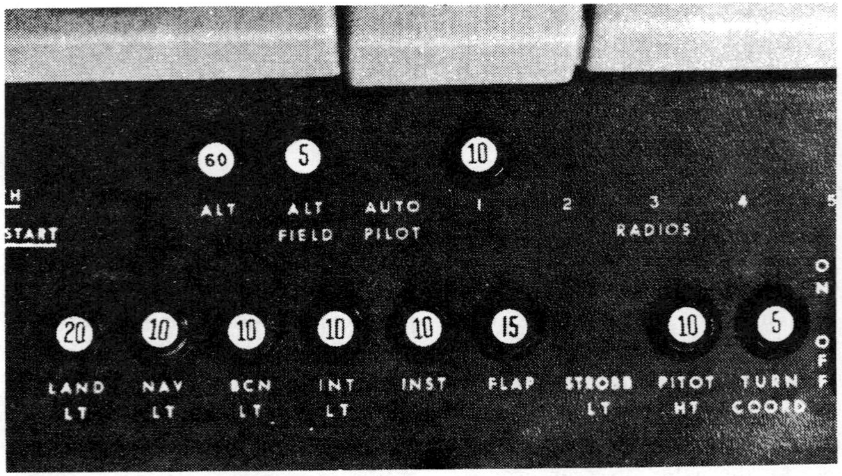

Figure 2-3. *Circuit breaker panel.*

A bus bar is used as a terminal in the airplane electrical system to connect the main electrical system to the equipment using electricity as a source

Figure 2-2. *Electrical system schematic.*

If the needle indicates a minus value, it means that the generator or alternator output is inadequate and energy is being drawn from the battery to supply the system. This could be caused by either a defective alternator/generator or by an overload in the system, or both. Full scale ammeter discharge or rapid fluctuation of the needle usually means generator/alternator malfunction. If this occurs, the pilot should cut the generator/alternator out of the system and conserve battery power by reducing the load on the electrical system.

A voltage regulator controls the rate of charge to the battery by stabilizing the generator or alternator electrical output. The generator/alternator voltage output is usually slightly higher than the battery voltage. For example, a 12-volt battery system would be fed by a generator/alternator system of approximately 14 volts. The difference in voltage keeps the battery charged.

An inverter is installed on some airplanes to change direct current to alternating current.

Figure 2–4. *Ammeter.*

Engine Operation

Knowledge of a few general principles of engine operation will help the pilot obtain increased dependability and efficiency from the engine and, in many instances, this knowledge will help in avoiding engine failure.

In this short chapter, it is impractical to discuss in detail the various types of engines and the finer points of operation which can be learned only through experience. Information from the manufacturer's instruction manual; familiarity with the operating limitations for the airplane engine; and specific advice from a flight instructor, combined with the information contained within this section, should provide adequate information to operate an airplane engine satisfactorily.

How an Engine Operates   Most light airplane engines are internal combustion of the reciprocating type which operate on the same principle as automobile engines. They are called reciprocating engines because certain parts move back and forth in contrast to a circular motion such as a turbine. Some smaller airplanes are equipped with turbine engines, but this type will not be discussed in this handbook. As shown in Fig. 2–5, the reciprocating engine consists of cylinders, pistons, connecting rods, and a crankshaft. One end of a connecting rod is attached to a piston and the other end to the crankshaft. This converts the straight-line motion of the piston to the rotary motion of the crankshaft, which turns the propeller. At the closed end of the cylinder there are normally two spark plugs which ignite the fuel, and two openings over which valves open and close. One valve (the intake valve) when open admits the mixture of fuel and air, and the other (the exhaust valve) when open permits the burned gases to escape. For the engine to complete one cycle, the piston must complete four strokes. This requires two revolutions of the crankshaft. The four strokes are the intake, compression, power, and exhaust. The following describes one cycle of engine operation.

Diagram A of Fig. 2–6 shows the piston moving away from the cylinder head. The intake valve is opened and the fuel/air mixture is drawn into the cylinder. *This is the intake stroke.*

Diagram B shows the piston returning to the top of the cylinder. Both valves are closed, and the fuel/air mixture is compressed. *This is the compression stroke.*

Diagram C shows that when the piston is approximately at the top of the cylinder head, a spark from the plugs ignites the mixture, which burns at a controlled rate. Expansion of the burning gas exerts pressure on the piston, forcing it downward. *This is the power stroke.*

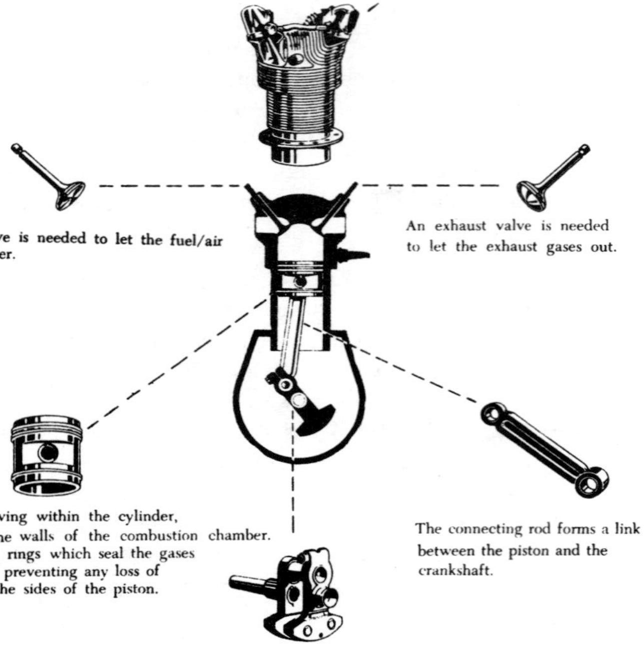

are added and the power strokes are timed to occur at successive intervals during the revolution of the crankshaft.

Aircraft engines are classified by the various ways the cylinders are arranged around the central crankcase. Most general aviation airplane engines are classed as the horizontally opposed, which have the cylinder banks arranged in two rows, directly opposite to each other and using the same crankshaft.

Larger and more powerful reciprocating engines are classed as radial engines. In these engines, the cylinders are placed in a circular pattern around the crankcase, which is placed in the center of the circle.

Other engine classifications are the in-line engine with the cylinders placed in one straight row, and the "vee" type with the cylinders placed in two rows forming a "V" similar to the V-8 engine used in automobiles.

Cooling System The burning fuel within the cylinders produces intense heat, most of which is expelled through the exhaust. Much of the remaining heat, however, must be removed to prevent the engine from overheating. In practically all automobile engines, excess heat is carried away by water circulating around the cylinder walls.

Most light airplane engines are air cooled. The cooling process is accomplished by cool air being forced into the engine compartment through openings in front of the engine cowl. This ram air is routed by baffles over fins attached to the engine cylinders, and other parts of the engine, where the air absorbs the engine heat. Expulsion of the hot air takes place through one or two openings at the rear bottom of the engine cowling.

Some airplanes are equipped with a device known as cowl flaps which are used to control engine temperatures during various flight operations. Cowl flaps are hinged covers which fit over the opening through which the hot air is expelled. By adjusting the cowl flap opening, the pilot can regulate the engine temperature during flight. If the engine temperature is low, the cowl flap can be closed, thereby restricting the flow of expelled hot air and increasing engine temperature. If the engine temperature is high, the cowl flaps can be opened to permit a greater flow of air through the system, thereby decreasing the engine temperature. Usually during low airspeed and high power operations such as takeoffs and climbs, the cowl flaps are opened. During higher speed and lower power operations such as cruising flight and descents, the cowl flaps are closed.

Under normal operating conditons in airplanes not equipped with cowl flaps, the engine temperature can be controlled by changing the airspeed or

Every internal combustion engine must have certain basic parts in order to change heat into mechanical energy.

The cylinder forms a part of the chamber in which the fuel is compressed and burned.

An exhaust valve is needed to let the exhaust gases out.

An intake valve is needed to let the fuel/air into the cylinder.

The piston, moving within the cylinder, forms one of the walls of the combustion chamber. The piston has rings which seal the gases in the cylinder, preventing any loss of power around the sides of the piston.

The connecting rod forms a link between the piston and the crankshaft.

The crankshaft and connecting rod change the straight line motion of the piston to a rotary turning motion. The crankshaft in an aircraft engine also absorbs the power or work from all the cylinders and transfers it to the propeller.

Figure 2–5. *Basic parts of a reciprocating engine.*

Diagram D shows that just before the piston completes the power stroke the exhaust valve starts to open, and the burned gases are forced out as the piston returns to the top of the cylinder. *This is the exhaust stroke.* The cycle is then ready to begin again as shown in Diagram A.

From this description, notice that each cylinder of the engine delivers power only once in every four strokes of the piston or every two revolutions of the crankshaft. The momentum of the crankshaft carries the piston through the other three strokes although the diagram shows the action of only one cylinder. To increase power and gain smoothness of operation, other cylinders

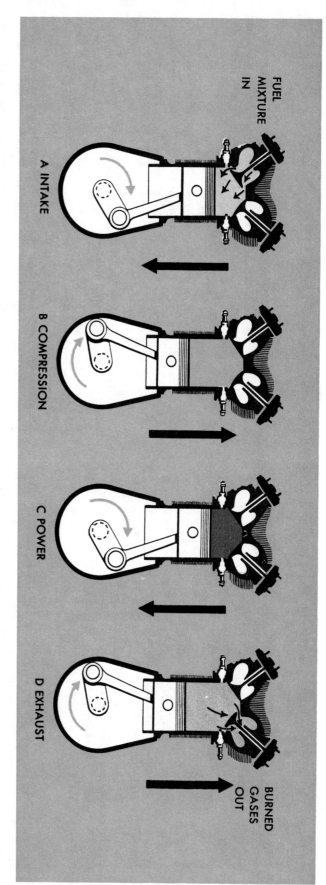

Figure 2-6. *Four strokes of the piston produce: (A) fuel mixture (light blue) is drawn into cylinder by downward stroke, (B) mixture (darker blue) is compressed by upward stroke, (C) spark ignites mixture (red) forcing piston downward and producing power that turns propeller, (D) burned gases (light red) pushed out of cylinder by upward stroke.*

the power output of the engine. High engine temperatures can be decreased by increasing the airspeed and/or reducing the power.

The oil temperature gauge indicates the temperature of the oil which is heated by the engine; therefore this gauge gives an indirect and delayed indication of rising engine temperatures. However, the oil temperature gauge should be used for determining engine temperature if this is the only means available.

Many airplanes are equipped with a cylinder head temperature gauge. This is an additional instrument which will indicate a direct and immediate engine temperature change. This instrument is calibrated in degrees Celsius or Fahrenheit, and is usually color coded with a green arc to indicate the normal operating range. A red line on the instrument indicates maximum allowable engine temperature.

To avoid excessive cylinder head temperatures a pilot can open the cowl flaps, increase airspeed, enrich the mixture, or reduce power. Any of these procedures will aid in reducing the engine temperature. Establishing a shallow climb increases the airflow through the cooling system, preventing excessively high engine temperatures.

When an airplane engine is operated on the ground, very little air flows past the cylinders (particularly if the engine is closely cowled) and overheating is likely to occur. Overheating may also occur during a prolonged climb, because the engine at this time is usually developing high power at relatively slow airspeed.

Operating the engine at higher than its designed temperature can cause loss of power, excessive oil consumption, and detonation. It will also lead to serious permanent damage, scoring the cylinder walls, damaging the pistons and rings, and burning and warping the valves. To aid the pilot in avoiding excessive temperatures, engine temperature instruments in the cockpit should be monitored in flight.

Ignition System The function of the ignition system is to provide a spark to ignite the fuel/air mixture in the cylinder. The magneto ignition system is used on most modern aircraft engines because it produces a hotter spark at high engine speeds than the battery system used in automobiles. Also, it does not depend upon an external source of energy such as the electrical system. Magnetos are self-contained engine driven units supplying ignition current. However, the magneto must be actuated by rotating the engine before current is supplied to the ignition system. The aircraft battery furnishes electrical power to operate the starter system; the starter system actuates the rotating element of the magneto; and the magneto then furnishes the spark to each cylinder to ignite the fuel/air mixture. After the engine starts, the starter system is disengaged, and the battery no longer has any part in the actual operation of the engine. If the battery (or master) switch is turned OFF, the engine will continue to run. However, this should not be done since battery power is necessary at low engine r.p.m. to operate other electrical equipment (radio, lights, etc.) and, when the generator or alternator is operating, the battery will be storing up a charge, if not already fully charged.

Most aircraft engines are equipped with a dual ignition system; that is, two magnetos to supply the electrical current to two spark plugs for each combustion chamber. One magneto system supplies the current to one set of plugs; the second magneto system supplies the current to the other set of plugs. This is the reason that the ignition switch has four positions: OFF, LEFT, RIGHT, and BOTH. With the switch in the "L" or "R" position, only one magneto is supplying current and only one set of spark plugs is firing. With the switch in the BOTH position, both magnetos are supplying current and both sets of spark plugs are firing. The main advantages of the dual system are:

1. Increased safety. In case one magneto system fails, the engine may be operated on the other system until a landing can be made.

NOTE: To ensure that both ignition systems are operating properly, each system is checked during the engine runup prior to flight. This check should be accomplished in accordance with the manufacturer's recommendations in the Aircraft Flight Manual or Pilot's Operating Handbook.

2. Improved burning and combustion of the mixture, and consequently improved performance.

It is important to turn the ignition switch to "BOTH" for flight and completely "OFF" when shutting down the engine after flight. Even with the electrical master switch "OFF" and the ignition switch on either "BOTH" or "LEFT" or "RIGHT" magnetos, the engine could fire if the propeller is moved from outside the airplane. Also, if the magneto switch ground wire is disconnected the magneto is "ON" even though the ignition switch is in the "OFF" position.

Fuel System The function of the fuel system is to provide a means of storing fuel in the airplane and transferring this fuel to the airplane engine. Fuel systems are classified according to the method used to furnish fuel to the engine from the fuel tanks. The two classifications are the "gravity feed" and the "fuel pump system."

The gravity feed system utilizes the force of gravity to transfer the fuel from the tanks to the engine. This system can be used on high-wing airplanes if the fuel tanks are installed in the wings. This places the fuel tanks above the carburetor and the fuel is gravity fed through the system and into the carburetor.

If the design of the airplane is such that gravity cannot be used to transfer fuel, fuel pumps are installed. This is true on low-wing airplanes where the fuel tanks in the wings are located below the carburetor.

Two fuel pump systems are used on most airplanes. The main pump system is engine driven and an auxiliary electric driven pump is provided for use in the event the engine pump fails. The auxiliary pump, commonly known as the "boost pump," provides added reliability to the fuel system, and is also used as an aid in engine starting. The electric auxiliary pump is controlled by a switch in the cockpit.

Because of variation in fuel system operating procedures, the pilot should consult the Aircraft Flight Manual or Pilot's Operating Handbook for specific operating procedures.

Fuel Tanks, Selectors, and Strainers Most airplanes are designed to use space in the wings to mount fuel tanks. All tanks have filler openings which are covered by a cap. This system also includes lines connecting to the engine, a fuel gauge, strainers, and vents which permit air to replace the fuel consumed during flight. Fuel overflow vents are provided to discharge fuel in the event the fuel expands because of high temperatures. Drain plugs or valves (sumps) are located at the bottom of the tanks from which water and other sediment can be drained from the tanks.

Fuel lines pass through a selector assembly located in the cockpit which provides a means for the pilot to turn the fuel "off," "on," or to select a particular tank from which to draw fuel. The fuel selector assembly may be a simple on/off valve, or a more complex arrangement which permits the pilot to select individual tanks or use all tanks at the same time.

Many airplanes are equipped with fuel strainers, called sumps, located at the low point in the fuel lines between the fuel selector and the carburetor. The sumps filter the fuel and trap water and sediment in a container which can be drained to remove foreign matter from the fuel.

Fuel Primer   A manual fuel primer is installed in some airplanes to aid in starting the engine, particularly when the weather is cold. Activating the primer draws fuel from the tanks and vaporizes the fuel directly into one or two of the cylinders through small fuel lines. When engines are cold and do not generate sufficient heat to vaporize the fuel, the primer is used not only to start the engine, but to keep the engine running until sufficient engine heat is generated.

Fuel Pressure Gauge   If a fuel pump is installed in the fuel system, a fuel pressure gauge is also included. This gauge indicates the pressure in the fuel lines. The normal operating pressure can be found in the airplane operating manual or on the gauge by color coding.

Induction, Carburetion, and Injection Systems   In reciprocating aircraft engines, the function of the induction system is to complete the process of taking in outside air, mixing it with fuel, and delivering this mixture to the cylinders. The system includes the air scoops and ducts, the carburetor or fuel injection system, the intake manifold, and (if installed) the turbo or super-chargers.

Two types of induction systems are commonly used in light airplane engines: (1) carburetor system, which mixes the fuel and air in the carburetor before this mixture enters the intake manifold, and (2) fuel injection system in which the fuel and air are mixed just prior to entering each cylinder. The fuel injection system does not utilize a carburetor.

The carburetor system uses one of two types of carburetor: (1) the float type carburetor which is generally installed in airplanes equipped with small horsepower engines, and (2) the pressure type, used in higher horsepower engines. The pressure type will not be discussed in this book, but many aspects of each are similar.

In the operation of the carburetor system the outside air first flows through an air filter, usually located at an air intake in the front part of the engine cowling. This filtered air flows into the carburetor and through a venturi, a narrow throat in the carburetor. When the air flows rapidly through the venturi, a low pressure area is created, which forces the fuel to flow through a main fuel jet located at the throat and into the airstream where it is mixed with the flowing air (Fig. 2–7).

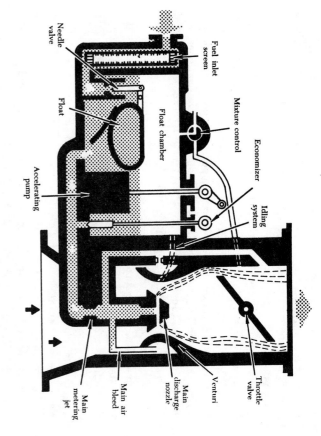

Figure 2–7. *A float-type carburetor.*

The fuel/air mixture is then drawn through the intake manifold and into the combustion chambers where it is ignited. The "float-type carburetor" acquires its name from a float which rests on fuel within the float chamber. A needle attached to the float opens and closes an opening in the fuel line. This meters the correct amount of fuel into the carburetor, depending upon the position of the float, which is controlled by the level of fuel in the float chamber. When the level of the fuel forces the float to rise, the needle closes the fuel opening and shuts off the fuel flow to the carburetor. It opens when the engine requires additional fuel.

Mixture Control   A "mixture control" in the cockpit is provided to change the fuel flow to the engine to compensate for varying air densities as the airplane changes altitude.

Carburetors are normally calibrated at sea level pressure to meter the correct amount of fuel with the mixture control in a "Full Rich" position. As altitude increases, air density decreases. This means that a given volume of air does not weigh as much at higher altitudes because it does not contain as many air molecules. As altitude increases, the weight of air decreases, even

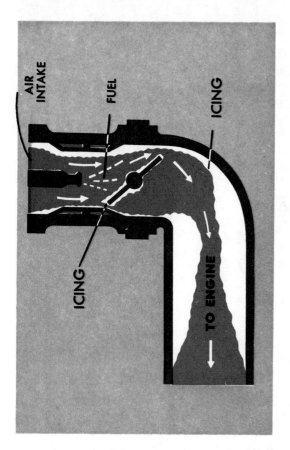

Figure 2–8. *Formation of ice (white) in the fuel intake system may reduce or block fuel flow (red) to the engine.*

though the volume of air entering the carburetor remains the same. To compensate for this difference the mixture control is used to adjust the ratio of fuel-to-air mixture entering the combustion chamber. This also regulates fuel consumption.

If the fuel/air mixture is too rich, i.e., too much fuel in terms of the weight of air, excessive fuel consumption, rough engine operation, and appreciable loss of power will occur. Because of excessive fuel, a cooling effect takes place which causes below normal temperatures in the combustion chambers. This cooling results in spark plug fouling. Conversely, operation with an excessively lean mixture, i.e., too little fuel in terms of the weight of air, will result in rough engine operation, detonation, overheating, and a loss of power.

To summarize, as the airplane climbs and the atmospheric pressure decreases, there is a corresponding decrease in the weight of air passing through the induction system. The volume of air, however, remains constant, and since it is the volume of airflow which determines the pressure drop at the throat of the venturi, the carburetor tends to meter the same amount of fuel to this thin air as to the dense air at sea level. Therefore, the mixture becomes richer as the airplane gains altitude. The mixture control prevents this by decreasing the rate of fuel discharge to compensate for the decrease in air density. However, the mixture must be enriched when descending from altitude.

Follow the manufacturer's recommendation for the particular airplane being flown to determine the proper leaning/enriching procedures.

Carburetor Icing. Carburetor icing is one cause of engine failure. The vaporization of fuel, combined with the expansion of air as it flows through the carburetor, causes a sudden cooling of the mixture. The temperature of the air passing through the carburetor may drop as much as 15° C. (60° F.) within a fraction of a second. Water vapor in the air is "squeezed out" by this cooling and, if the temperature in the carburetor reaches 0° C. (32° F.) or below, the moisture will be deposited as frost or ice inside the carburetor passages. Even a slight accumulation of this deposit will reduce power and may lead to complete engine failure, particularly when the throttle is partly or fully closed (Fig. 2–8).

Conditions Conducive to Carburetor Icing. On dry days, or when the temperature is well below freezing, the moisture in the air is not generally enough to cause trouble. But if the temperature is between −7° C. (20° F.) and 21° C. (70° F.), with visible moisture or high humidity, the pilot should be constantly on the alert for carburetor ice. During low or closed throttle settings, an engine is particularly susceptible to carburetor icing.

Indications of Carburetor Icing. For airplanes with fixed-pitch propellers, the first indication of carburetor icing is loss of r.p.m. For airplanes with controllable pitch (constant-speed) propellers, the first indication is usually a drop in manifold pressure. In both cases, a roughness in engine operation may develop later. There will be no reduction in r.p.m. in airplanes with constant-speed propellers, since propeller pitch is automatically adjusted to compensate for the loss of power, thus maintaining constant r.p.m.

Use of Carburetor Heat. The carburetor heater is an anti-icing device that preheats the air before it reaches the carburetor. This preheating can be used to melt any ice or snow entering the intake, to melt ice that forms in the carburetor passages (provided the accumulation is not too great), and to keep the fuel mixture above the freezing temperature to prevent formation of carburetor ice.

When conditions are conducive to carburetor icing during flight, periodic checks should be made to detect its presence. If detected, full carburetor heat should be applied immediately, and it should be left in the "on" position until the pilot is certain that all the ice has been removed. If ice was present, applying partial heat or leaving heat on for an insufficient time might aggravate the situation.

42

When heat is first applied there will be a drop in r.p.m. in airplanes equipped with fixed-pitch propellers; there will be a drop in manifold pressure in airplanes equipped with controllable-pitch propellers. If there is no carburetor ice present, there will be no further change in r.p.m. or manifold pressure until the carburetor heat is turned off, then the r.p.m. or manifold pressure will return to the original reading before heat was applied. If carburetor ice is present, there will normally be a rise in r.p.m. or manifold pressure after the initial drop (often accompanied by intermittent engine roughness); and then, when the carburetor heat is turned "off," the r.p.m. or manifold pressure will rise to a setting greater than that before application of the heat. The engine should also run more smoothly after the ice has been removed.

Whenever the throttle is closed during flight, the engine cools rapidly and vaporization of the fuel is less complete than if the engine is warm. Also, in this condition the engine is more susceptible to carburetor icing. Therefore, if the pilot suspects carburetor-icing conditions and anticipates closed-throttle operation, the carburetor heat should be turned to "full-on" before closing the throttle, and left on during the closed-throttle operation. The heat will aid in vaporizing the fuel and preventing carburetor ice. Periodically, however, the throttle should be opened smoothly for a few seconds to keep the engine warm, otherwise the carburetor heater may not provide enough heat to prevent icing.

Use of carburetor heat tends to reduce the output of the engine and also to increase the operating temperature. Therefore, the heat should not be used when full power is required (as during takeoff) or during normal engine operation except to check for the presence or removal of carburetor ice. In extreme cases of carburetor icing, after the ice has been removed it may be necessary to apply just enough carburetor heat to prevent further ice formation. However, this must be done with caution. Check the engine manufacturer's recommendations for the correct use of carburetor heat.

The carburetor heat should be checked during the preflight inspection. To properly perform this inspection, the manufacturer's recommendations should be followed.

Carburetor Air Temperature Gauge Some airplanes are equipped with a carburetor air temperature gauge which is useful in detecting potential icing conditions. Usually, the face of the gauge is calibrated in degrees Celsius (C.), with a yellow arc indicating the carburetor air temperatures at which icing may occur. This yellow arc ranges between −15° C. and +5° C. If the air temperature and moisture content of the air are such that the carburetor icing is improbable, the engine can be operated with the indicator in the yellow range with no adverse effects. However, if the atmospheric conditions are conducive to carburetor icing, the indicator must be kept outside the yellow arc by application of carburetor heat.

Certain carburetor air temperature gauges have a red radial which indicates the maximum permissible carburetor inlet air temperature recommended by the engine manufacturer; also, a green arc which indicates the normal operating range.

Outside Air Temperature Gauge (OAT) Most airplanes are equipped with an outside air temperature gauge calibrated in both degrees Celsius and Fahrenheit. It is used not only for obtaining the outside or ambient air temperature for calculating true airspeed, but also is useful in detecting potential icing conditions.

Fuel Injection System Fuel injectors have replaced carburetors in some airplanes. In this system, the fuel is normally injected either directly into the cylinders or just ahead of the intake valve. The fuel injection system is generally considered to be less susceptible to icing than the carburetor system. Impact icing of the air intake, however, is a possibility in either system. Impact icing occurs when ice forms on the exterior of the airplane and results in clogging openings such as the carburetor's air intake.

There are several types of fuel injection systems in use today. Although there are variations in design, the operational methods of each are generally similar. Most designs include an engine-driven fuel pump, a fuel/air control unit, fuel distributor, and discharge nozzles for each cylinder.

Some of the advantages of fuel injection are:

    a. Reduction in evaporative icing.
    b. Better fuel flow.
    c. Faster throttle response.
    d. Precise control of mixture.
    e. Better fuel distribution.
    f. Easier cold weather starts.

Disadvantages are usually associated with:

    a. Difficulty in starting a hot engine.
    b. Vapor locks during ground operations on hot days.
    c. Problems associated with restarting an engine that quits because of fuel starvation.

The air intake for the fuel injection system is somewhat similar to that used in the carburetor system. The fuel injection system, however, is equipped with an alternate air source located within the engine cowling. This source is used if the external air source is obstructed by ice or by other matter. The alternate air source is usually operated automatically with a backup manual system that can be used if the automatic feature malfunctions.

Proper Fuel is Essential   There are several grades of aviation fuel available; therefore, care must be exercised to assure that the correct aviation grade is being used for the specific type of engine. It can be harmful to the engine and dangerous to the flight if the wrong kind of fuel is used. It is the pilot's responsibility to obtain the proper grade of fuel. The proper grade is stated in the Aircraft Flight Manual or Pilot's Operating Handbook, on placards in the cockpit, and usually on the wing next to the filler caps.

The proper fuel for an engine will burn smoothly from the spark plug outward, exerting a smooth pressure downward on the piston. Using low-grade fuel or too lean a mixture can cause detonation. Detonation or knock is a sudden explosion or shock to a small area of the piston top, similar to striking it with a hammer (Fig. 2–9). Detonation produces extreme heat which often progresses into preignition, causing severe structural stresses on parts of the engine. Therefore, to prevent detonation the pilot should use the proper grade of fuel, maintain a sufficiently rich mixture, and maintain engine temperatures within the recommended limits.

Aviation gasolines are classified by octane and performance numbers (grades) which designate the antiknock value or knock resistance of the fuel mixture in the engine cylinder. The higher the grade of gasoline, the more pressure the fuel can withstand without detonating. For fuels that have two numbers, the first number indicates the lean-mixture rating and the second the rich-mixture rating. Thus, grade 100/130 fuel has a lean mixture rating of 100 and a rich mixture rating of 130.

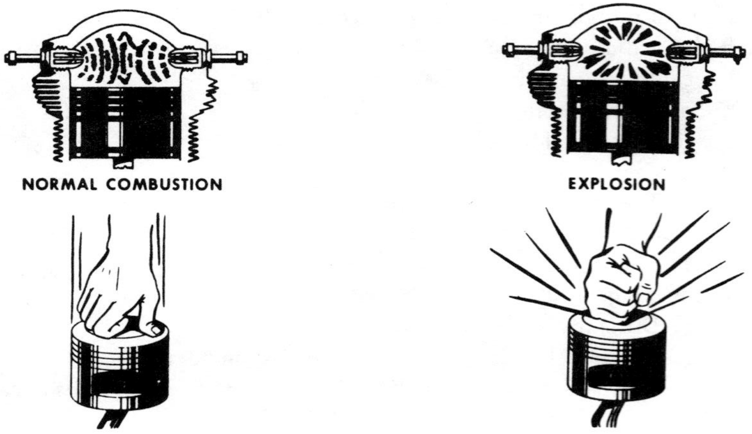

Two different scales are used to designate fuel grade. For fuels below grade 100 octane, numbers are used to designate the grade. For fuels of 100 octane and above, performance rating (knock-free power available) is represented by the numbers. Dyes are added to aviation fuels for easy identification of the fuel grade.

The grade of an aviation gasoline is not an indication of its fire hazard capabilities. Grade 91/96 gasoline is as easy to ignite as grade 115/145 and will explode with equal force. The grade indicates only the gasoline's performance in the airplane's engine.

Airplane engines are designed to operate using a specific grade of fuel as recommended by the manufacturer. Lower numbered octane fuel is used in lower compression engines because these fuels ignite at a lower temperature. Higher octane fuels are used in higher compression engines because they must ignite at higher temperatures but not prematurely. If the proper grade of fuel is not available it is possible, but not desirable, to use the next higher grade as a substitute.

A knock inhibitor is used to improve the antiknock qualities of a fuel. Such inhibitors must have a minimum of corrosive or other undesirable qualities. Perhaps the best available inhibitor in use at present is tetraethyl lead (TEL), which provides a thin layer of protective material to prevent undesirable corrosion.

The use of low-lead fuel in engines designed for higher leaded fuel or the use of higher leaded fuels in engines designed for low-lead fuels, may create problems.

Fuel Contamination   Water and dirt in fuel systems are dangerous; the pilot must either eliminate or prevent contamination. Of the accidents attributed to powerplant failure from fuel contamination, most have been traced to:

1. Inadequate preflight inspection by the pilot.
2. Servicing of aircraft with improperly filtered fuel from small tanks or drums.
3. Storing aircraft with partially filled fuel tanks.
4. Lack of proper maintenance.
   To help alleviate these problems, a substantial amount of fuel should be

NORMAL COMBUSTION

EXPLOSION

Figure 2–9. *Normal combustion and explosive combustion.*

drained from the fuel strainer (gascolator) quick drain and, if possible, from each fuel tank sump into a transparent container and checked for dirt and water. Quick-drain valves should be installed in aircraft fuel tanks if the tanks are not already equipped with these devices. Experiments have shown that when the fuel strainer is being drained, water in the tank may not appear until all the fuel has been drained from the lines leading to the tank. This indicates that the water remains in the tank and is not forcing the fuel out of the fuel lines leading to the fuel strainer. Therefore, drain enough fuel from the fuel strainer to be certain that fuel is being drained from the tank. The amount will depend on the length of fuel line from the tank to the drain. If water is found in the first sample, drain further samples until no trace appears.

Experiments have also shown that water will still remain in the fuel tanks after the drainage from the fuel strainer had ceased to show any trace of water. This residual water can be removed only by draining the fuel tank sumps.

The pilot should be able to identify suspended water droplets in the fuel from a cloudy appearance of the fuel; or the clear separation of water from the colored fuel which occurs after the water has settled to the bottom of the tank. Water is the principal contaminant of fuel, and to increase flight safety the fuel sumps should be drained during preflight.

In addition to the above measures the following should be considered. The fuel tanks should be filled after each flight, or at least after the last flight of the day. This will prevent moisture condensation within the tank since no air space will be left. If the pilot chooses to refuel with only the amount that can be carried on the next flight—perhaps a day later—there is an added risk of fuel contamination by moisture condensation within the tank. Each additional day may add to the amount of moisture condensation within the tank or tanks.

Another preventive measure the pilot can take is to avoid refueling from cans and drums. This practice introduces a major likelihood of fuel contamination.

As has been pointed out, the practice of using a funnel and chamois skin when refueling from cans or drums is hazardous under any condition, and should be discouraged. It is recognized, of course, that in remote areas or in emergency situations, there may be no alternative to refueling from sources with inadequate anticontamination systems, and a chamois skin and funnel may be the only possible means of filtering fuel.

In addition, it should be clearly understood that the use of a chamois will not always assure decontaminated fuel. Worn out chamois will not filter water; neither will a new, clean chamois that is already water-wet or damp. Most imitation chamois skins will not filter water. There are many filters available that are more effective than the old chamois and funnel system.

Refueling Procedures   Static electricity, formed by the friction of air passing over the surfaces of an airplane in flight and by the flow of fuel through the hose and nozzle, creates a fire hazard during refueling. To guard against the possibility of a spark igniting fuel fumes, a ground wire should be attached to the aircraft before the cap is removed from the tank. The refueling nozzle should be grounded to the aircraft before refueling is begun and throughout the refueling process. The fuel truck should also be grounded to the aircraft and the ground.

If fueling from drums or cans is necessary, proper bonding and grounding connections are extremely important, since there is an ever-present danger of static discharge and fuel vapor explosion. Nylon, dacron, or wool clothing are especially prone to accumulate and discharge static electricity from the person to the funnel or nozzle. Drums should be placed near grounding posts and the following sequence of connections observed:

1. Drum to ground.
2. Ground to aircraft.
3. Drum to aircraft.
4. Nozzle to aircraft before the aircraft tank cover is opened.
5. When disconnecting, reverse the order—4, 3, 2, 1.

The passage of fuel through a chamois increases the charge of static electricity and the danger of sparks. The aircraft must be properly grounded and the nozzle, chamois filter, and funnel bonded to the aircraft. If a can is used, it should be connected to either the grounding post or the funnel. Under no circumstances should a plastic bucket or similar nonconductive container be used in this operation.

Oil System   Proper lubrication of the engine is essential to the extension of engine life and prevention of excessive maintenance.

The oil system provides a means of storing and circulating oil throughout the internal components of a reciprocating engine. Lubricating oil serves two purposes: (1) it furnishes a coating of oil over the surfaces of the moving parts, preventing direct metal-to-metal contact and the generation of heat, and (2) it absorbs and dissipates, through the oil cooling system, part of the engine heat produced by the internal combustion process.

Usually the engine oil is stored in a sump at the bottom of the engine crankcase. An opening to the oil sump is provided through which oil can be

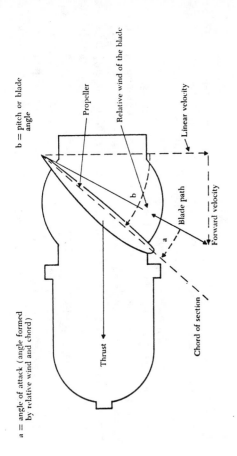

a = angle of attack (angle formed by relative wind and chord)

b = pitch or blade angle

Propeller

Relative wind of the blade

Thrust

Linear velocity

Blade path

Forward velocity

Chord of section

Figure 2-10. *Factors affecting propellers.*

Figure 2-11. *Changes in propeller blade angle from hub to tip.*

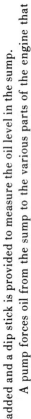

added and a dip stick is provided to measure the oil level in the sump.

A pump forces oil from the sump to the various parts of the engine that require lubrication. The oil then drains back to the sump for recirculation.

Each engine is equipped with an oil pressure gauge and an oil temperature gauge which are monitored to determine that the oil system is functioning properly.

Oil pressure gauges indicate pounds of pressure per square inch (PSI), and are color coded with a green arc to indicate the normal operating range. Also, at each end of the arc, some gauges have a red line to indicate high oil pressure, and another red line to indicate low oil pressure.

The oil pressure indication varies with the temperature of the oil. If the oil temperature is cold the pressure will be higher than if the oil is hot.

A loss of oil pressure is usually followed by engine failure. If this occurs while on the ground, the pilot must shut the engine down immediately; if in the air, land at a suitable emergency landing site.

The oil temperature gauge is calibrated in either Celsius or Fahrenheit and color coded in green to indicate the normal temperature operating range.

It is important that the pilot check the oil level before each flight. Starting a flight with an insufficient oil supply can lead to serious consequences. The airplane engine will burn off a certain amount of oil during operation, and beginning a flight when the oil level is low will usually result in an insufficient supply of oil before the flight terminates.

There are many different types of oil manufactured for aviation use. The engine manufacturer's recommendation should be followed to determine the type and weight of oil to use. This information can be found in the Aircraft Flight Manual or Pilot's Operating Handbook, or on placards on or near the oil filler cap.

Propeller · A detailed discussion of the propeller is quite complex and beyond the intended scope of this handbook. However, the following material is offered as an introduction to the function of the propeller.

A propeller is a rotating airfoil, and is subject to induced drag, stalls, and other aerodynamic principles that apply to any airfoil. It provides the necessary thrust to pull, or in some cases push, the airplane through the air. This is accomplished by using engine power to rotate the propeller which in turn generates thrust in much the same way as a wing produces lift. The propeller has an angle of attack which is the angle between the chord line of the propeller's airfoil and its relative wind (airflow opposite to the motion of the blade). Other factors affecting propellers are illustrated in Fig. 2-10.

A propeller blade is twisted. The blade angle changes from the hub to the tip with the greatest angle of incidence, or highest pitch, at the hub and the smallest at the tip (Fig. 2-11). The reason for the twist is to produce uniform lift from the hub to the tip. As the blade rotates there is a difference in the actual speed of the various portions of the blade. The tip of the blade travels faster than that part near the hub because the tip travels a greater distance than the hub in the same length of time (Fig. 2-12). Changing the angle of incidence (pitch) from the hub to the tip to correspond with the speed produces uniform lift throughout the length of the blade. If the propeller blade was designed with the same angle of incidence throughout its entire length, it would be extremely inefficient because as airspeed increases in flight the portion near the hub would have a negative angle of attack while the blade tip would be stalled.

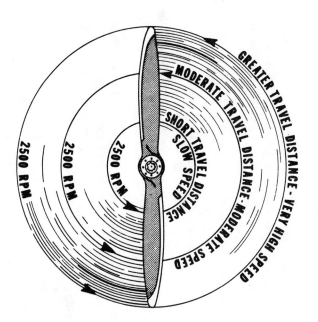

Figure 2-12. *Relationship of travel distance and speed of various portions of propeller blade.*

Geometric pitch is the distance in inches that the propeller would move forward in one revolution if it were rotated in a solid medium so as not to be affected by slippage as it is in the air. Effective pitch is the actual distance it moves forward through the air in one revolution. Propeller slip is the difference between the geometric pitch and effective pitch (Fig. 2–13). Pitch is proportional to the blade angle which is the angle between the chord line of the blade and the propeller's plane of rotation (Fig. 2-10).

Small airplanes are equipped with either one of two types of propellers. One is the fixed-pitch, and the other is the controllable pitch or constant-speed propeller.

Fixed-Pitch Propeller The pitch of this propeller is fixed by the manufacturer and cannot be changed by the pilot. There are two types of fixed-pitch propellers; the climb propeller and the cruise propeller. Whether the airplane has a climb or cruise propeller installed depends upon its intended use. The climb propeller has a lower pitch, therefore less drag. This results in the capability of higher r.p.m. and more horsepower being developed by the

engine. This increases performance during takeoffs and climbs, but decreases performance during cruising flight.

The cruise propeller has a higher pitch, therefore more drag. This results in lower r.p.m. and less horsepower capability. This decreases performance during takeoffs and climbs, but increases efficiency during cruising flight.

The propeller on a low-horsepower engine is usually mounted on a shaft which may be an extension of the engine crankshaft. In this case the revolutions per minute of the propeller would be the same as the engine r.p.m.

On higher horsepower engines the propeller is mounted on a shaft geared to the engine crankshaft. In this type, the r.p.m. of the propeller is different than that of the engine.

If the propeller is a fixed-pitch and the speed of the engine and propeller is the same, a tachometer is the only indicator of engine power.

A tachometer is calibrated in hundreds of revolutions per minute, and gives a direct indication of the engine and propeller r.p.m. The instrument is color coded with a green arc denoting the normal operating range and a red line denoting the maximum continuous operating r.p.m. Some tachometers have additional marking or interrupted arcs. Therefore the manufacturer's recommendations should be used as a reference to clarify any misunderstanding of tachometer markings.

The revolutions per minute are regulated by the throttle which controls the fuel/air flow to the engine. At a given altitude the higher the tachometer reading the higher the power output of the engine.

There is a condition under which the tachometer does not show correct power output of the engine. This occurs when operating altitude increases. For example, 2300 r.p.m. at 5,000 feet produce less horsepower than 2300

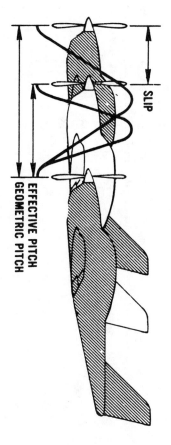

Figure 2-13. *Effective and geometric propeller pitch.*

r.p.m. at sea level. The reason for this is that air density decreases as altitude increases. Power output depends on air density, therefore decreasing the density decreases the power output of the engine. As altitude changes, the position of the throttle must be changed to maintain the same r.p.m. As altitude is increased, the throttle must be opened further to indicate the same r.p.m. as at a lower altitude.

Controllable-Pitch Propellers The pitch on these propellers can be changed in flight; therefore, they are referred to as controllable-pitch propellers. These propeller systems vary from a simple two-position propeller to more complex automatic constant-speed propellers.

The number of pitch positions at which the propeller can be set may be limited, such as a two-position propeller with only high or low pitch available. Many other propellers, however, are variable pitch, and can be adjusted to any pitch angle between a minimum and maximum pitch setting.

An airplane equipped with a controllable pitch propeller has two controls: (1) a throttle control and (2) a propeller control. The throttle controls the power output of the engine which is registered on the manifold pressure gauge. The manifold pressure gauge is a simple barometer that measure the air pressure in the engine intake manifold in inches of mercury. It is color coded with a green arc indicating the normal operating range.

The propeller control regulates the engine r.p.m. and in turn the propeller r.p.m. The r.p.m. is registered on the tachometer. The pilot can set the throttle control and propeller control at any desired manifold pressure and r.p.m. setting within the engine operating limitation.

Within a given power setting, when using a constant-speed propeller, the pilot can set the propeller control to a given r.p.m. and the propeller governor will automatically change the pitch (blade angle) to counteract any tendency for the engine to vary from this r.p.m. For example, if manifold pressure or engine power is increased, the propeller governor automatically increases the pitch of the blade (more propeller drag) to maintain the same r.p.m.

A controllable pitch propeller permits the pilot to select the blade angle that will result in the most efficient performance for a particular flight condition. A low blade angle or decreased pitch, reduces the propeller drag and allows more engine power for takeoffs. After airspeed is attained during cruising flight, the propeller blade is changed to a higher angle or increased pitch. Consequently the blade takes a larger bite of air at a lower power setting, and therefore increases the efficiency of the flight. This process is similar to shifting gears in an automobile from low gear to high gear.

For any given r.p.m. there is a manifold pressure that should not be exceeded. If manifold pressure is excessive for a given r.p.m., the pressure within the cylinders could be exceeded, thus placing undue stress on them. If repeated too frequently, this stress could weaken the cylinder components and eventually cause engine failure.

The pilot can avoid conditions that would possibly overstress the cylinders by being constantly aware of the r.p.m., especially when increasing the manifold pressure. Pilots should conform to the manufacturer's recommendations for power settings of a particular engine so as to maintain the proper relationship between manifold pressure and r.p.m. Remember, the combination to avoid is a high throttle setting (manifold pressure indication) and a low r.p.m. (tachometer indication).

When both manifold pressure and r.p.m. need to be changed, the pilot can further help avoid engine overstress by making power adjustments in the proper order. When power settings are being decreased, reduce manifold pressure before r.p.m. When power settings are being increased, reverse the order—increase r.p.m. first, then manifold pressure. If r.p.m. is reduced before manifold pressure, manifold pressure will automatically increase and possibly exceed manufacturer's tolerances.

Summarizing: In an airplane equipped with a controllable-pitch propeller, the throttle controls the manifold pressure and the propeller control is used to regulate the r.p.m. Avoid high manifold pressure settings with low r.p.m. The preceding is a standard procedure for most situations, but with unsupercharged engines it is sometimes modified to take advantage of auxiliary fuel metering devices in the carburetor. These devices function at full throttle settings, providing additional fuel flow. This additional fuel helps cool the engine during takeoffs and full-power climbs where engine overheating may be a problem. In such instances, a small reduction in r.p.m. is possible without overstressing the engine, even though the throttle is in the full-power position. If in doubt, the manufacturer's recommendations should be followed.

Starting the Engine

Before starting the engine, the airplane should be in an area where the propeller will not stir up gravel or dust so as to cause damage to the propeller or property. Rules of safety and courtesy should be strictly observed to avoid personal injury or annoyance. The wheels should be chocked and the brakes set to avoid hazards caused by unintentional movement.

Engines Equipped with a Starter The pilot should be familiar with the manufacturer's recommended starting procedures for the airplane being flown. This information can be found in the Aircraft Flight Manual or Pilot's Operating Handbook, or other sources. There are not only different procedures applicable to starting engines equipped with conventional carburetors and those equipped with fuel injection systems, but also between different systems of either carburetion or fuel injection. The pilot should always ascertain that no one is near the propeller, call "clear prop," and wait for a possible response before engaging the propeller. Continuous cranking beyond 30 seconds' duration may damage the starter. In addition, the starter motor should be allowed to cool at least 1 to 2 minutes between cranking periods. If the engine refuses to start under normal circumstances after a reasonable number of attempts, the possibility of problems with ignition or fuel flow should be investigated.

As soon as the engine starts, check for unintentional movement and set power to the recommended warmup r.p.m. The oil pressure should then be checked to determine that the oil system is functioning properly. If the gauge does not indicate oil pressure within 30 seconds, the engine should be stopped and a check should be made to determine what is causing the lack of oil pressure. If oil is not circulating properly, the engine can be seriously damaged in a short time. During cold weather there will be a much slower response in oil pressure indications than during warmer weather, because colder temperatures cause the oil to congeal (thicken) to a greater extent.

The engine must reach normal operating temperature before it will run smoothly and dependably. Temperature is indicated by the cylinder-head temperature gauge. If the airplane is not equipped with this gauge the oil temperature gauge must be used. Remember, in this case, that oil warms much slower in cold weather.

Before takeoff the pilot should perform all necessary checks for engine and airplane operation. Follow the manufacturer's recommendations when performing all checks. Use a checklist—*do not rely on memory.*

A word of caution to pilots operating airplanes with starters. If the battery is dead or so weak that it will not turn the engine, the battery should be replaced. Avoid turning the propeller by hand to start the engine. Under these conditions if the engine is started by hand the alternator or generator will generate electrical current into the dead battery and it is possible that the battery could overheat, resulting in a very hazardous operation.

Engines not Equipped with a Starter Because of the hazards involved in hand starting airplane engines, every precaution should be exercised. The safety measures previously mentioned should be adhered to, and it is *extremely* important that a competent pilot be at the controls in the cockpit. Also, the person turning the propeller should be thoroughly familiar with the technique. The following are additional suggestions to aid in increasing the safety factor while hand starting airplanes.

The person who turns the propeller is in charge, and calls out the commands, "gas on, switch off, throttle closed, brakes set." The pilot in the cockpit will check these items and repeat the phrase to assure that there is no misunderstanding. The person propping the airplane should push slightly on the airplane to assure that the brakes are set and are holding firmly. The switch and throttle must not be touched again until the person swinging the prop calls "contact." The pilot will repeat "contact" and then turn on the switch in that sequence—*never* turn the switch on and then call "contact."

For the person swinging the prop a few simple precautions will help avoid accidents.

When touching a propeller, always assume that the switch is on, even though the pilot may confirm the statement "switch off." The switches on many engine installations operate on the principle of short-circuiting the current. If the switch is faulty, as sometimes happens, it can be in the "off" position and still permit the current to flow to the spark plugs.

Be sure to stand on firm ground. Slippery grass, mud, grease, or loose gravel could cause a slip or fall into or under the propeller.

Never allow any portion of the body to get into the propeller arc of rotation. This applies even though the engine is not being cranked; occasionally, a hot engine will backfire after shutdown when the propeller has almost stopped rotating.

Stand close enough to the propeller to be able to step away as it is pulled down. Standing too far away from the propeller requires leaning forward to reach it. This is an off-balance position and it is possible to fall into the blades as the engine starts. Stepping away after cranking is also a safeguard in the event the brakes do not hold when the engine starts.

When swinging the propeller, always move the blade downward by pushing with the palms of the hands. If the blade is moved upward, or gripped tightly with the fingers and backfiring occurs, it could cause broken fingers or the body to be pulled into the path of the propeller blades.

When removing the chocks from in front of the wheels, it should be remembered that the propeller, when revolving, is almost invisible. There are cases on record where someone intending to remove the chocks walked directly into the propeller.

Unsupervised "hand propping" of an airplane should not be attempted by inexperienced persons. Regardless of the experience level, it should never be attempted by anyone without adhering to adequate safety measures. Uninformed or inexperienced persons or nonpilot passengers should never handle the throttle, brakes, or switches during starting procedures. The airplane should be securely chocked or tied down, and great care should be exercised in setting the throttle. It may be well to turn the fuel selector valve to the "off" position after properly priming the engine and prior to actually attempting the hand start. After it starts, the engine will usually run long enough with the fuel "off" to permit walking around the propeller and turning the fuel selector to the "on" position.

Idling the Engine During Flight

It might be well to briefly mention the potential problems that could be created by excessive idling of the engine during flight, particularly for long periods of time such as prolonged descents.

Whenever the throttle is closed during flight, the engine cools rapidly and vaporization of fuel is less complete. The airflow through the carburetor system under such conditions may not be of sufficient volume to assure a uniform mixture of fuel and air. Consequently, the engine may cease to operate because the mixture is too lean or too rich. Suddenly opening or closing the throttle could aggravate this condition, and the engine may cough once or twice, sputter, and stop.

Three precautions should be taken to prevent the engine from stopping while idling. First, make sure that the ground-idling speed is properly adjusted. Second, do not open or close the throttle abruptly. Third, keep the engine warm during glides by frequently opening the throttle for a few seconds.

Exhaust Gas Temperature (EGT) Gauge

Many airplanes are equipped with an exhaust gas temperature gauge. If properly used, this engine instrument can reduce fuel consumption by 10 percent because of its accuracy in indicating to the pilot the exact amount of fuel that should be metered to the engine.

An exhaust gas temperature gauge measures, in degrees Celsius or Fahrenheit, the temperature of the exhaust gases at the exhaust manifold. This temperature measurement varies with ratio of fuel to air entering the cylinders, and therefore can be used as a basis for regulating the fuel/air mixture. This is possible because this instrument is very sensitive to temperature changes.

Although the manufacturer's recommendation for leaning the mixture should be adhered to, the usual procedure for leaning the mixture on lower horsepower engines when an EGT is available is as follows:

The mixture is leaned slowly while observing the increase in exhaust gas temperature on the gauge. When the EGT reaches a peak, the mixture should be enriched until the EGT gauge indicates a decrease in temperature. The number of degrees drop is recommended by the engine manufacturer, usually approximately 25° to 75°. Engines equipped with carburetors will run rough when leaned to the peak EGT reading, but will run smooth after the mixture is enriched slightly.

Superchargers or Turbochargers

As an airplane climbs to altitude, the density of the air decreases. Any decrease in fuel or air will cause a decrease in engine power. Superchargers or turbochargers compress the air leading to the engine, thereby maintaining a constant air density as the airplane climbs.

Superchargers incorporate a blower driven by the engine crankshaft to raise the pressure of the air as it enters the engine.

Turbochargers incorporate a compressor unit driven by an impeller placed in the exhaust stream of the engine.

These units extend the engine's sea level power output to 20,000 feet or higher above mean sea level.

Aircraft Documents, Maintenance, and Inspections

Aircraft Owner Responsibilities An aircraft owner assumes responsibilities similar to those of an automobile owner. An automobile is usually registered in the state where the owner resides and state license plates must be obtained. The registered owner of an aircraft is also responsible for certain items such as:

1. Having a current Airworthiness Certificate and Aircraft Registration Certificate in the aircraft.
2. Maintaining the aircraft in an airworthy condition.
3. Assuring that maintenance is properly recorded.
4. Keeping abreast of current regulations concerning the operation and maintenance of the aircraft.
5. Notifying the FAA Aircraft Registry immediately of any change of permanent mailing address, or of the sale or export of the aircraft, or of the loss of U.S. citizenship.

Some states require that an automobile be inspected periodically (most states every 6 to 12 calendar months) to assure that it is in a safe operating condition. An aircraft must be inspected in accordance with an annual inspection program or with one of the five inspection programs outlined in Federal Aviation Regulations Part 91, Section 91.217, in order to maintain a current Airworthiness Certificate.

Some similarities between automobile and aircraft responsibilities are shown in the following chart:

Automobile/Aircraft Comparison Chart

Responsibility	Automobile	Aircraft
Registration	Yes	Yes
Inspection (most states)	Yes	Yes
Compulsory insurance (most states)	Yes	No
Reporting of accidents	Yes	Yes
Required maintenance records	No	Yes
Controlled maintenance	No	Yes
Maximum speed restrictions	Yes	Yes

Certificate of Aircraft Registration Before an aircraft can be legally flown it must be registered with the FAA Aircraft Registry and have within it a Certificate of Registration issued to the owner as evidence of the registration (Fig. 2-14). An aircraft is eligible for registration only if it is not registered under the laws of any foreign country.

The Certificate of Aircraft Registration will expire when:

1. The aircraft is registered under the laws of a foreign country.
2. The registration of the aircraft is cancelled at the written request of the owner.
3. The aircraft is totally destroyed or scrapped.
4. The ownership of the aircraft is transferred.
5. The holder of the certificate loses United States citizenship.
6. Thirty days have elapsed since the death of the holder of the certificate.

When the aircraft is destroyed, scrapped, or sold, the previous owner must notify the FAA by filling in the back of the Certificate of Aircraft Registration, and mailing it to the FAA Aircraft Registry.

When a U.S. civil aircraft is transferred to a person who is not a U.S. citizen, the U.S.-registered owner is required to remove the United States registration and nationality marks from the aircraft before the aircraft is delivered.

A Dealers Aircraft Registration Certificate is another form of registration certificate, but it is valid only for required flight tests by the manufacturer or in flights that are necessary for the sale of the aircraft by the manufacturer or a dealer. It must be removed by the dealer when the aircraft is sold.

The FAA does not issue any certificate of ownership or endorse any information with respect to ownership on a Certificate of Aircraft Registration.

NOTE: For any additional information concerning the Aircraft Registration Application or the Aircraft Bill of Sale, contact the nearest General Aviation District Office or Flight Standards District Office.

Airworthiness Certificate An Airworthiness Certificate is issued by the Federal Aviation Administration only after the aircraft has been inspected and it is found that it meets the requirements of the Federal Aviation Regulations (FARS), and is in a condition for safe operation. Under any circumstances, the aircraft must meet the requirements of the original type certificate. The certificate must be displayed in the aircraft so that it is legible to passengers or crew whenever the aircraft is operated. The Airworthiness Certificate may be transferred with the aircraft except when it is sold to a foreign purchaser.

The Standard Airworthiness Certificate, Fig. 2-15, is issued for aircraft type certificated in the normal, utility, acrobatic, and transport categories or for manned free balloons. An explanation of each item in the certificate follows:

Item 1. Nationality—The "N" indicates the aircraft is of United States registry. Registration Marks—the number, in this case 12345, is the registration number assigned to the aircraft.

Item 2. Indicates the make and model of the aircraft.

Item 3. Is the serial number assigned to the aircraft, as noted on the aircraft data plate.

Item 4. Indicates that the aircraft, in this case, must be operated in accordance with the limitations specified for the NORMAL category.

Item 5. Indicates the aircraft is considered in a condition for safe operation at the time of inspection and issuance of the certificate. Any exemptions from the applicable airworthiness standards are briefly noted here, and the exemption number. The word NONE will be entered if no exemption exists.

Item 6. Indicates the Airworthiness Certificate is in effect indefinitely if the aircraft is maintained in accordance with Parts 21, 43, and 91 of the

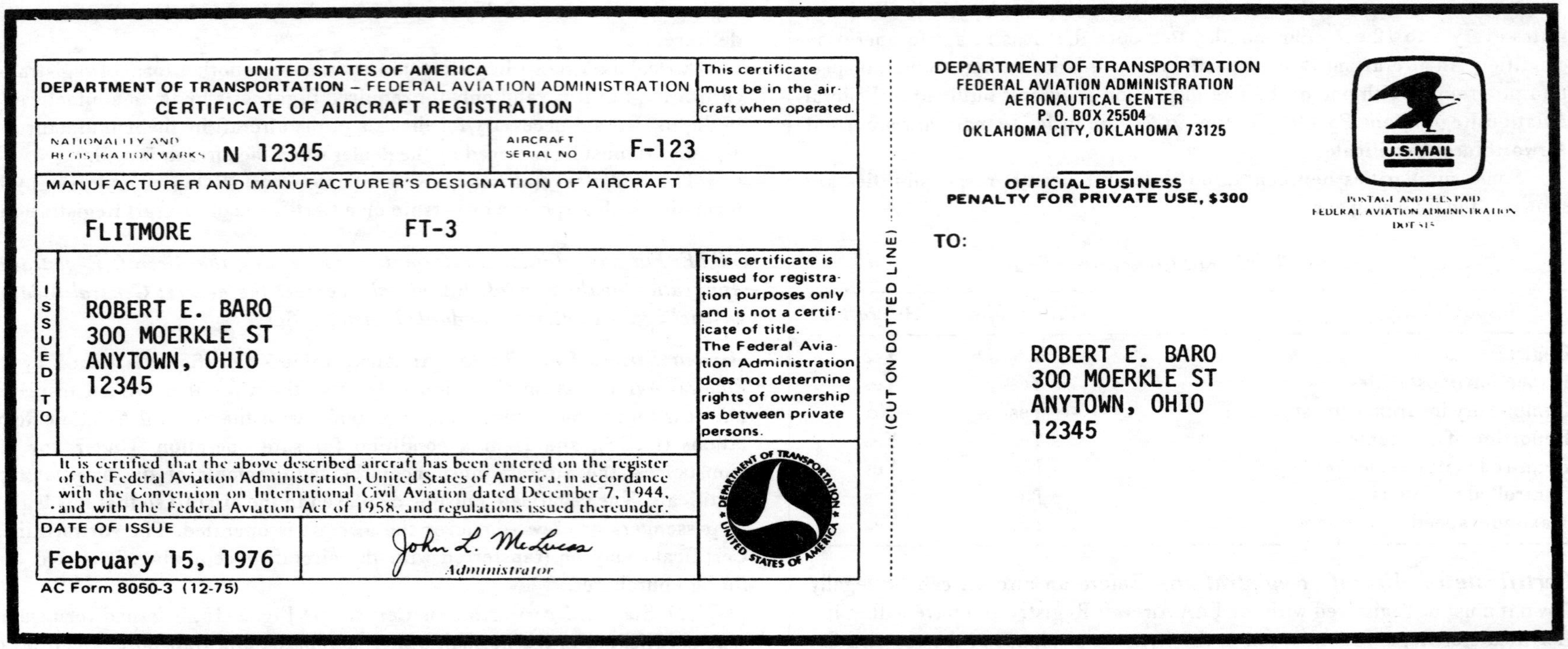

Figure 2-14. *Certificate of aircraft registration.*

Federal Aviation Regulations and the aircraft is registered in the United States. Also included are the date the certificate was issued and the signature and office identification of the FAA representative.

The Special Airworthiness Certificate is issued for all aircraft certificated in other than the Standard classifications (Experimental, Restricted, Limited, and Provisional).

In purchasing an aircraft classed as other than Standard, it is suggested that the local FAA General Aviation District Office be contacted for an explanation of the pertinent airworthiness requirements and the limitations of such a certificate.

In summary, the FAA initially determines that the aircraft is in a condition for safe operation and conforms to type design, then issues an Airworthiness Certificate. A Standard Airworthiness Certificate remains in effect

so long as the aircraft receives the required maintenance and is properly registered in the United States. Flight safety relies in part on the condition of the aircraft, which may be determined on inspection by mechanics, approved repair stations, or manufacturers who meet specific requirements of FAR Part 43.

Aircraft Maintenance Maintenance means the inspection, overhaul, and repair of aircraft, including the replacement of parts. A PROPERLY MAINTAINED AIRCRAFT IS A SAFE AIRCRAFT.

The purpose of maintenance is to ensure that the aircraft is kept to an acceptable standard of airworthiness throughout its operational life.

Although maintenance requirements will vary for different types of aircraft, experience shows that most aircraft will need some type of preventive

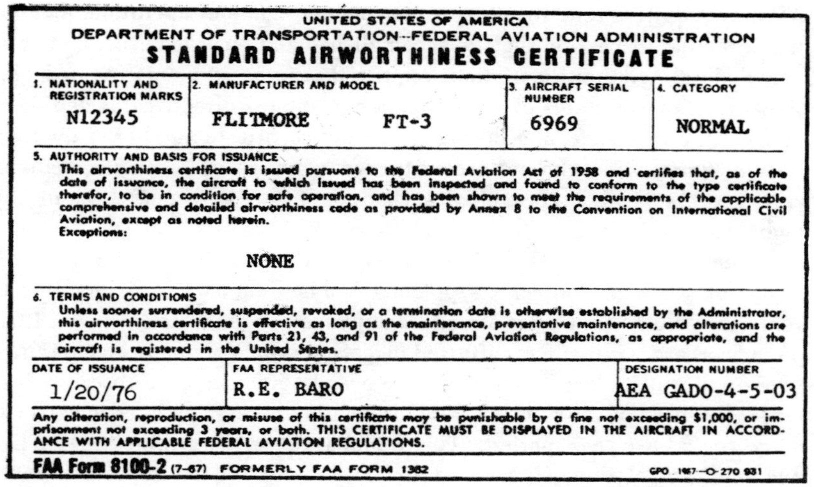

DEPARTMENT OF TRANSPORTATION—FEDERAL AVIATION ADMINISTRATION
UNITED STATES OF AMERICA
STANDARD AIRWORTHINESS CERTIFICATE

1. NATIONALITY AND REGISTRATION MARKS	2. MANUFACTURER AND MODEL	3. AIRCRAFT SERIAL NUMBER	4. CATEGORY
N12345	FLITMORE FT-3	6969	NORMAL

5. AUTHORITY AND BASIS FOR ISSUANCE
This airworthiness certificate is issued pursuant to the Federal Aviation Act of 1958 and certifies that, as of the date of issuance, the aircraft to which issued has been inspected and found to conform to the type certificate therefor, to be in condition for safe operation, and has been shown to meet the requirements of the applicable comprehensive and detailed airworthiness code as provided by Annex 8 to the Convention on International Civil Aviation, except as noted herein.

6. TERMS AND CONDITIONS
Unless sooner surrendered, suspended, revoked, or a termination date is otherwise established by the Administrator, this airworthiness certificate is effective as long as the maintenance, preventative maintenance, and alterations are performed in accordance with Parts 21, 43, and 91 of the Federal Aviation Regulation, as appropriate, and the aircraft is registered in the United States.

Any alteration, reproduction, or misuse of this certificate may be punishable by a fine not exceeding $1,000, or imprisonment not exceeding 3 years, or both. THIS CERTIFICATE MUST BE DISPLAYED IN THE AIRCRAFT IN ACCORDANCE WITH APPLICABLE FEDERAL AVIATION REGULATIONS.

NONE

DATE OF ISSUANCE	FAA REPRESENTATIVE	DESIGNATION NUMBER
1/20/76	R.E. BARO	AEA GADO-4-5-03

FAA Form 8100-2 (7-67) FORMERLY FAA FORM 1362

GPO 187-O-270 881

Figure 2-15. *Standard airworthiness certificate.*

maintenance every 25 hours of flying time or less, and minor maintenance at least every 100 hours. This is influenced by the kind of operation, climatic conditions, storage facilities, age, and construction of the aircraft. Most manufacturers supply service information which should be used in maintaining the aircraft.

Inspections FAR Part 91 places primary responsibility on the owner or operator for maintaining an aircraft in an airworthy condition. Certain inspections must be performed on the aircraft and the owner must maintain the airworthiness of the aircraft during the time between required inspections by having any unsafe defects corrected.

Federal Aviation Regulations require the inspection of all civil aircraft at specific intervals for the purpose of determining the overall condition. The interval depends generally upon the type of operations in which the aircraft is engaged. Some aircraft need to be inspected at least once each 12 calendar months, while inspection is required for others after each 100 hours of operation. In other instances, an aircraft may be inspected in accordance with an inspection system set up to provide for total inspection of the aircraft on the basis of time, time in service, number of system operations, or any combination of these.

To determine the specific inspection requirements and rules for the per-

formance of inspections, refer to the Federal Aviation Regulations which prescribe the requirements for various types of operations.

Annual Inspection. A reciprocating powered single-engine aircraft flown for pleasure is required to be inspected at least annually by a certificated airframe and powerplant mechanic holding an inspection authorization, or by a certificated repair station that is appropriately rated, or by the manufacturer of the aircraft. The aircraft may not be operated unless the annual inspection has been performed within the preceding 12 calendar months. A period of 12 calendar months extends from any day of any month to the last day of the same month the following year. However, an aircraft with the annual inspection overdue may be operated under a special flight permit for the purpose of flying the aircraft to a location where the annual inspection can be performed.

100-Hour Inspection. A reciprocating powered single-engine aircraft used to carry passengers or for flight instruction for hire must be inspected within each 100 hours of time in service by a certificated airframe and powerplant mechanic, a certificated repair station that is appropriately rated, or the aircraft manufacturer. An annual inspection is acceptable as a 100-hour inspection, but the reverse is not true.

Other Inspection Programs. The annual and 100-hour inspection requirements do not apply to large airplanes, turbojet or turbo-propeller-powered multiengine airplanes, or to airplanes for which the owner or operator complies with the progressive inspection requirements. Details of these requirements may be determined by reference to Parts 43 and 91 of the Federal Aviation Regulations and by inquiry at a local FAA General Aviation or Flight Standards District Office.

Preflight Inspection. Although not required by Federal Aviation Regulations, a careful pilot will always conduct a thorough preflight inspection before every flight to assure that the aircraft is safe for flight.

Preventive Maintenance Simple or minor preservation operations and the replacement of small standard parts, not involving complex assembly operations, are considered preventive maintenance. A certificated pilot may perform preventive maintenance on any aircraft, owned or operated by the pilot, that is not used in air carrier service. Typical preventive maintenance operations are found in FAR Part 43, Maintenance, Preventive Maintenance, Rebuilding, and Alteration. Part 43 also contains other rules to be followed in the maintenance of aircraft.

Repairs and Alterations Except as noted under "Preventive Maintenance," all repairs and alterations are classed as either major or minor. Major repairs or alterations must be approved for return to service by an appropriately rated certificated repair station, an airframe and powerplant mechanic holding an inspection authorization, or a representative of the Administrator. Minor repairs and alterations may be approved for return to service by a certificated airframe and powerplant mechanic or an appropriately certificated repair station.

Special Flight Permits   A special flight permit is an authorization to operate an aircraft that may not currently meet applicable airworthiness requirements, but is safe for a specific flight. Before the permit is issued, an FAA inspector may personally inspect the aircraft or require it to be inspected by a certificated airframe and powerplant mechanic or repair station to determine its safety for the intended flight. The inspection must be recorded in the aircraft records.

The special flight permit is issued to allow the aircraft to be flown to a base where repairs, alterations, or maintenance can be performed; for delivering or exporting the aircraft; or for evacuating an aircraft from an area of impending danger. A special flight permit may be issued to allow the operation of an overweight aircraft for flight beyond its normal range over water or land areas where adequate landing facilities or fuel is not available.

If a special flight permit is needed, assistance and the necessary forms may be obtained from the local Flight Standards or General Aviation District Office.

Airworthiness Directives   A primary safety function of the Federal Aviation Administration is to require correction of unsafe conditions found in an aircraft, aircraft engine, propeller, or applicance when such conditions exist and are likely to exist or develop in other products of the same design. The unsafe condition may exist because of a design defect, maintenance, or other causes. FAR Part 39, Airworthiness Directives, defines the authority and responsibility of the Administrator for requiring the necessary corrective action. The Airworthiness Directives (ADs) are the media used to notify aircraft owners and other interested persons of unsafe conditions and to prescribe the conditions under which the product may continue to be operated.

Airworthiness Directives may be divided into two categories: (1) those of an emergency nature requiring immediate compliance upon receipt, and (2) those of a less urgent nature requiring compliance within a relatively longer period of time.

Airworthiness Directives are Federal Aviation Regulations and must be complied with, unless specific exemption is granted. It is the aircraft owner or operator's responsibility to assure compliance with all pertinent ADs. This includes those ADs that require recurrent or continuing action. For example, an AD may require a repetitive inspection each 50 hours of operation; meaning, the particular inspection shall be accomplished and recorded every 50 hours.

Federal Aviation Regulations require that a record be maintained which shows the current status of applicable airworthiness directives, including the method of compliance, and the signature and certificate number of the repair station or mechanic who performed the work. For ready reference, many aircraft owners have a chronological listing of the pertinent ADs in their logbooks.

The Airworthiness Directives Summary contains all the valid ADs previously published and biweekly supplements. The Summary is divided into two volumes. Volume I includes directives applicable to small aircraft, i.e., 12,500 pounds or less maximum certificated takeoff weight. Volume II includes directives applicable to large aircraft—those over 12,500 pounds. Subscription service will consist of the summary and automatic biweekly updates to each summary for a two-year period. The Summary of Airworthiness Directives, Volume I and Volume II, are sold and distributed for the Superintendent of Documents by the FAA from Oklahoma City. For further Information on how to order Airworthiness Directives, order Advisory Circular 39-6E from

U.S. Department of Transportation
Publications Section, M-443.1
Washington, D.C. 20590

Preflight Inspection   The preflight inspection of the airplane is one of the pilot's most important duties. A number of serious airplane accidents have been traced directly to poor preflight inspection practices. The preflight inspection should be a thorough and systematic means by which the pilot determines that the airplane is ready for safe flight.

Many Aircraft Flight Manuals or Pilot's Operating Handbooks contain a section devoted to a systematic method of performing a preflight inspection that should be used by the pilot for guidance.

The following guide is a practical inspection checklist which can be used on almost any single-engine or light twin-engine aircraft, provided it is modified to suit the airplane type and the manufacturer's recommendations. The circled numbers on the inspection chart (Fig. 2-16), correspond to the numbers indicated on the itemized list. By following the numerically indicated route, as indicated by the arrows on the inspection chart, an effective and organized preflight inspection can be accomplished.

Before entering the airplane—Stand off and observe the general overall appearance of the airplane for obvious defects and discrepancies.

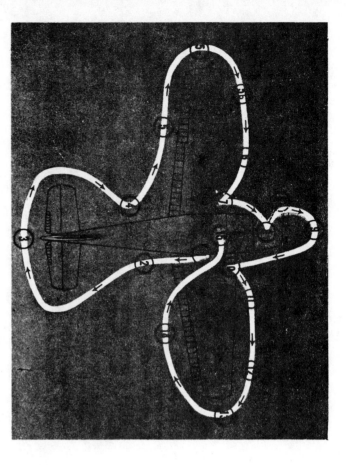

Figure 2-16. *Preflight inspection chart.*

1. Cockpit/Cabin:
Battery and ignition switches—"OFF."
Control locks—"REMOVE."
Landing gear switch—gear "DOWN" position.

2. Fuselage:
Baggage compartment—contents secure and door locked.
Airspeed static source—free from obstructions.
Condition of covering—missing or loose rivets, cracks, tears in fabric, etc.
Anticollision and navigation lights—condition and security.

3. Empennage:
Deicer boots—condition and security.
Control surface locks—"REMOVE."
Fixed and movable control surfaces—dents, cracks, excess play, hinge pins and bolts for security and condition.

Tailwheel—spring, steering arms and chains, tire inflation, and condition.
Lights—navigation anticollision lights for condition and security.

4. Fuselage:
Same as item 2.

5. Wing:
Control surface locks—"REMOVE."
Control surfaces, including flaps—dents, cracks, excess play, hinge pins and bolts for security and condition.
General condition of wings and covering—torn fabric, bulges or wrinkles, loose or missing rivets, "oil cans," etc.
Wingtip and navigation light—security and damage.
Deicer boots—general condition and security.
Landing light—condition, cleanliness, and security.
Stall warning vane—freedom of movement. Prior to inspection turn master switch "ON" so that stall warning signal can be checked when vane is deflected.

6. Landing gear:
Wheels and brakes—condition and security, indications of fluid leakage at fittings, fluid lines and adjacent area.
Tires—cuts, bruises, excessive wear, and proper inflation.
Oleos and shock struts—cleanliness and proper inflation.
Shock cords—general condition.
Wheel fairing—general condition and security. On streamline wheel fairing, look inside for accumulation of mud, ice, etc.
Limit and position switches—security, cleanliness, and condition.
Ground safety locks—"REMOVE."

7. Fuel tank:
Fuel quantity in tank.
Fuel tank filler cap and fairing covers—secure.
Fuel tank vents—obstructions.
When fuel tank is equipped with a quick or snap-type drain valve, drain a sufficient amount of fuel into a container to check for the presence of water and sediment.

8. Engine:
Engine oil quantity—secure filler cap.
General condition—check for fuel and oil leaks.
Cowling, access doors, and cowl flaps—condition and security.
Carburetor filter—cleanliness and security.

55

Drain a sufficient quantity of fuel from the main fuel sump drain to determine that there is no water or sediment remaining in the system.

Nose landing gear—wheel and tire—cuts, bruises, excessive wear, and proper inflation.

Oleo and shock strut—proper inflation and cleanliness.

Wheel well and fairing—general condition and security.

Limit and position switches—cleanliness, condition, and security.

Ground safety lock—"REMOVE."

9. Propeller:

Propeller and spinner—security, oil leakage, and condition. Be particularly observant for deep nicks and scratches.

Assure that ground area under propeller is free of loose stones, cinders, etc.

10. Fuel tank:

Same as item 7.

11. Landing gear:

Same as item 6.

12. Pitot:

Pitot cover—"REMOVE."

Pitot and static ports—remove obstructions.

General condition and alignment.

13. Wingtip and navigation lights.

Same as item 5.

14. Wing:

Same as item 5.

15. Cockpit:

Cleanliness and loose articles.

Windshield and windows—obvious defects and cleanliness.

Safety belt and shoulder harness—condition and security.

Adjust rudder pedals so full rudder travel may be assured.

Parking brake—"SET."

Landing gear and flap switches or levers in proper position.

Check all switches and controls.

Trim tabs—"SET."

Pilot's seat—"LOCKED."

Ground Runup and Functional Check It is desirable to have the airplane headed into the wind.

Navigation and communication equipment—"OFF."

Fuel tank selector valve—"ON."

Brakes—"ON."

Start engine in accordance with manufacturer's recommended procedure.

Check all instruments for proper operation and indication.

Check all powerplant controls:

At idle r.p.m., momentarily switch magnetos to "OFF," and check for proper "ground." If the engine continues to operate after the switch is turned to "OFF," it indicates a faulty ground circuit between the magneto and switch and should be corrected before further engine operation.

Check flight controls, including flaps, for free and smooth operation in proper direction.

Check radio receiver and tune to proper frequency.

Set altimeter and clock.

After storage—A thorough inspection is recommended for aircraft that have been tied down or stored for an extensive period of time. Inactive aircraft are frequently used for nesting by insects and animals. Birds' nests in air intake scoops impair airflow, resulting in excessively rich mixtures and may cause engine stoppage. Nests lodged between engine cylinders and engine baffles cause overheating, preignition, and detonation. Insect nests obstructing fuel tank vents cause lean mixtures and fuel starvation. Mice remove rib stitching, making wings unsafe. Excretions from rodents are highly corrosive to aluminum alloy metals and harmful to fabric and wood. Deterioration or excessive weather-checking of fuel, oil, hydraulic, or induction hoses may result in leaks and faulty operation.

Be sure to inspect:

Oil coolers   —intake scoops

Carburetors   —intake screens and passages

Fuel tank vents   —free of obstruction

Pitot tubes   —free of obstructions

Fuselage   —interior and baggage areas

Wings   —interior of wings and control surfaces

Static vents   —free of obstructions

Cold Weather—Aircraft having fuel tank caps installed flush with the wing upper surface are susceptible to collecting water in the filter overflow well. This water may freeze during cold weather operations, resulting in blockage of the fuel tank vent and engine failure. Partial obstruction of the vent may cause erratic engine operation and loss of power. In the case of aircraft using engine-driven fuel pumps, if the vent is clogged, the tank may collapse causing structural damage.

Check carburetor air scoop for obstructions and open the drain. Water accumulation in the air scoop may freeze and the engine will not develop full power on takeoff. Check carburetor air filter screens for obstructions to airflow from ice and snow accumulation.

Drain fuel tank sumps regularly. Water can form in the fuel tank due to condensation from rapid and extreme temperature changes and may freeze. This can result in restricted fuel flow and cracked lines and fuel strainer bowls.

Check for ice accumulation in the rear section of the fuselage and inside of the wings and control surfaces. Blowing and drifting snow may seep into the fuselage, wings, and control surfaces, eventually melting and accumulating in a low point where it may ultimately freeze. The weight of this ice may be great enough to seriously affect flight characteristics. Make sure that all frost, snow, and ice are removed from the aircraft, especially from the top of wings and other airfoil surfaces.

Check for proper operation of oil cooler shutters or use of covers as specified by the aircraft manufacturer. Determine that the grade of oil used conforms with the engine manufacturer's recommendation.

CHAPTER III—FLIGHT INSTRUMENTS

The use of instruments as an aid to flight enables the pilot to operate the airplane more precisely, and therefore, obtain maximum performance and enhanced safety. This is particularly true when flying greater distances. Manufacturers have provided the necessary flight instruments; however, it is the pilot's responsibility to gain the essential knowledge about how the instruments operate so that they can be used effectively.

This chapter covers the operational aspects of the pitot-static system and associated instruments: the vacuum system and associated instruments; and the magnetic compass.

The Pitot-Static System and Associated Instruments

There are two major parts of the pitot-static system; (1) the impact pressure chamber and lines, and (2) the static pressure chamber and lines. This system provides the source of air pressure for the operation of the altimeter, vertical speed indicator (vertical velocity indicator), and the airspeed indicator.

In older airplanes, the impact pressure and the static pressure were generated from one common source known as the pitot-static tube.

The installation in newer airplanes separates the pitot and static sources. In this system the impact air pressure (air striking the airplane because of its forward motion) is taken from a pitot tube, which is mounted either on the leading edge of the wing or on the nose, and aligned to the relative wind. On certain airplanes the pitot tube is located on the vertical stabilizer. These locations provide minimum disturbance or turbulence caused by the motion of the airplane through the air. The static pressure (pressure of the still air) is usually taken from the static line attached to a vent or vents mounted flush with the side of the fuselage. On most airplanes using a flush-type static source, there are two vents, one on each side of the fuselage. This compensates for any possible variation in static pressure due to erratic changes in airplane attitude.

The openings of both the pitot tube and the static vent should be checked during the preflight inspection to assure that they are free from obstructions. Clogged or partially clogged openings should be cleaned by a certificated mechanic. Blowing into these openings is not recommended because this could damage any of the three instruments (Fig. 3–1).

Briefly the operation of the pitot-static system is as follows: As the airplane moves through the air the impact pressure on the open pitot tube affects the pressure in the pitot chamber. Any change of pressure in the pitot

chamber is transmitted through a line connected to the airspeed indicator which utilizes impact pressure for its operation. The static chamber is vented through small holes to the free undisturbed air, and as the atmospheric pressure increases or decreases, the pressure in the static chamber changes accordingly. Again, this pressure change is transmitted through lines to the instruments which utilize static pressure as illustrated in Fig. 3–1.

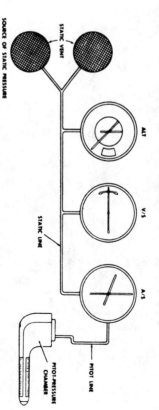

SOURCE OR STATIC PRESSURE

STATIC VENT

ALT   V/S   A/S

STATIC LINE

PITOT LINE

PITOT-PRESSURE CHAMBER

Flush-type static source

Figure 3–1. *Pitot-static system with instruments.*

An alternate source for static pressure is provided in some airplanes in the event the static ports become clogged. This source usually is vented to the pressure inside the cockpit. Because of the venturi effect of the flow of air over the cockpit, this alternate static pressure is usually lower than the pressure provided by the normal static air source. When the alternate static source is used, the following differences in the instrument indications usually occur: the altimeter will indicate higher than the actual altitude, the airspeed greater than the actual airspeed, and the vertical speed will indicate a climb while in level flight.

The Altimeter

The altimeter (Fig. 3–2) measures the height of the airplane above a given level. Since it is the only instrument that gives altitude information, the altimeter is one of the most important instruments in the airplane. To use the altimeter effectively, the pilot must thoroughly understand its principle of operation and the effect of atmospheric pressure and temperature on the altimeter.

The dial of a typical altimeter is graduated with numerals arranged clockwise from 0 to 9 inclusive as shown in Fig. 3-2. Movement of the aneroid element is transmitted through a gear train to the three hands which sweep the calibrated dial to indicate the altitude. The shortest hand indicates altitude in tens of thousands of feet; the intermediate hand, in thousands of feet; and the longest hand in hundreds of feet, subdivided into 20-foot increments.

This indicated altitude is correct, however, only if the sea level barometric pressure is standard (29.92 inches of mercury), the sea level free air temperature is standard (+15° C. or 59° F.), and furthermore, the pressure and temperature decrease at a standard rate with increase in altitude. Since atmospheric pressure continually changes, a means is provided to adjust the altimeter to compensate for nonstandard conditions. This is accomplished through a system by which the altimeter setting (local station barometric pressure reduced to sea level) is set to a barometric scale located on the face of the altimeter. Only after the altimeter is set properly will it indicate the correct altitude.

Effect of Nonstandard Pressure and Temperature. If no means were provided for adjusting altimeters to nonstandard pressure, flight could be hazardous. For example, if a flight is made from a high pressure area to a low pressure area without adjusting the altimeter, the actual altitude of the airplane will be lower than the indicated altitude, and when flying from a low pressure area to high pressure area, the actual altitude of the airplane will be higher than the indicated altitude. Fortunately, this error can be corrected by setting the altimeter properly.

Variations in air temperature also affect the altimeter. On a warm day the expanded air is lighter in weight per unit volume than on a cold day, and consequently the pressure levels are raised. For example, the pressure level where the altimeter indicates 10,000 ft. will be HIGHER on a warm day than under standard conditions. On a cold day the reverse is true, and the 10,000-foot level would be LOWER. The adjustment made by the pilot to compensate for nonstandard pressures does not compensate for nonstandard temperatures. Therefore, if terrain or obstacle clearance is a factor in the selection of a cruising altitude, particularly at higher altitudes, remember to anticipate that COLDER-THAN-STANDARD TEMPERATURE will place the aircraft LOWER than the altimeter indicates. Therefore, a higher altitude should be used to provide adequate terrain clearance.

Setting the Altimeter. To adjust the altimeter for variation in atmospheric pressure, the pressure scale in the altimeter setting window, calibrated in inches of mercury (Hg), is adjusted to correspond with the given altimeter

Figure 3-2. *Sensitive altimeter. The instrument is adjusted by the knob (lower left) so the current altimeter setting (30.34) appears in the window to the right.*

Principle of Operation. The pressure altimeter is simply an aneroid barometer that measures the pressure of the atmosphere at the level where the altimeter is located, and presents an altitude indication in feet. The altimeter uses static pressure as its source of operation. Air is more dense at the surface of the earth than aloft, therefore as altitude increases, atmospheric pressure decreases. This difference in pressure at various levels causes the altimeter to indicate changes in altitude.

The presentation of altitude varies considerably between different types of altimeters. Some have one pointer while others have more. Only the multipointer type will be discussed in this handbook.

setting. Altimeter settings can be defined as station pressure reduced to sea level, expressed in inches of mercury.

The station reporting the altimeter setting takes an hourly measurement of the station's atmospheric pressure and corrects this value to sea level pressure. These altimeter settings reflect height above sea level only in the vicinity of the reporting station. Therefore it is necessary to adjust the altimeter setting as the flight progresses from one station to the next.

FAA regulations provide the following concerning altimeter settings: The cruising altitude of an aircraft below 18,000 ft. MSL shall be maintained by reference to an altimeter that is set to the current reported altimeter setting of a station located along the route of flight and within 100 nautical miles of the aircraft. If there is no such station, the current reported altimeter setting of an appropriate available station shall be used. In an aircraft having no radio, the altimeter shall be set to the elevation of the departure airport or an appropriate altimeter setting available before departure.

Many pilots confidently expect that the current altimeter setting will compensate for irregularities in atmospheric pressure at all altitudes. This is not always true because the altimeter setting broadcast by ground stations is the station pressure corrected to mean sea level. The altimeter setting does not account for distortion at higher levels, particularly the effect of nonstandard temperature.

It should be pointed out, however, that if each pilot in a given area were to use the same altimeter setting, each altimeter will be equally affected by temperature and pressure variation errors, making it possible to maintain desired altitude separation between aircraft.

When flying over high mountainous terrain, certain atmospheric conditions can cause the altimeter to indicate an altitude of 1,000 ft., or more, HIGHER than the actual altitude. For this reason a generous margin of altitude should be allowed—not only for possible altimeter error, but also for possible downdrafts which are particularly prevalent if high winds are encountered.

To illustrate the use of the altimeter setting system, follow a flight from Love Field, Dallas, Texas, to Abilene Municipal Airport, Abilene, Texas, via the Mineral Wells VOR. Before takeoff from Love Field, the pilot receives a current altimeter setting of 29.85 from the control tower or Automatic Terminal Information Service. This value is set in the altimeter setting window of the altimeter. The altimeter indication should then be compared with the known Dallas field elevation of 485 ft. If the altimeter is calibrated properly, it should indicate the field elevation of 485 ft. Since most altimeters are

not perfectly calibrated, an error may exist. If an altimeter indication varies from the field elevation more than 75 feet, the accuracy of the instrument is questionable and it should be referred to an instrument technician for recalibration.

When over the Mineral Wells VOR, assume the pilot receives a current altimeter setting of 29.94 from the Mineral Wells FAA Flight Service Station, and applies this setting to the altimeter. Before entering the traffic pattern at Abilene Municipal Airport, further assume a new altimeter setting of 29.69 is received from the Abilene Control Tower, and applied to the altimeter. If the pilot desires to fly the traffic pattern at approximately 800 ft. above terrain—the field elevation at Abilene is 1,778 ft.—an indicated altitude of approximately 2,600 ft. should be maintained (1,778 ft. + 800 ft. = 2,578 ft.). Upon landing the altimeter should indicate the field elevation at Abilene Municipal (1,778 ft.).

The importance of properly setting and reading the altimeter cannot be overemphasized.

Let's assume that the pilot neglected to adjust the altimeter at Abilene to the current setting, and uses the Mineral Wells altimeter setting of 29.94. If this occurred, the airplane, when in the Abilene traffic pattern, would be approximately 250 ft. below the proper traffic pattern altitude while indicating an altitude of 2,600 ft., and the altimeter would indicate approximately 250 ft. more than the field elevation or 2,028 ft. upon landing.

Actual altimeter setting 29.94
Proper altimeter setting 29.69
Difference . . . . .25 of an inch

(1 inch of pressure is equal to approximately 1,000 ft. of altitude—.25 × 1,000 ft. = 250 ft.)

The above calculation may be confusing, particularly in determining whether to add or subtract the amount of altimeter error. The following additional explanation is offered and can be helpful in finding the solution to this type of problem.

There are two means by which the altimeter pointers can be moved. One utilizes changes in air pressure while the other utilizes the mechanical makeup of the altimeter setting system.

When the airplane altitude is changed, the changing pressure within the altimeter case expands or contracts the aneroid barometer which through linkage rotates the pointers. A decrease in pressure causes the altimeter to indicate an increase in altitude and an increase in pressure causes the altimeter to indicate a decrease in altitude. It is obvious then that if the airplane is

29.92 inches of mercury. Pressure altitude is used for computer solutions to determine density altitude, true altitude, true airspeed, etc.

TRUE ALTITUDE—The true vertical distance of the aircraft above sea level—the actual altitude. (Often expressed in this manner; 10,900 ft. MSL.) Airport, terrain, and obstacle elevations found on aeronautical charts are true altitudes.

DENSITY ALTITUDE—This altitude is pressure altitude corrected for nonstandard temperature variations. When conditions are standard, pressure altitude and density altitude are the same. Consequently, if the temperature is above standard, the density altitude will be higher than pressure altitude. If the temperature is below standard, the density altitude will be lower than pressure altitude. This is an important altitude because it is directly related to the aircraft's takeoff and climb performance.

Vertical Speed Indicator

The vertical speed or vertical velocity indicator (Fig. 3–3) indicates whether the aircraft is climbing, descending, or in level flight. The rate of climb or descent is indicated in feet per minute. If properly calibrated, this indicator will register zero in level flight.

Principle of Operation. Although the vertical speed indicator operates solely from static pressure, it is a differential pressure instrument. The case of the instrument is airtight except for a small connection through a restricted passage to the static line of the pitot static system.

A diaphragm with connecting linkage and gearing to the indicator

Figure 3–3. *Vertical speed indicator.*

flown from a pressure level of 28.75 inches of Hg to a pressure level of 29.75 inches of Hg, the altimeter would show a decrease of approximately 1,000 feet in altitude.

The other method of moving the pointers does not rely on changing air pressure but the mechanical construction of the altimeter. When the knob on the altimeter is rotated, the altimeter setting pressure scale moves simultaneously with the altimeter pointers. This may be confusing because the numerical values of pressure indicated in the window increase while the altimeter indicates an increase in altitude; or decrease while the altimeter indicates a decrease in altitude. This is contrary to the reaction on the pointers when air pressure changes, and is based solely on the mechanical makeup of the altimeter. To further explain this point, assume that the proper altimeter setting is 29.50 and the actual setting is 30.00 or a .50 difference. This would cause a 500-foot error in altitude. In this case if the altimeter setting is adjusted from 30.00 to 29.50, the numerical value decreases; but the altimeter indicates a decrease of 500 ft. in altitude. Before the correction was made the airplane was flying at an altitude 500 feet higher than it should have been with the correct altimeter setting.

Types of Altitude

Knowing the aircraft's altitude is vitally important to the pilot for several reasons. The pilot must be sure that the airplane is flying high enough to clear the highest terrain or obstruction along the intended route; this is especially important when visibility is restricted. To keep above mountain peaks, the pilot must note the altitude of the aircraft and elevation of the surrounding terrain at all times. To reduce the possibility of a midair collision, the pilot must maintain altitudes in accordance with air traffic rules. Often certain altitudes are selected to take advantage of favorable winds and weather conditions. Also, a knowledge of the altitude is necessary to calculate true airspeeds.

Altitude is vertical distance above some point or level used as a reference. There may be as many kinds of altitude as there are reference levels from which altitude is measured and each may be used for specific reasons. Pilots are usually concerned, however, with five types of altitudes:

ABSOLUTE ALTITUDE—The vertical distance of an aircraft above the terrain.

INDICATED ALTITUDE—That altitude read directly from the altimeter (uncorrected) after it is set to the current altimeter setting.

PRESSURE ALTITUDE—The altitude indicated when the altimeter setting window is adjusted to 29.92. This is the standard datum plane, a theoretical plane where air pressure (corrected to 15° C.) is equal to

pointer is located inside the sealed case. Both the diaphragm and the case receive air from the static line at existing atmospheric pressure. When the aircraft is on the ground or in level flight, the pressures inside the diaphragm and the instrument case remain the same and the pointer is at the zero indication. When the aircraft climbs or descends the pressure inside the diaphragm changes immediately but, due to the metering action of the restricted passage, the case pressure will remain higher or lower for a short time causing the diaphragm to contract or expand. This causes a differential pressure which is indicated on the instrument needle as a climb or descent.

The Airspeed Indicator. The airspeed indicator (Fig. 3–4) is a sensitive, differential pressure gauge which measures and shows promptly the difference between (1) pitot, or impact pressure, and (2) static pressure, the un-

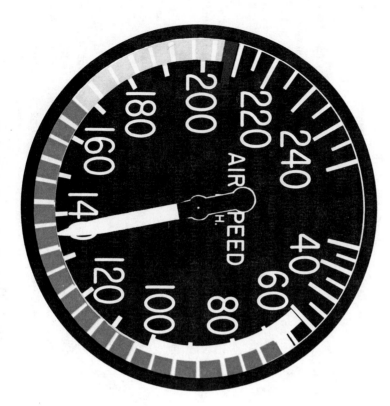

Figure 3–4. *Airspeed indicator.*

disturbed atmospheric pressure at flight level. These two pressures will be equal when the aircraft is parked on the ground in calm air. When the aircraft moves through the air, the pressure on the pitot line becomes greater than the pressure in the static lines. This difference in pressure is registered by the airspeed pointer on the face of the instrument, which is calibrated in miles per hour, knots, or both.

Kinds of Airspeed. There are three kinds of airspeed that the pilot should understand: (1) indicated airspeed; (2) calibrated airspeed; and (3) true airspeed.

Indicated Airspeed. The direct instrument reading obtained from the airspeed indicator, uncorrected for variations in atmospheric density, installation error, or instrument error.

Calibrated Airspeed. Calibrated airspeed (CAS) is indicated airspeed corrected for installation error and instrument error. Although manufacturers attempt to keep airspeed errors to a minimum, it is not possible to eliminate all errors throughout the airspeed operating range. At certain airspeeds and with certain flap settings, the installation and instrument error may be several miles per hour. This error is generally greatest at low airspeeds. In the cruising and higher airspeed ranges, indicated airspeed and calibrated airspeed are approximately the same.

It may be important to refer to the airspeed calibration chart to correct for possible airspeed errors because airspeed limitations such as those found on the color-coded face of the airspeed indicator, on placards in the cockpit, or in the Airplane Flight Manual or Owner's Handbook are usually calibrated airspeeds. Some manufacturers use indicated rather than calibrated airspeed to denote the airspeed limitations mentioned.

The airspeed indicator should be calibrated periodically because leaks may develop or moisture may collect in the tubing. Dirt, dust, ice, or snow collecting at the mouth of the tube may obstruct air passage and prevent correct indications, and also vibrations may destroy the sensitivity of the diaphragm.

True Airspeed. The airspeed indicator is calibrated to indicate true airspeed under standard sea level conditions—that is, 29.92 inches of Hg and 15° C. Because air density decreases with an increase in altitude, the airplane has to be flown faster at higher altitudes to cause the same pressure difference between pitot impact pressure and static pressure. Therefore, for a given true airspeed, indicated airspeed decreases as altitude increases or for a given indicated airspeed, true airspeed increases with an increase in altitude.

A pilot can find true airspeed by two methods. The first method, which is more accurate, involves using a computer. In this method, the calibrated airspeed is corrected for temperature and pressure variation by using the airspeed correction scale on the computer.

A second method, which is a "rule of thumb," can be used to compute the approximate true airspeed. This is done by adding to the indicated airspeed 2% of the indicated airspeed for each 1,000 ft. of altitude.

Sample Problem:

Given:

IAS. . . . . 140 m.p.h.

Altitude . . . . 6,000 ft.

Find: True Airspeed (TAS)

Solution:

$2\% \times 6 = 12\%$ (.12)

$140 \times .12 = 16.8$

$140 + 16.8 = 156.8$ m.p.h. (TAS)

The Airspeed Indicator Markings. Airplanes weighing 12,500 lbs. or less manufactured after 1945 and certificated by FAA, are required to have airspeed indicators that conform in a standard color-coded marking system. This system of color-coded markings, pictured in Fig 3–5, enables the pilot to determine at a glance certain airspeed limitations which are important to the safe operation of the aircraft. For example, if during the execution of a maneuver, the pilot notes that the airspeed needle is in the yellow arc and is rapidly approaching the red line, immediate corrective action to reduce the airspeed should be taken. It is essential at high airspeed that the pilot use smooth control pressures to avoid severe stresses upon the aircraft structure.

The following is a description of the standard color-code markings on airspeed indicators used on single-engine light airplanes:

FLAP OPERATING RANGE (the white arc).

POWER-OFF STALLING SPEED WITH THE WING FLAPS AND LANDING GEAR IN THE LANDING POSITION (the lower limit of the white arc).

MAXIMUM FLAPS EXTENDED SPEED (the upper limit of the white arc). This is the highest airspeed at which the pilot should extend full flaps. If flaps are operated at higher airspeeds, severe strain or structural failure could result.

NORMAL OPERATING RANGE (the green arc).

POWER-OFF STALLING SPEED WITH THE WING FLAPS AND LANDING GEAR RETRACTED (the lower limit of the green arc).

MAXIMUM STRUCTURAL CRUISING SPEED (the upper limit of the green arc). This is the maximum speed for normal operation.

CAUTION RANGE (the yellow arc). The pilot should avoid this area unless in smooth air.

NEVER-EXCEED SPEED (the red line). This is the maximum speed at which the airplane can be operated in smooth air. This speed should never be exceeded intentionally.

Other Airspeed Limitations. There are other important airspeed limitations *not* marked on the face of the airspeed indicator. These speeds are generally found on placards in view of the pilot and in the Airplane Flight Manual or Owner's Handbook.

One example is the MANEUVERING SPEED. This is the "rough air" speed and the maximum speed for abrupt maneuvers. If during flight, rough air or severe turbulence is encountered, the airspeed should be reduced to maneuvering speed or less to minimize the stress on the airplane structure.

Other important airspeeds include LANDING GEAR OPERATING SPEED, the maximum speed for extending or retracting the landing gear if using aircraft equipped with retractable landing gear; the BEST ANGLE OF CLIMB SPEED, important when a short field takeoff to clear an obstacle is required; and the BEST RATE OF CLIMB SPEED, the airspeed that will give

Figure 3–5. *Airspeed indicator showing color-coded marking system.*

the pilot the most altitude in a given period of time. The pilot who flies the increasingly popular light twin-engine aircraft must know the aircraft's MINIMUM CONTROL SPEED, the minimum flight speed at which the aircraft is satisfactorily controllable when an engine is suddenly made inoperative with the remaining engine at takeoff power. The last two airspeeds are now marked either on the face of the airspeed indicator or on the instrument panel of recently manufactured airplanes.

Description of these airspeed limitations are, through choice, limited to layman's language.

The following are abbreviations for performance speeds:

"V_a" —design maneuvering speed.
"V_c" —design cruising speed.
"V_f" —design flap speed.
"V_{fe}" —maximum flap extended speed.
"V_{le}" —maximum landing gear extended speed.
"V_{lo}" —maximum landing gear operating speed.
"V_{lof}" —lift-off speed.
"V_{ne}" —never-exceed speed.
"V_r" —rotation speed.
"V_s" —the stalling speed or the minimum steady flight speed at which the airplane is controllable.
"V_{so}" —the stalling speed or the minimum steady flight speed in the landing configuration.
"V_{s1}" —the stalling speed or the minimum steady flight speed obtained in a specified configuration.
"V_x" —speed for best angle of climb.
"V_y" —speed for best rate of climb.

Sources of Power for Gyroscopic Operation The gyroscopic instruments can be operated either by a vacuum or an electrical system. In some airplanes, all the gyros are either vacuum or electrically operated; in others, vacuum systems provide the power for the heading and attitude indicators, while the electrical system provides the power to drive the gyroscope of the turn needle.

Gyroscopic Flight Instruments Several flight instruments utilize the properties of a gyroscope for their operation. The most common instruments containing gyroscopes are the turn indicator, turn coordinator, heading indicator, and the attitude indicator. To understand how these instruments operate requires a knowledge of the instrument power systems, gyroscopic principles, and the operating principles of each instrument.

Vacuum or Pressure System. The vacuum or pressure system spins the gyro by drawing a stream of air against the rotor vanes to spin the rotor at high speeds essentially the same as a water wheel or turbine operates. Either a venturi tube or a vacuum pump can be used to provide the vacuum required to spin the rotors.

The amount of vacuum required for instrument operation is usually between 3.5" to 4.5" Hg. The turn-and-bank indicators used in some installations require a lower vacuum setting. These values are obtained by using regulating valves in the individual instrument supply line.

Venturi-Tube Systems. The venturi tube was used as a means to operate the gyroscopes in many airplanes manufactured years ago. The advantages of the venturi tube as a suction source are its relatively low cost and simplicity of installation and operation.

The venturi tube is mounted on the exterior of the aircraft fuselage. Throughout the normal operating airspeed range the velocity of the air through the venturi creates sufficient suction to spin the gyro.

The limitations of the venturi system should be noted. The venturi is designed to produce the desired vacuum at approximately 100 m.p.h. under standard sea level conditions. Wide variations in airspeed or air density, or restriction to airflow by ice accretion, will affect the pressure at the venturi throat and thus the vacuum driving the gyro rotor. Since the rotor does not reach normal operating speed until after takeoff, preflight operational checks of venturi-powered gyro instruments are impossible. For this reason the system is adequate only for light aircraft instrument training and limited flying under instrument weather conditions. Aircraft flown throughout a wider range of speed, altitude, and weather conditions require a more effective source of power which is independent of airspeed and less susceptible to adverse atmospheric conditions.

Engine Driven Vacuum Pump. One source of vacuum for the gyros installed in light aircraft is the vane-type engine-driven pump which is mounted on the accessory drive shaft of the engine, and is connected to the engine lubrication system to seal, cool, and lubricate the pump. Another type of pump is mounted on the side of the engine block forward of the engine cylinders and is driven from the engine crankshaft by a pulley and V-belt arrangement. Pump capacity and pump size vary in different aircraft, depending on the number of gyros to be operated.

A typical vacuum system (Fig. 3-6) consists of an engine-driven vacuum pump, an air/oil separator, a vacuum regulator, a relief valve, an air filter, and tubing and manifolds necessary to complete the connections. A suction gauge

Rigidity in space can best be explained by applying Newton's first Law of Motion which states, "a body at rest will remain at rest; or if in motion in a straight line, it will continue in a straight line unless acted upon by an outside force." An example of this law is the rotor of a universally mounted gyro. When the wheel is spinning, it exhibits the ability to remain in its original plane of rotation regardless of how the base is moved. However, since it is impossible to design bearings without some friction present, there will be some deflective force upon the wheel.

The flight instruments using the gyroscopic property of rigidity for their operation are the attitude indicator and the heading indicator; therefore, their rotors must be freely or universally mounted.

The second property of a gyroscope—precession—is the resultant action or deflection of a spinning wheel when a deflective force is applied to its rim (Fig. 3-7). When a deflective force is applied to the rim of a rotating wheel,

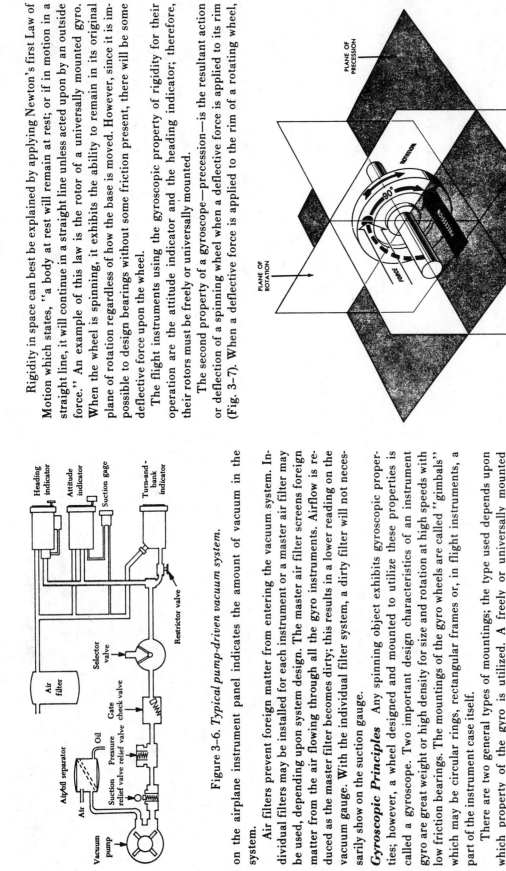

Figure 3-7. *Precession of a gyroscope resulting from an applied deflective force.*

Figure 3-6. *Typical pump-driven vacuum system.*

on the airplane instrument panel indicates the amount of vacuum in the system.

Air filters prevent foreign matter from entering the vacuum system. Individual filters may be installed for each instrument or a master air filter may be used, depending upon system design. The master air filter screens foreign matter from the air flowing through all the gyro instruments. Airflow is reduced as the master filter becomes dirty; this results in a lower reading on the vacuum gauge. With the individual filter system, a dirty filter will not necessarily show on the suction gauge.

Gyroscopic Principles Any spinning object exhibits gyroscopic properties; however, a wheel designed and mounted to utilize these properties is called a gyroscope. Two important design characteristics of an instrument gyro are great weight or high density for size and rotation at high speeds with low friction bearings. The mountings of the gyro wheels are called "gimbals" which may be circular rings, rectangular frames or, in flight instruments, a part of the instrument case itself.

There are two general types of mountings; the type used depends upon which property of the gyro is utilized. A freely or universally mounted gyroscope is free to rotate in any direction about its center of gravity. Such a wheel is said to have three planes of freedom. The wheel or rotor is free to rotate in any plane in relation to the base and is so balanced that with the gyro wheel at rest, it will remain in the position in which it is placed. Restricted or semirigidly mounted gyroscopes are those mounted so that one of the planes of freedom is held fixed in relation to the base.

There are two fundamental properties of gyroscopic action; rigidity in space, and precession.

66

the resultant force is 90° ahead in the direction of rotation and in the direction of the applied force. The rate at which the wheel precesses is inversely proportional to the speed of the rotor and proportional to the deflective force. The force with which the wheel precesses is the same as the deflective force applied (minus the friction in the bearings). If too great a deflective force is applied for the amount of rigidity in the wheel, the wheel precesses and topples over at the same time.

Turn-and-Slip Indicator

The turn and slip indicator was one of the first instruments used for controlling an airplane without visual reference to the ground or horizon (Fig. 3-8). Its principal uses in airplanes are to indicate turn and to serve as an emergency source of bank information in the event the attitude indicator fails.

The turn and slip indicator is actually a combination of two instruments: the turn needle and the ball or inclinometer. The needle is gyro operated to show rate of turn, and the ball reacts to gravity and/or centrifugal force to indicate the need for directional trim.

As stated previously, the turn needle is operated by a gyro, driven by either vacuum or electricity. Semirigid mounting of the gyro permits it to rotate freely about the lateral and longitudinal axes of the airplane but is restricted in rotation about the vertical axis. When the airplane is turned or rotated around the vertical axis a deflective force is set up causing the gyro to precess, which results in tilting the gyro. This tilting is transmitted to the turn needle through linkage. As the rate of turn increases, the precession of the gyro increases, resulting in an increased rate of turn being indicated. A spring

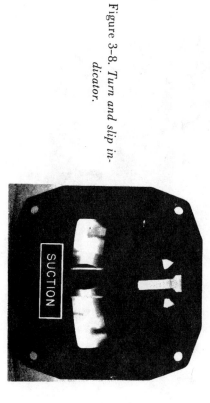

Figure 3–8. *Turn and slip indicator.*

assembly attached to the gyro keeps the gyro upright when a deflective force is not applied. This spring is also adjustable to calibrate the instrument for a given rate or turn. In addition, a dampening mechanism prevents excessive oscillation of the turn needle.

The Turn Needle. The turn needle indicates the rate (number of degrees per second) at which the aircraft is turning about its vertical axis. Unlike the attitude indicator, it does not give a direct indication of the banked attitude of the aircraft. For any given airspeed, however, there is a specific angle of bank necessary to maintain a coordinated turn at a given rate. The faster the airspeed, the greater the angle of bank required to obtain a given rate of turn. Thus, the turn needle gives only an indirect indication of the aircraft's banking attitude or angle of bank.

Types of Turn Needles. There are two types of turn needles—the "2 minute" turn needle and the "4 minute" turn needle. When using a 2-minute turn needle, a 360° turn made at a rate indicated by a one-needle width deflection would require 2 minutes to complete. In this case, the aircraft would be turning at a rate of 3° per second, which is considered a standard rate turn. When using a 4-minute turn needle, a 360° turn made at a rate indicated by a one-needle width deflection would require 4 minutes to complete. In this case, the aircraft is turning at a rate of 1.5° per second. A standard rate turn of 3° per second would be indicated on this type of turn needle by a two-needle width deflection.

The Ball. This part of the instrument is a simple inclinometer consisting of a sealed, curved glass tube containing kerosene and a black agate or steel ball bearing which is free to move inside the tube. The fluid provides a dampening action which ensures smooth and easy movement of the ball. The tube is curved so that when in a horizontal position the ball tends to seek the lowest point. Two reference markers are provided as an aid in determining when the ball is in the center of the tube. During coordinated straight-and-level flight, the force of gravity causes the ball to rest in the lowest part of the tube, centered between the reference lines. During a coordinated turn, turning forces are balanced, which causes the ball to remain centered in the tube. If turning forces are unbalanced, such as during a slip or skid, the ball moves away from the center of the tube in the direction of the excessive force.

The ball then is actually a balance indicator, and is used as a visual aid to determine coordinated use of the aileron and rudder control. During a turn it indicates the relationship between the angle of bank and rate of turn. It indicates the "quality" of the turn or whether the aircraft has the correct angle of bank for the rate of turn.

Figure 3–9. *Indications of the ball in various types of turns.*

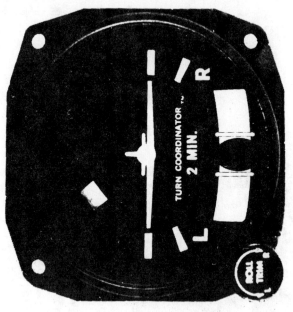

Figure 3–10. *Turn coordinator.*

In a coordinated turn, the ball assumes a position between the reference markers (Fig. 3–9).

In a skid, the rate of turn is too great for the angle of bank, and the excessive centrifugal force moves the ball to the outside of the turn (Fig. 3–9). To achieve coordinated flight from a skid requires that the bank be increased or the rate of turn decreased, or a combination of both.

In a slip, the rate of turn is too slow for the angle of bank, and the lack of centrifugal force moves the ball to the inside of the turn (Fig. 3–9). To achieve coordinated flight from a slip requires the bank be decreased or the rate of turn increased, or a combination of both.

It is very important for the pilot to understand that the ball should be kept in the center at all times during flight except for certain maneuvers such as intentional slips. If the ball is not centered, it means that abnormal forces are being created which could cause the airplane to unexpectedly stall or spin, which is dangerous, particularly close to the ground.

Turn Coordinator In recent years a new type of turn indicator has been developed and is used quite extensively. This instrument is referred to as a "Turn Coordinator" (Fig. 3–10). In place of the turn needle indication, this instrument shows the movement of the airplane about the longitudinal axis by displaying a miniature airplane on the instrument. The movement of the miniature airplane on the instrument is proportional to the roll rate of the airplane. When the roll rate is reduced to zero, or in other words the bank is held constant, the instrument provides an indication of the rate of turn. This

The new design features a realignment of the gyro in such a manner that it senses airplane movement about the yaw and roll axis and pictorially displays the resultant motion as described above. The conventional inclinometer (ball) is also incorporated in this instrument.

The Heading Indicator. The heading indicator (or directional gyro) is fundamentally a mechanical instrument designed to facilitate the use of the magnetic compass. Errors in the magnetic compass are numerous, making straight flight and precision turns to headings difficult to accomplish, particularly in turbulent air. Heading indicators (Figs. 3-11 and 3-12), however, are not affected by the forces that make the magnetic compass difficult to interpret.

The operation of the heading indicator depends upon the principle of rigidity in space. The rotor turns in a vertical plane, and fixed to the rotor is a compass card. Since the rotor remains rigid in space, the points on the card hold the same position in space relative to the vertical plane. As the instrument case and the airplane revolve around the vertical axis, the card provides clear and accurate heading information.

Heading information is displayed by one of two means. Fig. 3-11 shows the older type heading indicator which displays headings on a card mounted horizontally around the gyro mechanism. Only a small portion of the heading indicator can be seen through the window. Fig. 3-12 shows the more modern heading indicator which displays headings on a dial mounted vertically to the instrument. All headings can be seen on the latter instrument. Both are calibrated in five degree increments.

Because of precession, caused chiefly by bearing friction, the heading indicator will creep or drift from a heading to which it is set. Among other fac-

Figure 3-11. *Heading indicator.*

Figure 3-12. *Heading indicator.*

tors the amount of drift depends largely upon the condition of the instrument. If the bearings are worn, dirty, or improperly lubricated, the drift may be excessive.

Bear in mind that the heading indicator is not direction-seeking, as is the magnetic compass. It is important to check the indications frequently and reset the heading indicator to align it with the magnetic compass when required. Adjusting the heading indicator to the magnetic compass heading should be done only when the airplane is in wings-level unaccelerated flight; otherwise erroneous magnetic compass readings may be obtained.

The bank and pitch limits of the heading indicator vary with the particular design and make of instrument. On some heading indicators found in light airplanes, the limits are approximately 55° of pitch and 55° of bank. When either of these attitude limits is exceeded, the instrument "tumbles" or "spills" and no longer gives the correct indication until reset. After spilling, it may be reset with the caging knob. Many of the modern instruments used are designed in such a manner that they will not tumble.

The Attitude Indicator. The attitude indicator, with its miniature aircraft and horizon bar, displays a picture of the attitude of the airplane (Fig 3-13). The relationship of the miniature aircraft to the horizon bar is the same as the relationship of the real aircraft to the actual horizon. The instrument gives an instantaneous indication of even the smallest changes in attitude.

The gyro in the attitude indicator is mounted on a horizontal plane and depends upon rigidity in space for its operation. The horizon bar represents the true horizon. This bar is fixed to the gyro and remains in a horizontal plane as the airplane is pitched or banked about its lateral or longitudinal axis, indicating the attitude of the airplane relative to the true horizon.

An adjustment knob is provided with which the pilot may move the miniature airplane up or down to align the miniature airplane with the horizon bar to suit the pilot's line of vision. Normally, the miniature airplane is adjusted so that the wings overlap the horizon bar when the airplane is in straight-and-level cruising flight.

Certain attitude indicators are equipped with a caging mechanism. If the instrument is so equipped, it should be uncaged only in straight-and-level flight; otherwise, it will be unreliable.

The pitch and bank limits depend upon the make and model of the instrument. Limits in the banking plane are usually from 100° to 110°, and the pitch limits are usually from 60° to 70°. If either limit is exceeded, the instrument will tumble or spill and will give incorrect indications until reset. A number of modern attitude indicators will not tumble.

on the instrument panel. Its indications are very close approximations of the actual attitude of the airplane.

Figure 3-13. Attitude indicator.

Every pilot should be able to interpret the banking scale indicators on the top of the instrument move in the opposite banking scale (Fig. 3–14). Most direction from that in which the airplane is actually banked. This may confuse the pilot if the indicator is used to determine the direction of bank. This scale should be used only to control the degree of desired bank. The relationship of the miniature airplane to the horizon bar should be used for an indication of the direction of bank. An attitude indicator is now available with a banking scale indicator that moves in the same direction as the bank.

The attitude indicator is reliable and the most realistic flight instrument

Magnetic Compass   Since the magnetic compass works on the principle of magnetism, it is well for the pilot to have at least a basic understanding of magnetism. A simple bar magnet has two centers of magnetism which are called poles. Lines of magnetic force flow out from each pole in all directions, eventually bending around and returning to the other pole. The area through which these lines of force flow is called the field of the magnet (Fig. 3–15). For the purpose of this discussion, the poles are designated "north" and "south." If two-bar magnets are placed near each other the north pole of one will attract the south pole of the other. There is evidence that there is a magnetic field surrounding the earth, and this theory is applied in the design of the magnetic compass. It acts very much as though there were a huge bar magnet running along the axis of the earth which ends several hundred miles below the surface.

The lines of force have a vertical component (or pull) which is zero at the equator but builds to 100 per cent of the total force at the poles. If magnetic needles, such as the airplane magnetic compass bars, are held along these lines of force, the vertical component causes one end of the needle to dip or deflect downward. The amount of dip increases as the needles are moved closer and closer to the poles. It is this deflection or dip which causes some of the larger compass errors.

The magnetic compass (Fig. 3–16), which is the only direction seeking in-

Level flight

Climbing turn to the left

Descending turn to the left

strument in the airplane, is simple in construction. It contains two steel magnetized needles fastened to a float around which is mounted a compass card. The needles are parallel, with their north seeking ends pointed in the same direction. The compass card has letters for cardinal headings, and each 30° interval is represented by a number, the last zero of which is omitted. For example, 30° would appear as a 3 and 300° would appear as 30. Between these numbers, the card is graduated for each 5°.

The float assembly is housed in a bowl filled with acid-free white kerosene. The purposes of the liquid are to dampen out excessive oscillations of the compass card and relieve by buoyancy part of the weight of the float from the bearings. Jewel bearings are used to mount the float assembly on top of a pedestal. A line (called the lubber line) is mounted behind the glass of the instrument that can be used for a reference line when aligning the headings on the compass card.

Compass Errors *Variation.* Although the magnetic field of the earth lies roughly north and south, the earth's magnetic poles do not coincide with its geographic poles, which are used in the construction of aeronautical charts. Consequently, at most places on the earth's surface, the direction-sensitive steel needles which seek the earth's magnetic field will not point to True North but to Magnetic North. Furthermore, local magnetic fields from mineral deposits and other conditions may distort the earth's magnetic field and cause an additional error in position of the compass' north-seeking magnetized needles with reference to True North. The angular difference between True North and the direction indicated by the magnetic compass—excluding deviation error—is variation. Variation is different for different points on the earth's surface and is shown on the aeronautical charts as

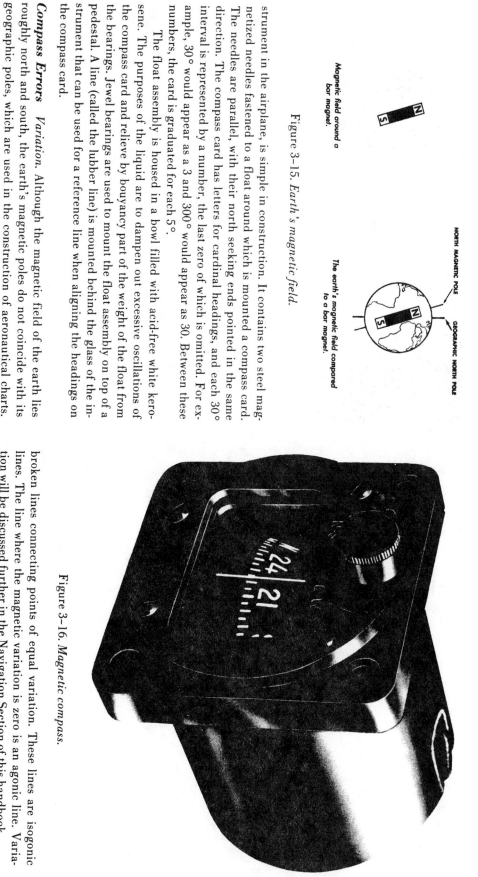

Magnetic field around a bar magnet.

The earth's magnetic field compared to a bar magnet.

NORTH MAGNETIC POLE

GEOGRAPHIC NORTH POLE

Figure 3-15. *Earth's magnetic field.*

broken lines connecting points of equal variation. These lines are isogonic lines. The line where the magnetic variation is zero is an agonic line. Variation will be discussed further in the Navigation Section of this handbook.

Deviation. Actually, a compass is very rarely influenced solely by the earth's magnetic lines of force. Magnetic disturbances from magnetic fields produced by metals and electrical accessories in an aircraft disturb the compass needles and produce an additional error. The difference between the direction indicated by a magnetic compass not installed in an airplane and one installed in an airplane, is deviation.

If an aircraft changes heading, the compass' direction-sensitive, magnetized needles will continue to point in about the same direction while the aircraft turns with relation to it. As the aircraft turns, metallic and electrical

Figure 3-16. *Magnetic compass.*

equipment in the aircraft change their position relative to the steel needles; hence, their influence on the compass needle changes and deviation changes. Thus, deviation depends, in part, on the heading of the aircraft. Although compensating magnets on the compass are adjusted to reduce this deviation on most headings, it is impossible to eliminate this error entirely on all headings. Therefore, a deviation card, installed in the cockpit in view of the pilot, enables the pilot to maintain the desired magnetic headings. Deviation will be discussed further in the Navigation Section of this handbook.

Using the Magnetic Compass   Since the magnetic compass is the only direction-seeking instrument in most airplanes, the pilot must be able to turn the airplane to a magnetic compass heading and maintain this heading. It will help to remember the following characteristics of the magnetic compass which are caused by magnetic dip:

1. If on a northerly heading and a turn is made toward east or west, the initial indication of the compass lags or indicates a turn in the opposite direction. This lag diminishes as the turn progresses toward east or west where there is no turn error.

2. If on a southerly heading and a turn is made toward the east or west, the initial indication of the compass needle will indicate a greater amount of turn than is actually made. This lead also diminishes as the turn progresses toward east or west where there is no turn error.

3. If a turn is made to a northerly heading from any direction, the compass indication when approaching northerly headings leads or is ahead of the turn. Therefore, the rollout of the turn is made before the desired heading is reached. If a turn is made to a southerly heading from any direction, the compass indication when approaching southerly headings lags behind the turn. Therefore, the rollout is made after the desired heading is passed. The amount of lead or lag is maximum on the north-south headings and depends upon the angle of bank used and geographic position of the airplane with regard to latitude.

4. When on an east or west heading, no error is apparent while entering a turn to north or south; however, an increase in airspeed or acceleration will cause the compass to indicate a turn toward north; a decrease in airspeed or acceleration will cause the compass to indicate a turn toward south.

5. If on a north or south heading, no error will be apparent because of acceleration or deceleration.

The magnetic compass should be read only when the aircraft is flying straight and level at a constant speed. This will help reduce errors to the minimum.

If the pilot thoroughly understands the errors and characteristics of the magnetic compass, this instrument can become the most reliable means of determining heading.

CHAPTER IV—AIRPLANE PERFORMANCE

Performance can be defined as the ability to operate or function, usually with regard to effectiveness. Airplane performance is the capability of the airplane, if operated within its limitations, to accomplish maneuvers which serve a specific purpose. For example, most present-day airplanes are designed clean and sleek, which results in greater range, speed, and payload with decreased operating costs. This type of airplane is used for cross-country flights. Airplanes used for short flights and carrying heavy loads, such as those used in certain agricultural operations, are designed differently, but still exhibit good performance for that purpose. Some of the factors which represent good performance are short takeoff and landing distance, increased climb capability, and greater speeds using less fuel.

Because of its effect on performance, airplane weight and balance calculations are included in this chapter. Also included is an introduction to determining takeoff, cruise, and landing performance. For information relating to weight and balance, takeoff, cruise, and landing performance for a specific make and model of airplane, reference should be made to that Aircraft's Flight Manual or Pilot's Operating Handbook.

Weight Control

Weight (in the context of this discussion) is the force with which gravity attracts a body toward the center of the earth. It is a product of the mass of a body and the acceleration acting on the body. Weight is a major problem in airplane construction and operation, and demands respect from all pilots.

The force of gravity continually attempts to pull the airplane down toward earth. The force of lift is the only force that counteracts weight and sustains the airplane in flight. However, the amount of lift produced by an airfoil is limited by the airfoil design, angle of attack, airspeed, and air density. Therefore, to assure that the lift generated is sufficient to counteract weight, loading the airplane beyond the manufacturer's recommended weight must be avoided. If the weight is greater than the lift generated, altitude cannot be maintained.

Effects of Weight

Any item aboard the airplane which increases the total weight significantly is undesirable as far as flight is concerned. Manufacturers attempt to make the airplane as light as possible without sacrificing strength or safety.

The pilot of an airplane should always be aware of the consequences of overloading. An overloaded airplane may not be able to leave the ground, or if it does become airborne, it may exhibit unexpected and unusually poor flight characteristics. Each airplane has limits which if exceeded result in inferior operation and possible disaster. If an airplane is not properly loaded, the initial indication of poor performance usually takes place during takeoff, which is the most unfortunate place for the airplane and the pilot to be in trouble.

Excessive weight reduces the flight performance of an airplane in almost every respect. The most important performance deficiencies of the overloaded airplane are:

Higher takeoff speed.
Longer takeoff run.
Reduced rate and angle of climb.
Lower maximum altitude.
Shorter range.
Reduced cruising speed.
Reduced maneuverability.
Higher stalling speed.
Higher landing speed.
Longer landing roll.
Excessive weight on the nosewheel.

The pilot must be knowledgeable in the effect of weight on the performance of the particular airplane being flown. Preflight planning should include a check of performance charts to determine if the airplane weight may contribute to hazardous flight operations. Pilots should recognize and avoid such airplane performance-reducing factors as: high-density altitude, frost on the wings, low engine power, and severe or uncoordinated maneuvers. Excessive weight in itself reduces the safety margins available to the pilot, and becomes even more hazardous when other performance-reducing factors mentioned are combined with overweight. The pilot must also consider the consequences of an overweight airplane if an emergency condition arises. If an engine fails on takeoff or ice forms at low altitude, it is usually too late to reduce the airplane weight to keep it in the air.

Weight Changes   The weight of the airplane can be changed by altering the fuel load. Gasoline has considerable weight—6 lbs. per gallon—30 gallons may weigh more than one passenger. But it must be remembered that if weight is lowered by reducing fuel, the range of the airplane is shortened. During flight, fuel burn is normally the only weight change that takes place. As fuel is used, the airplane becomes lighter and performance is improved; this is one of the few advantages about the consumption of the fuel supply.

Changes of fixed equipment can have a major effect upon the weight of the airplane. An airplane can be overloaded by the installation of extra radios or instruments. Repairs or modifications usually add to the weight of the airplane; it is a rare exception when a structural or equipment change results in a reduction of weight. When an airplane ages, it tends to put on weight. The total effect of this gain is referred to as "Service Weight Pickup." Most service weight pickup is the known weight of actual parts installed in repair, overhaul, and modification. In addition, an unknown weight pickup results from the collection of trash and hardware, moisture absorption of soundproofing, and the accumulation of dirt and grease. This pickup can only be determined by weighing the airplane.

Balance, Stability, and Center of Gravity   Balance refers to the location of the center of gravity (c.g.) of an airplane, and is important to airplane stability and safety in flight. An airplane should not be flown if the pilot is unsatisfied with its loading and the resulting weight and balance condition. The c.g. is a point at which an airplane would balance if it were suspended at that point.

The prime concern of airplane balancing is the fore and aft location of the c.g. along the longitudinal axis. Location of the c.g. with reference to the lateral axis is also important. For each item of weight existing to the left of the fuselage centerline, there is generally an equal weight existing at a corresponding location on the right. This may be upset, however, by unbalanced lateral loading. The position of the lateral c.g. is not computed, but the pilot must be aware that adverse effects will certainly arise as a result of a laterally unbalanced condition. Lateral unbalance will occur if the fuel load is mismanaged by supplying the engine(s) unevenly from tanks on one side of the airplane (Fig. 4-1). The pilot can compensate for the resulting wing-heavy condition by adjusting the aileron trim tab or by holding a constant aileron control pressure. However, this places the airplane controls in an out-of-

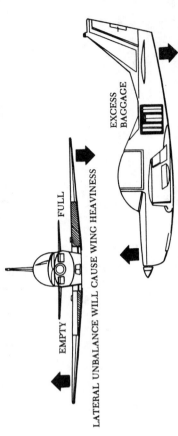

LATERAL UNBALANCE WILL CAUSE WING HEAVINESS

LONGITUDINAL UNBALANCE WILL CAUSE NOSE OR TAIL HEAVINESS

Figure 4-1. *Lateral or longitudinal unbalance.*

streamline condition, increases drag, and results in decreased operating efficiency. Since lateral balance is relatively easy to control and longitudinal balance is more critical, further reference to balance in this handbook will mean longitudinal location of the center of gravity.

The c.g. is not necessarily a fixed point; its location depends on the distribution of weight in the airplane. As variable load items are shifted or expended, there is a resultant shift in c.g. location. The pilot should realize that if the c.g. of an airplane is displaced too far forward on the longitudinal axis, a nose-heavy condition will result. Conversely, if the c.g. is displaced too far aft on the longitudinal axis, a tail-heavy condition will result (Fig. 4-1). It is possible that an unfavorable location of the c.g. could produce such an unstable condition that the pilot could not control the airplane.

In any event, flying an airplane which is out of balance can produce increased pilot fatigue with obvious effects on the safety and efficiency of flight. The pilot's natural correction for longitudinal unbalance is a change of trim to remove the excessive control pressure. Excessive trim, however, has the effect of not only reducing aerodynamic efficiency but also reducing primary control travel distance in the direction the trim is applied.

Effects of Adverse Balance   Adverse balance conditions affect the airplane flight characteristics in much the same manner as those mentioned for an excess weight condition. In addition, there are two essential airplane characteristics which may be seriously affected by improper balance; these are stability and control. Loading in a nose-heavy condition causes problems in controlling and raising the nose, especially during takeoff and landing. Loading in a tail-heavy condition has a most serious effect upon longitudinal stability, and can reduce the airplane's capability to recover from stalls and spins. Another undesirable characteristic produced from tail-heavy loading is that it produces very light stick forces. This makes it easy for the pilot to inadvertently overstress the airplane.

Limits for the location of the airplane's c.g. are established by the manufacturer. These are the fore and aft limits beyond which the c.g. should not be located for flight. These limits are published for each airplane in the "FAA Aircraft Specifications." If, after loading, the c.g. is not within the allowable limits, it will be necessary to relocate some items within the airplane before flight is attempted.

The forward c.g. limit is often established at a location which is determined by the landing characteristics of the airplane. It may be possible to maintain stable and safe cruising flight if the c.g. is located ahead of the prescribed forward limit, but during landing which is one of the most critical phases of flight, exceeding the forward c.g. limit may cause problems. Manufacturers purposely place the forward c.g. limit as far rearward as possible to aid pilots in avoiding damage to the airplane when landing.

A restricted forward c.g. limit is also specified to assure that sufficient elevator deflection is available at minimum airspeed. When structural limitations or large stick forces do not limit the forward c.g. position, it is located at the position where full-up elevator is required to obtain a high angle of attack for landing.

The aft c.g. limit is the most rearward position at which the c.g. can be located for the most critical maneuver or operation. As the c.g. moves aft, a less stable condition occurs which decreases the ability of the airplane to right itself after maneuvering or after disturbances by gusts.

For some airplanes the c.g. limits, both fore and aft, may be specified to vary as gross weight changes. They may also be changed for certain operations such as acrobatic flight, retraction of the landing gear, or the installation of special loads and devices which change the flight characteristics.

The actual location of the c.g. can be altered by many variable factors and are usually controlled by the pilot. Placement of baggage and cargo items determine the c.g. location. The assignment of seats to passengers can also be used as a means of obtaining a favorable balance. If the airplane is tail-heavy, it is only logical to place heavy passengers in a front seat.

The loading and selective use of fuel from various tank locations can affect airplane balance. Large airplanes must have fuel loaded in a manner determined by the total load, and during flight fuel use must be managed in a sequence that will keep the load in balance.

Management of Weight and Balance Control   Weight and balance control should be a matter of concern to all pilots. The pilot has to assume a large share of the management burden, such as control over the loading and fuel management (the two variable factors which can change both total weight and c.g. location); and weight and balance information available in the form of airplane records and operating handbooks. Loading information is available in the form of placards in baggage compartments and on tank caps.

The airplane owner or operator should make certain that up-to-date information is available in the airplane for the pilot's use, and should ensure that appropriate entries are made in the airplane records when repairs or modifications have been accomplished. Weight changes must be accounted for and proper notations made in weight and balance records. Without such information the pilot has no foundation upon which to base the necessary calculations and decisions.

The airplane manufacturer and the **Federal Aviation Administration** have major roles in designing and certificating the airplane with a safe and workable means of controlling weight and balance.

Terms and Definitions   The pilot needs to be familiar with terms used in working the problems related to weight and balance. The following list of terms and their definitions is fairly well standardized, and knowledge of these terms will aid the pilot to better understand weight and balance calculations of any airplane.

1. *Arm (moment arm)*—is the horizontal distance in inches from the reference datum line to the center of gravity of an item. The algebraic sign is plus (+) if measured aft of the datum, and minus (−) if measured forward of the datum.

2. *Center of gravity (c.g.)*—is the point about which an airplane would balance if it were possible to suspend it at that point. It is the mass center of the airplane, or the theoretical point at which the entire weight of the airplane is assumed to be concentrated. It may be expressed in percent of MAC (mean aerodynamic chord) or in inches from the reference datum.

3. *Center of gravity limits*—are the specified forward and aft points within which the c.g. must be located during flight. These limits are indicated on pertinent airplane specifications.

4. *Center of gravity range*—is the distance between the forward and aft c.g. limits indicated on pertinent airplane specifications.

Aircraft Weight Nomenclature

(General Aviation Aircraft)

Term	Example (pounds)	Notes
Empty weight	2,905	Includes: Airframe, engines, all fixed and permanent operating equipment, and unusable fuel and oil.*
+ Useful load	1,695	Includes: Pilot, copilot, passengers, baggage, fuel, and oil.
= Takeoff weight	4,600	
− Fuel used	460	Includes: Fuel burned.
= Landing weight	4,140	

NOTE: *The weights above are used for illustration only. The actual values will vary for each aircraft and each flight. * Some aircraft include all oil in empty weight.*

Control of Loading—General Aviation Airplanes Before any flight, the pilot should determine the weight and balance condition of the airplane. Too frequently airplanes are loaded by guess and intuition with occasional grim results. There is no excuse for following this method. Simple and orderly procedures, based on sound principles, have been devised by airplane manufacturers for the determination of loading conditions. The pilot, however, must use these procedures and exercise good judgment. In many modern airplanes it is not possible to fill all seats, baggage compartments, and fuel tanks, and still remain within the approved weight and balance limits. If the maximum passenger load is carried, the pilot must often reduce the fuel load or reduce the amount of baggage.

Basic Principles of Weight and Balance Computations It might be advantageous at this point to review and discuss some of the basic principles of how weight and balance can be determined. The following method of computation can be applied to any object or vehicle where weight and balance information is essential, but to fulfill the purpose of this handbook it is directed primarily toward the airplane.

By determining the weight of the empty airplane and adding the weight of everything loaded on the airplane, a total weight can be determined. This is

5. *Datum (reference datum)*—is an imaginary vertical plane or line from which all measurements of arm are taken. The datum is established by the manufacturer. Once the datum has been selected, all moment arms and the location of c.g. range are measured from this point.

6. *Delta*—is a Greek letter expressed by the symbol Δ to indicate a change of values. As an example, Δ c.g. indicates a change (or movement) of the c.g.

7. *Fuel load*—is the expendable part of the load of the airplane. It includes only usable fuel, not fuel required to fill the lines or that which remains trapped in the tank sumps.

8. *Moment*—is the product of the weight of an item multiplied by its arm. Moments are expressed in pound-inches (lb.-in.). Total moment is the weight of the airplane multiplied by the distance between the datum and the c.g.

9. *Moment index (or index)*—is a moment divided by a constant such as 100, 1,000, or 10,000. The purpose of using a moment index is to simplify weight and balance computations of airplanes where heavy items and long arms result in large, unmanageable numbers.

10. *Mean aerodynamic chord (MAC)*—is the average distance from the leading edge to the trailing edge of the wing.

11. *Standard weights*—have been established for numerous items involved in weight and balance computations. These weights should not be used if actual weights are available. Some of the standard weights are:

General aviation—crew and passenger	170 lbs. each
Gasoline	6 lbs./U.S. gal.
Oil	7.5 lbs./U.S. gal.
Water	8.35 lbs./U.S. gal.

12. *Station*—is a location in the airplane which is identified by a number designating its distance in inches from the datum. The datum is, therefore, identified as station zero. An item located at station +50 would have an arm of 50 inches.

13. *Useful load*—is the weight of the pilot, copilot, passengers, baggage, usable fuel, and drainable oil. It is the empty weight subtracted from the maximum allowable gross weight. This term applies to general aviation aircraft only.

14. *Weight, empty*—consists of the airframe, engines, and all items of operating equipment that have fixed locations and are permanently installed in the airplane. It includes optional and special equipment, fixed ballast, hydraulic fluid, unusable (residual) fuel, and undrainable (residual) oil.

quite simple, but to distribute this weight in such a manner that the entire mass of the loaded airplane is balanced around a point (c.g.) which must be located within specified limits presents a greater problem, particularly if the basic principles of weight and balance are not understood.

The point where the airplane will balance can be determined by locating the center of gravity, which is, as stated in the definitions of terms, the imaginary point where all the weight is concentrated. To provide the necessary balance between longitudinal stability and elevator control, the center of gravity is usually located slightly forward of the center of lift. This loading condition causes a nose-down tendency in flight, which is desirable during flight at a high angle of attack and slow speeds.

A safe zone within which the balance point (c.g.) must fall is called the c.g. range. The extremities of the range are called the forward c.g. limits and aft c.g. limits. These limits are usually specified in inches, along the longitudinal axis of the airplane, measured from a datum reference. The datum is an arbitrary point, established by airplane designers, which may vary in location between different airplanes (Fig. 4–2).

The distance from the datum to any component part of the airplane, or any object loaded on the airplane, is called the arm. When the object or component is located aft of the datum it is measured in positive inches; if located

forward of the datum it is measured as negative inches, or minus inches (Fig. 4–2). The location of the object or part is often referred to as the station. If the weight of any object or component is multiplied by the distance from the datum (arm), the product is the moment. The moment is the measurement of the gravitational force which causes a tendency of the weight to rotate about a point or axis and is expressed in pound-inches.

To illustrate, assume a weight of 50 pounds is placed on the board at a station or point 100 inches from the datum (Fig. 4–3). The downward force of the weight can be determined by multiplying 50 pounds by 100 inches, which produces a moment of 5,000 pounds-inches.

To establish a balance, a total of 5,000 pound-inches must be applied to the other end of the board. Any combination of weight and distance which, when multiplied, produces 5,000 pound-inches moment will balance the

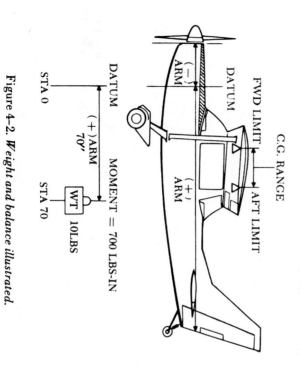

Figure 4–2. *Weight and balance illustrated.*

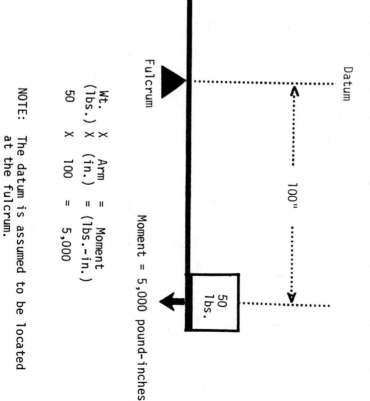

Figure 4–3. *Determining moments.*

Item	Weight (lbs.)	Arm (in.)	Positive Moments	Negative Moments
Airplane Empty Weight	1,000.0	6	6000.0	
Pilot	150.0	11	1650.0	
Baggage	40.0	32	1280.0	
Oil—4 qts. (7.5 lb./gal.)	7.5	−4		−30.0
Fuel—20 gal. (6 lb./gal.)	120.0	16	1920.0	
TOTAL WEIGHT	1,317.5		10850.0	
			−30.0	
TOTAL MOMENTS			10820.0	

NOTE: Remember the positive moments must be added and the total negative moments must be subtracted from the total positive moments.

In the preceding example, if the maximum allowable gross weight is 1,400 lbs., the loaded weight is 82.5 pounds below the maximum. Another 82.5 pounds could be added without exceeding the maximum allowable gross weight. Also, with this information the loaded center of gravity or the balance point of the loaded airplane can be determined. This can be accomplished by dividing the total moments by the total weight as follows:

$$c.g. = \frac{\text{Total moments}}{\text{Total weight}}$$

or

$$c.g. = \frac{10820.0}{1317.5}$$

$$c.g. = 8.21$$

If the center of gravity limit ranges from 7.50″ fore to 8.50″ aft it can be seen that actual loaded center of gravity falls between these two limits. Therefore, the loading on this airplane is both within the weight limits and the balance limits. A number of other methods of computing weight and balance will now be discussed.

Useful Load Check A simple weight check should always be made by pilots before flight to determine if the useful load is exceeded. The check may

board. For example, as illustrated in Fig. 4–4, if a 100-pound weight is placed at a point (station) 25 inches from the datum, and another 50-pound weight is placed at a point (station) 50 inches from the datum, the sum of the product of the two weights and their distances will total a moment of 5,000 pound-inches which will balance the board.

The following is intended to simply illustrate how this method of determining balance is applied to the airplane. However, bear in mind the complexity of the problems increases to some extent in different types of airplanes, particularly in respect to size and the number of items that the airplane is designed to carry.

In Fig. 4–5, the fulcrum is at the empty weight c.g. location, but the datum reference is moved to a convenient place to the left of the fulcrum. Also in this illustration the airplane empty weight and empty center of gravity are included in the calculations. This is required in all calculations of weight and balance, but for the sake of simplicity it was intentionally omitted when determining the weight and balance of the board previously discussed. Note that a table is constructed following the illustration, which aids in simplifying the problem.

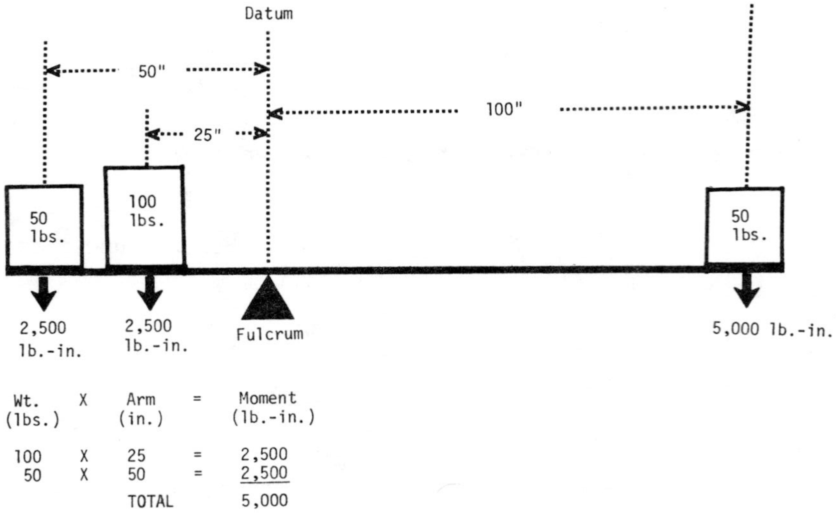

Figure 4–4. *Establishing a balance.*

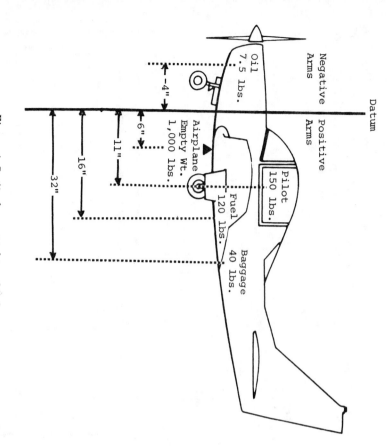

Figure 4–5. Airplane weight and balance.

be a mental calculation if the pilot is familiar with the airplane's limits and knows that unusually heavy loads are not aboard. But when all seats are being occupied, fuel tanks are full, and some baggage is aboard, the pilot should calculate the weight and balance carefully.

The pilot must know the useful load limit of the airplane. This information may be found in the latest weight and balance report, in a logbook, or on a major repair and alterations form. If useful load is not stated directly, simply subtract empty weight from maximum takeoff weight.

The check is simple. Merely add the weight of the items included in the useful load—then check this total against the limit. The calculations might look like this:

Example:

	Pounds
M. A. Jones (instructor)	175
Pilot	180
Fuel—30 gal. (6 lbs./gal.)	180
Oil—8 qts. (7.5 lbs./gal.)	15
Baggage	5
Total	555

Useful load limit is 575 lbs.
The calculations show that the useful load is not exceeded.

Now suppose that Mr. Jones is replaced by an instructor weighing 210 pounds. A useful load check will show that the useful load limit is exceeded, and the load must be reduced to or below the specified useful load limit. In some airplanes there is no alternative but to reduce the fuel load, even if all the baggage is removed.

Initial weight and balance calculations for airplanes and some examples in handbooks and manuals make the assumption that the pilot and passengers weigh a standard 170 pounds. Heavy pilots or passengers can seriously overload a small airplane. A student and instructor may easily weigh 220 pounds each, especially if dressed in winter clothing; this represents a potential overload of 100 pounds. The baggage compartment should not be overloaded. The maximum baggage compartment weight limits should be observed and followed. Frequently, a restriction is placed on rear seat occupancy with the maximum baggage aboard.

Weight and Balance Restrictions

The airplane's weight and balance restrictions should be closely followed. The loading conditions and empty weight of a particular airplane (Fig. 4–6) may differ from that found in the Aircraft Flight Manual or Pilot's Operating Handbook because modifications or equipment changes may have been made. Sample loading problems in the Aircraft Flight Manual or Pilot's Operating Handbook are intended for guidance only; therefore, each airplane must be treated separately. Although an airplane is certified for a specified maximum gross takeoff weight, it will not safely take off with this load under all conditions. Conditions which affect takeoff and climb performance such as high elevations, high temperatures, and high humidity (high-density altitudes), may require a reduction in weight before flight is attempted. Other factors to consider prior to takeoff are runway length, runway surface, runway slope, surface wind, and the presence of obstacles. These factors may require a reduction in weight prior to flight.

Some airplanes are designed so that it is impossible to load them in a manner that will place the c.g. outside the fore or aft limits. These are usually small airplanes having the seats, fuel, and baggage compartments located very near the c.g. limits. Nevertheless, these airplanes can be overloaded.

Some airplanes, however, can be loaded in such a manner that the c.g. will be beyond limits, even though the useful load is not exceeded. An out-of-balance condition is serious from the standpoint of stability and control. The pilot can determine if the load is within limits by using a loading schedule. If available, this schedule may be found in the weight and balance report, the aircraft logbook, the Aircraft Flight Manual or Pilot's Operating Handbook or may be posted in the airplane in the form of a placard. A typical placard is shown in Fig. 4-7.

The loading schedule should be used as a suggested loading plan only. A

LOADING SCHEDULE

FUEL	PASSENGERS	BAGGAGE
FULL	2 REAR	100 LBS
39 GAL	1 FRONT AND 2 REAR	NONE
FULL	1 FRONT AND 1 REAR	FULL

INCLUDES PILOT AND FULL OIL

Figure 4–7. *Loading schedule placard.*

WEIGHT & BALANCE DATA

AIRCRAFT SERIAL NO. 15556480   FAA REGISTRATION NO. N3248X

ITEM	WEIGHT	× ARM =	MOMENT
STANDARD AIRPLANE	975.0	32.0	31200.0
OPTIONAL EQUIPMENT	89.0	26.1	2322.9
PAINT	15.5	85.3	1322.2
UNUSABLE FUEL	20.0	43.0	860.0
LICENSED EMPTY WEIGHT	1099.5	32.5	35705.1

(GROSS WT) − (LICENSED EMPTY WT) = USEFUL LOAD

(1800 LB) − (1099.5 LB) = 700.5 LBS

IT IS THE RESPONSIBILITY OF THE OWNER AND PILOT TO ENSURE THAT THE AIRPLANE IS PROPERLY LOADED. THE DATA ABOVE INDICATES THE EMPTY WEIGHT, C.G. AND USEFUL LOAD WHEN THE AIRPLANE WAS RELEASED FROM THE FACTORY. REFER TO THE LATEST WEIGHT AND BALANCE RECORD WHEN ALTERATIONS HAVE BEEN MADE.

SAMPLE LOADING PROBLEM

ITEM	WEIGHT (LBS)	ARM (IN)	MOMENT (LB-IN/1000)
LICENSED EMPTY WEIGHT	1099.5		35.7
OIL	12	−15.0	−0.2
PILOT & PASSENGER	340	40.0	13.6
FUEL	188.5	43.0	8.1
BAGGAGE	160	65.0	10.4
TOTAL LOADED AIRPLANE	1800		67.6

Figure 4–6. *Weight and balance data.*

check by means of weight and balance calculations should be made to determine if limitations are being exceeded if any doubt arises. The loading schedule assumes that each passenger weight is approximately 170 pounds. It is obvious that passenger weights can vary considerably from the assumed standard.

Aircraft Flight Manual or Pilot's Operating Handbook. Each airplane is furnished with the Aircraft Flight Manual or Pilot's Operating Handbook. Airplanes weighing less than 6,000 pounds may have information furnished in the form of placards, markings, or manuals. When an Aircraft Flight Manual or Pilot's Operating Handbook is furnished, the following is included:

a. Limitations and data:
1. The maximum weight.
2. The empty weight and c.g. location.
3. The useful load.
4. The composition of the useful load, including the total weight of fuel and oil with full tanks.

b. Load distribution:
The established c.g. limits are furnished in the Aircraft Flight Manual or Pilot's Operating Handbook. If the available loading space is placarded or arranged so that reasonable distribution of the useful load will not result in placing the c.g. outside of the stated limits, the Aircraft Flight Manual or Pilot's Operating Handbook may not include any information other than the statement of c.g. limits. In other cases, the manual includes enough information to assure that any loading combinations will result in keeping the c.g. within established limits.

Light Single-Engine Airplane Loading Problems Airplane manufacturers use one of several available systems to provide loading information. The following weight and balance problems will illustrate how to determine if the maximum weight limit is exceeded or the c.g. is located beyond limits.

Assume a flight is planned in a single-engine, four-place airplane. The load consists of the pilot, one front seat passenger and two rear seat passengers, full fuel and oil and 60 pounds of baggage (Fig. 4–8).

Here is how this weight and balance problem is solved by using one of two different methods:

Example:
Solution by index table—
1. From the manual or weight and balance report, determine the empty weight and empty weight c.g. (arm) of the airplane.
2. Determine the arms for all useful load items.
3. Determine the maximum weight and c.g. range. (For this problem the maximum takeoff gross weight = 2,400 lb., and the c.g. range = Sta. 35.6 to 45.8)
4. Calculate the actual weights for the useful load items.
5. Construct a table as follows, and enter the appropriate values. Multiply each individual weight and arm to obtain moments.

	Weight	$\times$ Arm =	Moment (lb.-in.)
Airplane—empty	1,340	38.5	51,590
Oil—(8 qts.)	15	−20.0	−300
Pilot & front passenger	320	35.0	11,200
Fuel (40 gal.)	240	48.0	11,520
Rear passengers	300	72.0	21,600
Baggage	60	92.0	5,520
Total	2,275		101,130

NOTE: The oil tank for this airplane is located forward of the datum, and has a negative value; therefore, subtract the oil moment when totaling the moment column.

6. Adding the weights produces a total of 2,275 lb. and adding the moments produces a total of 101,130 lb.-in. The c.g. is calculated by dividing the total moment by the total weight:

$$\frac{101,130}{2,275} = 44.4 \text{ in. aft of datum.}$$

7. The total weight of 2,275 lb. does not exceed the maximum weight of 2,400 lb. and the computed c.g. of 44.4 falls within the allowable c.g. range of 35.6 to 45.8 in. aft of datum.

The other method makes use of graphic aids. Weight and balance computations are greatly simplified by two graphic aids—the loading graph and the center of gravity moment envelope. The loading graph (Fig. 4–9), is typical of those found in Pilot's Operating Handbooks.

This graph, in effect, multiplies weight by the arm, resulting in the moment. Note that the moment has been divided by a reduction factor (1000), resulting in an index number. Weight values appear along the left side of the

Now project a line vertically downward to the index number. For example, the index number of a pilot weighing 170 lb. is 6.1. The c.g. moment envelope (Fig. 4–10) eliminates the division to compute the c.g. It gives an acceptable range of index numbers for any airplane weight from minimum to maximum.

If the lines from total weight and total moment intersect within the envelope, the airplane is within weight and balance limits. In solving the sample problem, follow this procedure:

Example—
1. Determine the airplane empty weight and the empty weight index from the weight and balance report.
2. Construct a table such as the one that follows. In the left column, enter the actual weights of the empty airplane, oil, pilot and front seat passenger, fuel, rear seat passenger, and baggage. In the right column, enter the airplane empty weight index (moment/1,000).
3. From the loading graph (Fig. 4–9) determine the index number (moment/1,000) of each item and enter it in the table.
4. Add the two columns to determine the total weight and total moments.
5. Refer to the c.g. moment envelope (Fig. 4–10) and find the point of

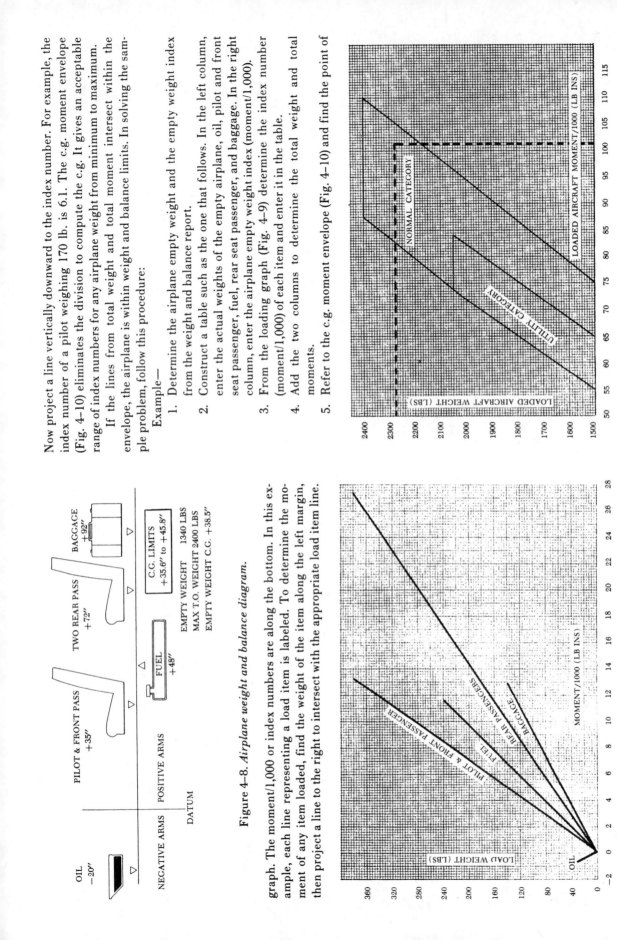

Figure 4–10. *C.G. moment envelope.*

Figure 4–8. *Airplane weight and balance diagram.*

OIL −20″

PILOT & FRONT PASS +35″

TWO REAR PASS +72″

BAGGAGE +92″

FUEL +48″

C.G. LIMITS +35.6″ to +45.8″

EMPTY WEIGHT 1340 LBS
MAX T.O. WEIGHT 2400 LBS
EMPTY WEIGHT C.G. +38.5″

NEGATIVE ARMS   POSITIVE ARMS

DATUM

graph. The moment/1,000 or index numbers are along the bottom. In this example, each line representing a load item is labeled. To determine the moment of any item loaded, find the weight of the item along the left margin, then project a line to the right to intersect with the appropriate load item line.

Figure 4–9. *Loading graph.*

intersection of a line projected right from total weight (2,275 lbs.) and a line projected upward from total moment/1,000 [101.2].

6. If the point of intersection falls within the envelope, as it does in this problem, the weight and c.g. are within limits.

Sample Loading Problem

Item	Weight	Moment/1,000
1. Empty airplane weight.	1,340	51.6
2. Oil.	15	-0.3
3. Pilot and front passenger.	320	11.2
4. Fuel.	240	11.6
5. Rear seat passengers.	300	21.6
6. Baggage.	60	5.5
7. Total airplane weight	2,275	101.2

Change of Weight. In many instances, the pilot must be able to solve problems which involve the shifting, addition, or removal of weight. For example, the airplane may be loaded within the allowable takeoff weight limit, but the c.g. limit may be exceeded. The most satisfactory solution to this problem is to relocate baggage, or passengers, or both. The pilot should be able to determine the minimum amount of load that needs to be shifted to make the airplane safe for flight. The pilot should also be able to determine if shifting the load to a new location will correct an out-of-limit condition. There are some standardized and simple calculations which can help make these determinations.

Weight Shifting. When weight is shifted from one location to another, the total weight of the airplane does not change. The total moments, however, do change in relation and proportion to the direction and distance the weight is moved. When weight is moved forward, total moments decrease; when weight is moved aft, total moments increase. The amount of moment change is proportional to the amount of weight moved. Since many airplanes have forward and aft baggage compartments, weight may be shifted from one to the other to change the c.g. If the airplane weight, c.g., and total moments are known, the new c.g. (after the weight shift) can be determined by dividing the new total moments by the total airplane weight.

Example:

To determine the new total moments, find out how many moments are gained or lost when the weight is shifted.

The weight shift conditions indicated for the airplane illustrated in Fig.

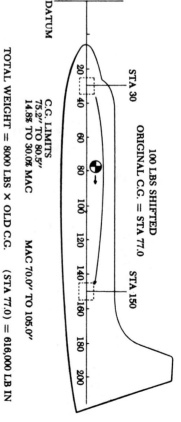

TOTAL WEIGHT = 8000 LBS × OLD C.G. (STA 77.0) = 616,000 LB IN

MAC 70.0″ TO 105.0″

Figure 4–11. *Weight shifting diagram.*

4–11, show that 100 pounds have been shifted from Sta. 30 to Sta. 150. This movement increases the total moments of the airplane by 12,000 lb.-in.

Baggage moment when at Sta. 150 = 100 lb. × 150 in. = 15,000 lb.-in.
Baggage moment when at Sta. 30 = 100 lb. × 30 in. = 3,000 lb.-in.
Moment Change = 12,000 lb.-in.

By adding the moment change to the original moment (or subtracting if the weight has been moved forward instead of aft), the new total moments can be determined. The new c.g. can then be determined by dividing the new total moments by the total weight.

Total moments = 616,000 + 12,000 = 628,000

$$\text{c.g.} = \frac{628,000}{8,000} = 78.5 \text{ in.}$$

The shift of the baggage has caused the c.g. to shift to Sta. 78.5.

A simple solution may be obtained by using the computer or electronic calculator. This can be done because the c.g. will shift a distance which is proportional to the distance the weight has shifted. The solution to the proportion problem is shown in Fig. 4–12.

Example:

$$1. \frac{\text{Weight shifted}}{\text{Total Weight}} = \frac{\Delta \text{c.g. (change of c.g.)}}{\text{Distance weight is shifted}}$$

$$\frac{100}{8,000} = \frac{\Delta \text{c.g.}}{120}$$

$$\Delta \text{c.g.} = 1.5 \text{ in.}$$

2. The change of c.g. is added to (or subtracted from) the original c.g. to determine the new c.g.:

$$77 + 1.5 = 78.5 \text{ in. aft of datum}$$

If the products are the same, the decimal is properly placed. If not the same, the wrong decimal location for the Δ c.g. has been selected and the decimal should be relocated accordingly.

The shifting weight proportion formula can also be used to determine how much weight must be shifted to achieve a desired shift of the c.g. The following problem illustrates a solution to this type of problem.

Example:

Given—

Airplane total weight . . . . 7,800 lb.
c.g. . . . . Sta. 81.5 in.
Aft c.g. limit . . . . 80.5 in.

Find—

How much weight must be shifted from the aft cargo compartment at Sta. 150 to the forward cargo compartment at Sta. 30 to move the c.g. to exactly the aft limit?

Solution—

1. Use the shifting weight proportion:

$$\frac{\text{Weight to be shifted}}{\text{Total Weight}} = \frac{\Delta \text{ c.g.}}{\text{Dist. wt. shifted}}$$

$$\frac{\text{Weight to be shifted}}{7,800} = \frac{1.0 \text{ in.}}{120 \text{ in.}}$$

Weight to be shifted = 65 lb.

2. Cross multiply to check for correct location of decimal point.

$$7,800 \times 1.0 = 7,800$$
$$65 \times 120 = 7,800$$

A computer or electronic calculator can be used to determine the solution, similar to the previous problem.

Weight Addition or Removal. In many instances the weight and balance of the airplane will change by adding or removing weight. When this is done a new c.g. must be calculated and checked against the limitations to determine that the new c.g. is within limits. This type of weight and balance problem is commonly encountered when the airplane burns fuel in flight, thereby reducing the fuel weight. Many airplanes are designed with the fuel tanks located close to the c.g., therefore the consumption of fuel does not affect the c.g. to any great extent. Certain airplane fuel tanks, however, are

---

Values Needed:

Weight to be Shifted	=	100 lbs.
Total Weight	=	8,000 lbs.
Change in CG (Unknown)	=	ΔCG
Distance Weight Moved	=	120 in.
Old CG Location	=	77 in.

Set Up Proportion:

$$\frac{\text{Weight to be Shifted}}{\text{Total Weight}} : \frac{\text{Change in CG}}{\text{Distance Weight Moved}}$$

$$\frac{100}{8,000} : \frac{\Delta CG}{120}$$

Cross Multiply:

$$100 \times 120 = 12,000$$
$$8,000 \times \Delta CG = 8,000 \Delta CG$$
$$8,000 \, \Delta CG = 12,000$$

Divide:

$$12,000 \div 8,000 = 1.5 \text{ in.}$$
$$\Delta CG = 1.5 \text{ in.}$$

(Add if weight shifted rearward. Subtract if weight shifted forward.)
In this Problem Add:

Old CG	=	77.0 in.
ΔCG	=	1.5 in.
New CG	=	78.5 in.

Figure 4-12. *Solution to proportion problem.*

Difficulty may arise in the computer-type solution when locating the decimal point in the answer. To make certain that Δ c.g. in the above problem is not .15 in. or 15.0 in., insert the answer obtained into the formula, or in other words substitute the answer 1.5 for the unknown Δ c.g., and cross multiply as follows:

$$1.5 \times 8,000 = 12,000$$
$$100 \times 120 = 12,000$$

located so that careful planning is required to prevent the c.g. from shifting out of limits during flight. If this weight condition is possible, it is advisable to calculate the weight and balance twice before flight. First determine the weight and balance with all items loaded except fuel; and second, determine the weight and balance, including the fuel. This will provide an indication of how fuel consumption affects balance.

The addition or removal of cargo causes a c.g. change, which should be calculated before flight. These problems can be solved by calculations involving total moments. However, a shortcut formula which can be adapted to the aeronautical computer may be used to simplify computations:

$$\frac{\text{Weight added (or removed)}}{\text{New total weight}} = \frac{\Delta \text{ c.g.}}{\text{Distance between wt. and old c.g.}}$$

In this formula, the terms "new" and "old" refer to conditions before and after the weight change.

It may be more convenient to use another form of this formula to find the weight change needed to accomplish a particular c.g. change (Δ c.g.). In this case use:

$$\frac{\text{Weight added or removed}}{\text{Old total weight}} = \frac{\Delta \text{ c.g.}}{\text{Distance between wt. and new c.g.}}$$

Notice that the terms "new" and "old" are found on both sides of the equation in either of the above proportions. If the "new" total weight is used the distance must be calculated from the "old" c.g. Just the opposite is true if the "old" total weight is used.

A typical problem involves the calculation of a new c.g. for an airplane which has been loaded and is ready for flight, and receives additional cargo or passengers just before departure time.

Example:

Given—
  Airplane total weight—6,860 lb.
  c.g. —Sta. 80.0
Find—What is the location of the c.g. if 140 lb. of baggage are added to station 150?

Solution—
  1. Use the added weight formula:

$$\frac{\text{Added weight}}{\text{New total weight}} = \frac{\Delta \text{ c.g.}}{\text{Dist. between wt. and old c.g.}}$$

$$\frac{140}{6,860 + 140} = \frac{\Delta \text{ c.g.}}{150 - 80}$$

$$\frac{140}{7,000} = \frac{\Delta \text{ c.g.}}{70}$$

  2. Add c.g. to the old c.g.;
     New c.g. = 80.0 in. + 1.4 in. = 81.4

Example:

Given—
  Airplane total weight — 6,100 lb.
  c.g. —Sta. 78
Find—What is the location of the c.g. if 100 lb. are removed from station 150?

Solution—
  1. Use the removed weight formula:

$$\frac{\text{Weight removed}}{\text{New total weight}} = \frac{\Delta \text{ c.g.}}{\text{Dist. between wt. and old c.g.}}$$

$$\frac{100}{6,100 - 100} = \frac{\Delta \text{ c.g.}}{150 - 78}$$

$$\frac{100}{6,000} = \frac{\Delta \text{ c.g.}}{72}$$

  Δ c.g. = 1.2 in. forward
  2. Subtract Δ c.g. from old c.g.:
     New c.g. = 78 in. − 1.2 in. = 76.8 in.

NOTE: *In the above two examples, the Δ c.g. is either added to or subtracted from the old c.g. Deciding which way the c.g. will shift for the particular weight change. If calculating which way the c.g. will shift to accomplish is best handled by mentally the c.g. is shifting aft, the Δ c.g. is added to the old c.g.; if it is shifting forward, the Δ c.g. is subtracted from the old c.g.*

To summarize c.g. movement:

Weight added fwd. of old c.g. }
Weight removed aft of old c.g. } c.g. moves fwd.
Weight added aft of old c.g. }
Weight removed fwd. of old c.g. } c.g. moves aft.

Example:

Given—

Airplane total weight — 7,000 lb.

c.g. — Sta. 79.0

Rear c.g. limit — Sta. 80.5

Find—How far aft can additional baggage weighing 200 lb. be placed without exceeding the rear c.g. limit?

Solution—

1. Use the added weight formula:

$$\frac{\text{Added weight}}{\text{New total weight}} = \frac{\Delta\,\text{c.g.}}{\text{Dist. between wt. and old c.g.}}$$

$$\frac{200}{7,200} = \frac{1.5}{\text{Dist. between wt. and old c.g.}}$$

Distance between wt. and old c.g. = 54 in.

2. Add to old c.g.:

79 in. + 54 in. = 133 in. aft of datum.

When the 200 lb. are located at Sta. 133, the new c.g. will be exactly on the aft limit; if the weight is located any further to the rear, the aft c.g. limit will be exceeded.

Example:

Given—

Airplane total weight—6,400 lb.

c.g. —Sta. 80.0

Aft c.g. limit —Sta. 80.5

Find—How much baggage can be located in the aft baggage compartment at station 150 without exceeding the aft c.g. limit?

Solution—

Use the added weight formula.

NOTE: In this problem, the new total weight is not given; therefore, it is more convenient to use the version of the formula which makes use of the old total weight.

$$\frac{\text{Added weight}}{\text{Old total weight}} = \frac{\Delta\,\text{c.g.}}{\text{Dist. between wt. and new c.g.}}$$

$$\frac{\text{Added weight}}{6,400} = \frac{.5}{150 - 80.5}$$

Added weight = 46 lb.

Airplane Performance Many accidents have occurred because pilots have failed to understand the effect of varying conditions on airplane performance. In addition to the effects of weight and balance previously discussed, other factors such as density altitude, humidity, winds, runway gradient, and runway surface conditions all have a profound effect in changing airplane performance.

Density Altitude. Air density is perhaps the single most important factor affecting airplane performance. It has a direct bearing on the power output of the engine, efficiency of the propeller, and the lift generated by the wings.

As previously discussed in this handbook, when the air temperature increases, the density of the air decreases. Also, as altitude increases, the density of the air decreases. The density of the air can be described by referring to a corresponding altitude; therefore, the term used to describe air density is density altitude. To avoid confusion remember that a decrease in air density means a high density altitude, and an increase in air density, means a lower density altitude. Density altitude is determined by first finding pressure altitude, and then correcting this altitude for nonstandard temperature variations. It is important to remember that as air density decreases (higher density altitude) airplane performance decreases, and as air density increases (lower density altitude) airplane performance increases.

Humidity. Because of evaporation, the atmosphere always contains some moisture in the form of water vapor. This water vapor replaces molecules of dry air and because water vapor weighs less than dry air, any given volume of moist air weighs less—is less dense—than an equal volume of dry air.

Usually during the operation of small airplanes the effect of humidity is not considered when determining density altitude, but keep in mind that high humidity will decrease airplane performance which, among other things, results in longer takeoff distances and decreased angle of climb.

Density Altitude Effect on Engine Power and Propeller Efficiency. An increase in air temperature or humidity, or decrease in air pressure resulting in a higher density altitude, significantly decreases power output and propeller efficiency.

The engine produces power in proportion to the weight or density of the air. Therefore, as air density decreases the power output of the engine decreases. This is true of all engines that are not equipped with a supercharger or turbocharger. Also, the propeller produces thrust in proportion to the mass of air being accelerated through the rotating blades. If the air is less dense, propeller efficiency is decreased.

The problem of high-density altitude operation is compounded by the fact that when the air is less dense, more engine power and increased propeller efficiency are needed to overcome the decreased lift efficiency of the airplane wing. This additional power and propeller efficiency are not available under high-density altitude conditions; consequently, airplane performance decreases considerably.

Effect of Wind on Airplane Performance. Surface winds during takeoffs and landings have, in a sense, an opposite effect on airplane performance to winds aloft during flight.

During takeoff, a headwind will shorten the takeoff run and increase the angle of climb. This increases performance and helps to compensate for lost performance due to high density altitude. A headwind during flight, however, has an opposite effect on performance because it decreases groundspeed, and consequently increases the total amount of fuel consumed for that flight.

During takeoff, a tailwind will increase the takeoff run and decrease the angle of climb. This decrease in performance should be carefully considered by the pilot before a downwind takeoff is attempted; otherwise the results could be disastrous. A tailwind during flight, however, increases the groundspeed, and conserves fuel.

During landing, a headwind will steepen the approach angle and shorten the landing roll, while a tailwind will decrease the approach angle and increase the landing roll, and again, downwind operations should be considered very carefully by the pilot before being attempted.

Runway Surface Condition and Gradient. The takeoff distance is affected by the surface condition of the runway. If the runway is muddy, wet, soft, rough, or covered with tall grass, these conditions will act as a retarding force and increase the takeoff distance. Some of these surface conditions may decrease landing roll, but there are certain conditions such as ice or snow covering the surface that will affect braking action and increase the landing roll considerably. The upslope or downslope of the runway (runway gradient) is quite important when runway length and takeoff distance are critical. Upslope provides a retarding force which impedes acceleration, resulting in a longer ground run on takeoff.

Landing uphill usually results in a shorter landing roll. Downhill operations will usually have the reverse effect of shortening the takeoff distance and increasing the landing roll.

Ground Effect. When an airplane is flown at approximately one wing span or less above the surface, the vertical component of airflow is restricted and modified, and changes occur in the normal pattern of the airflow around the wing and from the wingtips. This change alters the direction of the rela-

tive wind in a manner that produces a smaller angle of attack. This means that a wing operating in ground effect with a given angle of attack will generate less induced drag than a wing out of ground effect. Therefore, it is more efficient. While this may be useful in specific situations, it can also trap the unwary into expecting greater climb performance than the airplane is capable of sustaining. In other words, an airplane can take off, and while in ground effect establish a climb angle and/or rate that cannot be maintained once the airplane reaches an altitude where ground effect can no longer influence performance. Conversely, on a landing, ground effect may produce "floating," and result in overshooting, particularly at fast approach speeds.

Use of Performance Charts

Most airplane manufacturers provide adequate information from which the pilot can determine airplane performance. This information can be found in Aircraft Flight Manuals, Pilot's Operating Handbooks, or by other means.

Two commonly used methods of depicting performance data are: (1) tables, which are compact arrangements of conditions and performance values placed in an orderly sequence—usually arranged in rows and columns, and (2) graphs, which are pictorial presentations consisting of straight lines, curves, broken lines, or a series of bars representing the successive changes in the value of a variable quantity or quantities. Airplane performance graphs are usually the straight-line or curved-line types. The straight-line graph is a result of two values that vary at a constant rate, while the curved-line graph is a result of two values that vary at a changing rate.

Because all values are not listed on the tables or graphs, interpolation is often required to determine intermediate values for a particular flight condition or performance situation. Interpolation will be discussed later in this chapter.

It should be kept in mind that the information on airplane performance charts is based on flight tests conducted under normal operating conditions, using average piloting skills, with the airplane and engine in good operating condition. Any deviation from the above conditions will affect airplane performance.

The performance data extracted from performance charts are accurate. To attain this accuracy, reasonable care must be exercised when computing performance information. It is a good safety practice to consider the performance of the airplane flown to be less than that predicted by the performance charts.

Standard atmospheric conditions (temperature 59° Fahrenheit [15° Celsius], zero relative humidity, and a pressure of 29.92 inches of mercury at sea level) are used in the development of performance charts. This provides a base from which to evaluate performance when actual atmospheric conditions change.

Interpolation. To interpolate means to compute intermediate values between a series of given values. In many instances when performance is critical, an accurate determination of the performance values is the only acceptable means to enhance safe flight. Guessing to determine these values should be avoided.

Interpolation is simple to perform if the method is understood. The following are examples of how to interpolate or accurately determine the intermediate values between a series of given values.

The numbers in column A range from 10 to 30 and the numbers in column B range from 50 to 100. Determine the intermediate numerical value in column B that would correspond with an intermediate value of 20 placed in column A.

A	B
10	50
20	X = unknown
30	100

It can be visualized that 20 is halfway between 10 and 30; therefore, the corresponding value of the unknown number in column B would be halfway between 50 and 100, or 75.

Many interpolation problems are more difficult to visualize than the preceding example; therefore, a systematic method must be used to determine the required intermediate value. The following describes one method that can be used.

The numbers in column A range from 10 to 30 with intermediate values of 15, 20, and 25. Determine the intermediate numerical value in column B that would correspond with 15 in column A.

A	B
10	50
15	
20	
25	
30	100

First, in column A, determine the relationship of 15 to the range between 10 and 30 as follows:

$$\frac{15 - 10}{30 - 10} = \frac{5}{20} \text{ or } \frac{1}{4}$$

We find that 15 is 1/4 of the range between 10 and 30.

Now determine 1/4 of the range of column B between 50 and 100 as follows:

$$100 - 50 = 50$$
$$1/4 \text{ of } 50 = 12.5$$

The answer 12.5 represents the number of units, but to arrive at the correct value, 12.5 must be added to the lower number in column B as follows:

$$50 + 12.5 = 62.5$$

The interpolation has been completed and 62.5 is the actual value which is 1/4 of the range of column B.

Another method of interpolation is shown below:

Using the same numbers as in the previous example, a proportion problem based on the relationship of the numbers can be set up.

```
   A                    B
  ┌10                  ┌50
 5┤                 x─┤50
  └15                  └ ?
20┤                  50┤
  20
  25
  30                   100
```

Proportion: $\dfrac{5}{20} = \dfrac{x}{50}$

Cross multiply: $\dfrac{5}{20} = \dfrac{x}{50}$

$$20x = 250$$
$$x = 12.5$$

The answer 12.5 must be added to 50 to arrive at the actual value of 62.5.

The following example illustrates the use of interpolation applied to a problem dealing with one aspect of airplane performance:

Temperature (°F)	Takeoff Distance (ft.)
70	1,173
80	1,356

If a distance of 1,173 feet is required for takeoff when the temperature is 70° F. and 1,356 feet for 80° F., what distance is required when the temperature is 75°? The solution to the problem can be determined as follows:

Performance Charts. Following are descriptions of various performance charts. The information on these charts is not intended for operational use, but rather for familiarization and study. Because performance charts are developed for each specific make, model, and type of airplane, care must be exercised by pilots to assure that the chart developed for the specific airplane flown is used when seeking performance data.

Density Altitude Charts. There are various methods that can be used to determine density altitude, one of which is charts. Fig. 4–13 illustrates a typical density altitude chart, and includes a sample problem that will aid in becoming familiar with the use of this chart.

PRESSURE ALTITUDE AND DENSITY CHART

A practice problem – find the Density Altitude with these existing conditions:

Airport elevation 2,545 feet, OAT 70° F., and Altimeter Setting 29.70.

Altimeter Setting in Hg.	Altitude Addition For Obtaining Pressure Altitude
28.3	1,533
28.4	1,436
28.5	1,340
28.6	1,244
28.7	1,148
28.8	1,053
28.9	957
29.0	863
29.1	768
29.2	673
29.3	579
29.4	485
29.5	392
29.6	298
29.7	205
29.8	112
29.9	20
29.92	0
30.0	-73
30.1	-165
30.2	-257
30.3	-348
30.4	-440
30.5	-531
30.6	-622
30.7	-712
30.8	-803

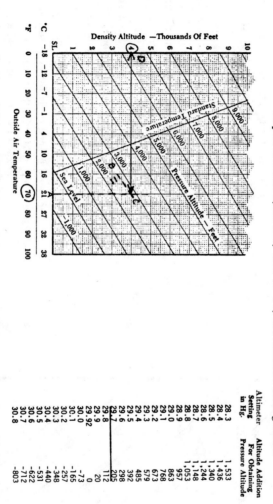

SOLUTION: The chart requires Pressure Altitude which is determined from the conversion table at the right of the graph. 2,545 + 205 = 2,750 feet Pressure Altitude.

Step 1. Draw a line parallel to the vertical lines from the 70° on the Fahrenheit Scale (A) to about the diagonal 3,000 feet Pressure Altitude line.

Step 2. Draw line B representing a value of 2,750 feet (interpolate 3/4 of distance from 2,000 to 3,000) parallel to the pressure altitude lines so that it intersects the line drawn in step 1.

Step 3. The intersection of these two lines (C) lies on the 4,000 foot value of the Density Altitude scale (D). THE DENSITY ALTITUDE IS 4,000 FEET.

Figure 4–13. *Pressure altitude and density altitude chart.*

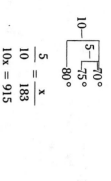

$$\frac{5}{10} = \frac{x}{183}$$
$$10x = 915$$
$$x = 91.5$$

1,173 + 91.5 = 1,264.5 ft., which is the distance required for takeoff with a temperature of 75° F.

Another type of chart used for determining density altitude is shown in Fig. 4–14, and includes a sample problem provided for study.

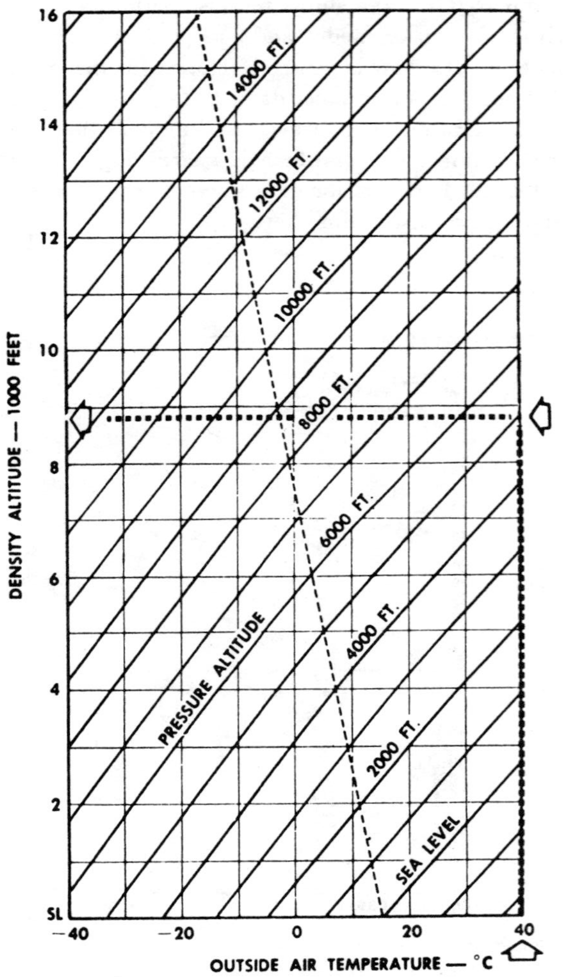

Takeoff Data Chart. Takeoff data charts are found in many Aircraft Flight Manuals or Pilot's Operating Handbooks. From this chart the pilot can determine (1) the length of the takeoff ground-run, and (2) the total distance required to clear a 50-foot obstacle under various airplane weights, head-winds, pressure altitudes, and temperatures. Charts for different airplanes will be different. Fig. 4–15 shows one such chart.

The first column of the chart illustrated gives three possible gross weights (3,100 lbs., 2,400 lbs., and 2,650 lbs.). The second column lists three windspeeds (0, 15, and 30 m.p.h.) opposite each gross weight. The remainder of the chart consists of pairs of columns, each pair having a main heading of a pressure altitude and temperature standard for that altitude (sea level, 59° F.; 2,500 ft., 50° F.; 5,000 ft., 41° F.; and 7,500 ft., 32° F.). The first column of each pair is headed "ground-run"; the second "to clear a 50-foot obstacle."

At the bottom of the chart is this note: "Increase distance 10 percent for each 25° F. above standard temperature for particular altitudes."

To determine the takeoff ground-run for a given set of conditions, the following procedure should be used:

(1) Locate the computed gross weight in the first column.
(2) Locate the existing headwind in the second column and on the same row as the computed gross weight in (1).
(3) Follow the headwind row out to the first column (headed by "ground-run") of the pair of columns headed by the flight altitude. The number at the intersection of this row and column is the length of the ground-run in feet for the given set of conditions, provided the temperature is standard for the altitude.
(4) Increase the number found in (3) by 10 percent for each 25° F. of temperature above standard (for that altitude). The resulting figure is the length of the ground-run.

The same procedure is followed to find the distance to clear a 50-foot obstacle except that in (3) the headwind row would be followed out to the second column (headed by "to clear a 50-foot obstacle") of the pair of columns headed by the altitude. To find distances based on conditions in between those listed in the chart, interpolation must be used.

Sample Problem. What will be the takeoff ground-run distance with the following conditions?

Gross weight . . . . 2,100 lb.
Pressure altitude . . . . 2,500 ft.
Temperature . . . . 75° F.
Headwind . . . . 15 m.p.h.

At an elevation of 5,000 ft. (assuming pressure altitude and elevation are identical) and a temperature of 40° C (104° F) the density altitude is approximately 8,750 ft.

Figure 4–14. *Determining density altitude.*

Solution—Applying steps (1), (2), and (3) to the performance chart, we obtain a figure of 225 ft. Since the temperature is 25° above standard, step (4) must also be applied. Ten percent of 225 is 22.5, or approximately 23. Adding 23 to 225 gives a total of 248 ft. for the takeoff ground-run. Putting this in tabular form, we have:

Ft.

Basic distance exclusive of correction for above standard temperature.............................. 225

Correction for above standard temperature (225 × 0.10).. 23

Approximate takeoff distance required.................... 248

Sample. What will be the distance required to take off and clear a 50-foot obstacle with the same airplane and with the following conditions?

Gross weight .... 2,650 lbs.

Pressure altitude .... 5,000 ft.

Temperature .... 91° F.

Headwind .... Calm

Solution—Following the four-step procedure, except using the "to clear a 50-foot obstacle" column, the solution of this problem gives these results:

Ft.

Basic distance exclusive of correction for above standard temperature 1,500

Correction for above standard temperature (1,500 × 0.20)....................................... 300

Approximate distance required to take off and clear a 50-foot obstacle........................... 1,800

TAKE-OFF DATA

TAKE-OFF DISTANCE WITH 20° FLAPS FROM HARD SURFACE RUNWAY.

GROSS WEIGHT LBS.	HEAD WIND MPH	AT SEA LEVEL & 59°F.		AT 2500 FT. & 50°F.		AT 5000 FT. & 41°F.		AT 7500 FT. & 32°F.	
		GROUND RUN	TO CLEAR 50' OBSTACLE	GROUND RUN	TO CLEAR 50' OBSTACLE	GROUND RUN	TO CLEAR 50' OBSTACLE	GROUND RUN	TO CLEAR 50' OBSTACLE
2100	0	335	715	390	810	465	935	560	1100
	15	185	465	225	540	270	625	330	745
	30	75	260	95	305	125	365	160	450
2400	0	440	895	525	1040	630	1210	770	1465
	15	255	600	310	700	380	835	475	1020
	30	115	350	150	420	190	510	245	640
2650	0	555	1080	665	1260	790	1500	965	1835
	15	330	735	405	865	490	1050	655	1345
	30	160	445	205	535	255	665	335	845

Note: Increase distances 10% for each 25°F above standard temperature for particular altitude.

Figure 4–15. *Takeoff performance data chart.*

CRUISE PERFORMANCE CHART

Altitude	RPM	M.P.	BHP	%BHP	TAS MPH	Gal/Hr.
2500	2450	23 21 20	175 166 157 148	76 72 68 63	158 154 151 148	14.2 13.4 12.7 12.0
	2300	23 22 21 20	164 153 143 135	71 67 62 59	154 149 145 142	13.1 12.2 11.5 11.0
	2200	23 22 21 20	153 144 135 126	67 63 59 55	149 146 142 138	12.1 11.4 10.8 10.2
Maximum Range Settings	2000	20 19 18 17	107 99 89 81	47 43 39 35	126 121 113 105	8.7 8.2 7.5 7.0
5000	2450	23 22 21 20	179 169 161 150	78 73 70 65	163 159 156 151	14.5 13.6 13.0 12.2
	2300	23 22 21 20	167 158 148 139	73 69 64 60	158 155 151 146	13.4 12.6 11.9 11.2
	2200	23 22 21 20	157 148 138 131	68 64 60 57	155 151 146 143	12.4 11.7 11.0 10.5
Maximum Range Settings	2000	19 18 17 16	103 94 86 79	45 41 37 34	126 118 111 103	8.5 7.9 7.3 6.8
7500	2450	21 20 19 18	163 153 143 133	71 67 62 58	161 157 152 147	13.1 12.4 11.7 11.0
	2300	21 20 19 18	151 142 133 125	66 62 58 54	156 151 147 142	12.2 11.6 11.0 10.5
	2200	21 20 19 18	143 134 126 118	62 58 54 51	152 148 143 138	11.4 10.7 10.2 9.7
Maximum Range Settings	2000	19 18 17 16	101 98 90 82	47 43 39 36	131 123 116 107	8.7 8.1 7.6 7.0

Data based on lean mixture, standard conditions, and maximum gross weight

Figure 4–16. *A cruise performance chart.*

Exercise: Find the takeoff ground-run distance and the distance necessary to clear a 50-foot obstacle under each of the following sets of conditions:

Gross weight (lbs.)	Headwind (m.p.h.)	Pressure Altitude (ft.)	Temperature (F°.)
1. 2,100	30	Sea level	59
2. 2,650	Calm	7,500	57
3. 2,400	15	2,500	50
4. 2,650	Calm	Sea level	109
5. 2,250	15	5,000	41

NOTE: The correct answers are given below:

	Ground Run (feet)	To Clear 50-foot (feet)
1.	75	260
2.	1,062	2,019
3.	310	700
4.	666	1,296
5.	325	730

Some airplanes require the use of partial flaps for best takeoff performance; others use no flaps because the additional drag caused by flaps more than offsets the lift advantage acquired from their use. The takeoff flap setting recommended in the Aircraft Flight Manual or Pilot's Operating Handbook should be used.

Cruise Performance Data. Cruise performance charts are compiled from actual tests and are a valuable aid in planning cross-country flights. Fuel consumption depends largely on altitude, power setting (manifold pressure and propeller r.p.m.), and mixture setting.

The following problem will show how to use the cruise performance chart in Fig. 4-16.

Sample Problem. How many flight hours of fuel remain under the following conditions?

Altitude.....................................5,000 ft.
Propeller r.p.m..........................2,300 r.p.m.
Manifold pressure (MP)..................22" Hg
Mixture..Lean
Fuel remaining..............................40 gal.

Solution.—

(1) Locate the altitude (5,000 ft.) in the altitude column (first column).

(2) Locate the r.p.m. (2300) in the r.p.m. column (second column) opposite the altitude (5,000 ft.) just found in (1).

(3) Locate the manifold pressure (22 inches of mercury) in the MP column (third column) opposite the r.p.m. (2300) just located in (2).

(4) Follow this manifold-pressure row out to the column headed by "Gal./Hr.," where the figure 12.6 is read. This is the rate of fuel consumption in gallons per hour.

(5) Divide fuel remaining (40 gallons) by rate of fuel consumption just found (12.6 gal./hr.). The result is 3.17, the number of flight hours remaining. The 3.17 hours are equivalent to 3 hours 10 minutes (multiply 0.17 by 60 minutes).

NOTE: The true airspeed (TAS) with this power setting would be 155 m.p.h. (next to last column).

Sample Problem. If in the preceding sample problem a power setting of 18 inches of manifold pressure and 2000 r.p.m. were used, how much more flight time would be available?

Solution—

(1) Following the same steps as in the preceding problem (except using the new r.p.m. and MP), a fuel consumption rate of 7.9 gallons per hour is found.

(2) Dividing the fuel remaining (40 gallons) by 7.9 gives a total remaining flight time of 5.06 hours. When converted, this is equivalent to 5 hours 4 minutes.

(3) Subtracting 3 hours 10 minutes from 5 hours 4 minutes gives an added flight time (endurance time) of 1 hour 54 minutes.

Sample Problems. Find the true airspeed (TAS), rate of fuel consumption, and total flight time available under the following conditions:

Altitude (ft.)	R.P.M.	Manifold Pressure	Fuel Available (gals.)
1. 2,500	2450	23	55
2. 5,000	2200	22	45
3. 7,500	2000	16	25
4. 2,500	2000	17	25
5. 5,000	2300	23	50

NOTE: For correct answers, see below:

	TAS	Gal./hr.	Flight Time
1.	158	14.2	3:52
2.	151	11.7	3:51
3.	107	7.0	3:34
4.	105	7.0	3:34
5.	158	13.4	3:44

Other types of cruise performance charts are shown in Figs. 4–17 and 4–19.

Some of the information that can be obtained from this chart includes recommended power settings at various altitudes, percent of brake horsepower at these settings, rate of fuel consumption (gal./hr.), true airspeed, hours of endurance with full tanks, and range in miles under standard conditions and zero wind. Not all of these values are contained in all charts. For example:

Refer to Fig. 4–17. At 5,000 ft., 2,300 r.p.m., and 21 inches of manifold pressure, 64% rated power is obtained, and approximately 151 m.p.h. true airspeed. Approximately 11.9 gal./hr. of fuel are consumed, which will give an endurance of 4.6 hrs. and a range of 700 miles under standard conditions, zero wind, and full fuel tanks.

CRUISE AND RANGE PERFORMANCE

Altitude	RPM	M.P.	BHP	%BHP	TAS MPH	Gal./Hr.	End. Hours	Mi. Gal.	Range Miles
5000	2450	23	179	78	163	14.5	3.8	11.2	615
		22	169	73	159	13.6	4.0	11.7	640
		21	161	70	156	13.0	4.2	11.7	660
		20	150	65	151	12.2	4.5	12.5	685
	2300	23	167	73	158	13.4	4.1	11.8	650
		22	158	69	155	12.6	4.4	12.2	675
		21	148	64	151	11.9	4.6	12.7	700
		20	139	60	146	11.2	4.9	13.1	720
	2200	23	157	68	155	12.4	4.4	12.5	685
		22	148	64	151	11.7	4.7	12.9	710
		21	138	60	146	11.0	5.0	13.3	730
		20	131	57	143	10.5	5.2	13.6	750

Cruise performance shown is based on standard conditions, zero wind, lean mixture, 55 gallons of fuel, no fuel reserve, and 2650 pounds gross weight.

Figure 4–17. *Cruise and range performance chart.*

Refer to Fig. 4–18. At 8,000 ft., 55% rated power and 10.3 gal./hr. fuel consumption can be obtained by using 2,200 r.p.m. and 19 inches of manifold pressure.

CRUISE PERFORMANCE

ALT.	RPM	% BHP	TAS MPH	58.8 Gal Endurance Hours	58.8 Gal Range Miles
2500	2500	75	130	6.0	773
	2350	63	118	7.1	832
	2200	53	107	8.4	894
3500	2525	75	131	6.0	775
	2400	65	121	6.9	827
	2250	55	110	8.0	874
4500	2550	75	132	6.0	780
	2400	63	120	7.0	841
	2250	53	109	8.3	905
5500 ⟵	2600	77	135	5.8	775
	2450 ⟵	65	123 ⟵	6.8 ⟵	837 ⟵
	2300	55	112	8.0	887

Figure 4-19. *Cruise performance chart.*

Power Setting Table —

Press. Alt. 1000 Feet	Std. Alt. Temp. °F	138 HP — 55% Rated Approx. Fuel 10.3 Gal./Hr. RPM AND MAN. PRESS.				163 HP — 65% Rated Approx. Fuel 12.3 Gal./Hr. RPM AND MAN. PRESS.			
		2100	2200	2300	2400	2100	2200	2300	2400
SL	59	21.6	20.8	20.2	19.6	24.2	23.3	22.6	22.0
1	55	21.4	20.6	20.0	19.3	23.9	23.0	22.4	21.8
2	52	21.1	20.4	19.7	19.1	23.7	22.8	22.2	21.5
3	48	20.9	20.1	19.5	18.9	23.4	22.5	21.9	21.3
4	45	20.6	19.9	19.3	18.7	23.1	22.3	21.7	21.0
5	41	20.4	19.7	19.1	18.5	22.9	22.0	21.4	20.8
6	38	20.1	19.5	18.9	18.3	22.6	21.8	21.2	20.6
7	34	19.9	19.2	18.6	18.0	22.3	21.5	21.0	20.4
9	27	19.4	18.8	18.2	17.6	—	—	20.5	19.9
8 ⟵	31	19.6	19.0 ⟵	18.4	17.8	—	21.3	20.7	20.1
10	23	19.1	18.6	18.0	17.4	—	—	—	19.6

Figure 4-18. *Power setting table.*

Refer to Fig. 4-19. An altitude of 5,500 ft. with 2,450 r.p.m., should result in 65% rated power, approximately 123 m.p.h. true airspeed, an endurance of 6.8 hrs., and a range of 837 miles.

Climb Performance. The rate of climb under various conditions can be determined by climb performance charts such as depicted in Figs. 4-20, and 4-21.

The information from these charts becomes exceedingly important when crossing high terrain or mountain ranges relatively soon after takeoff. Some charts also give the best climb airspeed and fuel consumed during the climb.

Refer to Fig. 4-20. At 5,000 ft., 41° F., and 2,100 lbs. gross weight, the rate of climb is 1,200 ft./min.; best climb speed is 82 m.p.h.; and fuel used to climb from sea level to 5,000 ft. is 2.8 gal. At a gross weight of 2,650 lbs. under the same conditions, the rate of climb is 795 ft./min.

Refer to Fig. 4-21. At a pressure altitude of 5,000 ft., 86° F., and 2,900 lbs. gross weight, the rate of climb is approximately 810 ft./min. Note that the pressure altitude and temperature must be converted to a density altitude. Refer to Fig. 4-14. The density altitude at this pressure altitude and temperature is approximately 7,750 ft. (86° F. = 30° C.) Enter this density altitude in the chart illustrated in Fig. 4-21 and find the rate of climb of approximately 810 ft./min.

Maximum Glide. These charts are available for some types of airplanes (Fig. 4-22). Note the stated conditions under which the chart values are determined. Various conditions will change these values and this must be considered when using this chart.

Crosswind and Headwind Component Charts. Takeoffs and landings in certain crosswind conditions are inadvisable or even dangerous. If the crosswind is strong enough to warrant an extreme drift correction, a hazardous landing condition may result. Therefore, always consider the takeoff or landing capabilities with respect to the reported surface wind conditions and the available landing directions.

Before an airplane is *type certificated* by the FAA, it must be flight tested to meet certain requirements. Among these is the demonstration of being

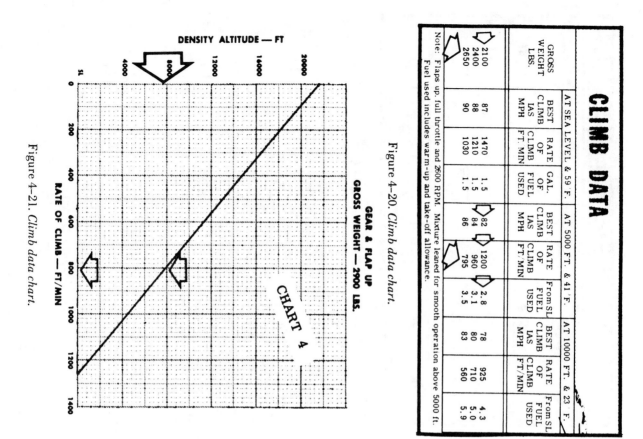

CLIMB DATA

GROSS WEIGHT LBS.	AT SEA LEVEL & 59° F.			AT 5000 FT. & 41° F.			AT 10000 FT. & 23° F.		
	BEST CLIMB IAS MPH	RATE OF CLIMB FT./MIN	GAL. OF FUEL USED	BEST CLIMB IAS MPH	RATE OF CLIMB FT./MIN	From SL FUEL USED	BEST CLIMB IAS MPH	RATE OF CLIMB FT./MIN	From SL FUEL USED
2100	87	1470	1.5	82	1200	2.8	78	925	4.3
2400	88	1210	1.5	84	960	3.1	80	710	5.0
2650	90	1030	1.5	86	795	3.5	83	560	5.9

Note: Flaps up, full throttle and 2600 RPM. Mixture leaned for smooth operation above 5000 ft. Fuel used includes warm-up and take-off allowance.

Figure 4-20. *Climb data chart.*

Figure 4-21. *Climb data chart.*

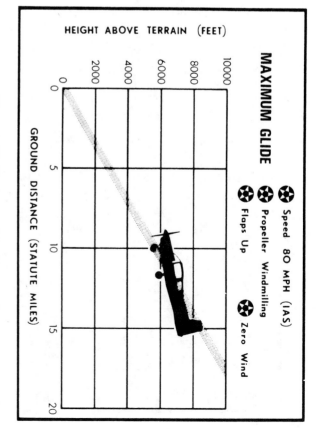

Figure 4-22. *Maximum glide distance chart.*

satisfactorily controllable with no exceptional degree of skill or alertness on the part of the pilot in 90° crosswinds up to a velocity equal to 0.2 V$_{so}$. This means a windspeed of two-tenths of the airplane's stalling speed with power off and gear and flaps down. (If the stalling speed is 60 knots, then the airplane must be capable of being landed in a 12 knot 90° crosswind.) To inform the pilot of the airplane's capability, regulations require that the demonstrated crosswind velocity be made available. Certain Airplane Owner's Manuals provide a chart for determining the maximum safe wind velocities for various degrees of crosswind for that particular airplane. The chart, Fig. 4-23, with the example included, will familiarize pilots with a method of determining crosswind components. The angle between the wind and nose is considered to be the same as the angle between the wind and the takeoff or landing runway.

Stall Speed Charts. Fig. 4-24 is a typical example of a Stall Speed Chart. Note the wide variation in stall speed between straight-and-level flight and various angles of bank. Note that the stall speed in a 60° bank with flaps up and power off (102 m.p.h.) is almost double the stall speed in straight-and-level

STALL SPEEDS IAS

CONFIGURATION	0°	ANGLE OF BANK 20°	40°	60°
Flaps Up — Power Off	72 mph	74 mph	82 mph	102 mph
Flaps Up — Power On	69 mph	71 mph	79 mph	98 mph
Flaps Down (30°) — Power Off	64 mph	66 mph	73 mph	91 mph
Flaps Down (30°) — Power On	55 mph	57 mph	63 mph	78 mph

Figure 4-24. *Stall speed chart.*

flight with flaps down and power on (55 m.p.h.). Even with power on in the 60° bank, the stall speed is reduced only 4 m.p.h. to 98 m.p.h. Study this chart and be aware of its significance, especially during traffic patterns and landings. Similar charts can be found in any airplane flight manual.

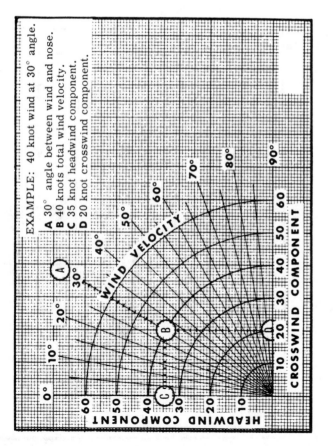

Figure 4-23. *Crosswind and headwind component chart.*

= Power Off = STALLING SPEEDS MPH = CAS

Gross Weight 1600 lbs.

CONDITION	ANGLE OF BANK 0°	20°	40°	60°
Flaps UP	55	57	63	78
Flaps 20°	49	51	56	70
Flaps 40°	48	49	54	67

Figure 4-25. *Stall speed chart.*

Another example of the Stall Speed Chart is shown in Fig. 4-25.

Landing Performance Data. Variables similar to those discussed under "Factors Affecting Takeoff Distance" also affect landing distances, although generally to a lesser extent. Consult your Aircraft Flight Manual or Pilot's Operating Handbook for landing distance data, recommended flap settings, and recommended approach airspeeds.

Sample Problem.—With a power off approach speed of 68 m.p.h. and 40° of flaps, approximately what ground roll will be required under the following conditions? (Refer to Fig. 4-26.)

Elevation . . . . Sea level
Gross weight . . . . 2,300 lbs.
Temperature . . . . 59° F.
Headwind . . . . Calm

Solution. Approximately 415 ft.

Sample Problem.—With power off approach speed of 72 m.p.h., and 40° of flaps, approximately what total landing distance (including ground roll) would be required to clear a 50-foot obstacle and land under the following conditions?

		AT SEA LEVEL & 59°F		AT 2500 FT & 50°F		AT 5000 FT & 41°F		AT 7500 FT & 32°F	
GROSS WEIGHT LBS.	APPROACH IAS MPH	GROUND ROLL	TO CLEAR 50' OBSTACLE	GROUND ROLL	TO CLEAR 50' OBSTACLE	GROUND ROLL	TO CLEAR 50' OBSTACLE	GROUND ROLL	TO CLEAR 50' OBSTACLE
2300	68	415	1015	445	1070	480	1130	520	1190
2600	72	470	1105	505	1165	545	1230	590	1300
2900	76	520	1190	560	1260	605	1330	655	1405

LANDING DISTANCE TABLE

NOTE: REDUCE LANDING DISTANCES 10% FOR EACH 6 MPH HEADWIND.   FLAPS 40° AND POWER OFF.

Figure 4-26. *A landing performance data chart.*

Elevation.................... 2,500 ft.
Gross weight................ 2,600 lb.
Temperature................. 50° F.
Headwind.................... 12 m.p.h.

Solution.

Basic landing distance before headwind correction...... 1,165 ft.
Correction for headwind (1,165 × .20)................ −233 ft.
Approximate landing distance................... 932 ft.

Combined Graphs. Some Aircraft Performance Charts incorporate two or more graphs into one when an aircraft flight performance involves several conditions. A simple combination of graphs is illustrated in Fig. 4–27. It requires three functions to solve for takeoff distance with adjustments for air density, gross weight, and headwind conditions. The first function converts pressure altitude to density altitude. The right margin of this portion of the graph, even though it is not numbered, represents density altitude and starts the second function, the effect of gross weight on takeoff distance. The right margin of this section represents takeoff distance with no wind and starts the final phase of correcting for effect of headwind.

NORMAL TAKE-OFF

ASSOCIATED CONDITIONS:

POWER	TAKE-OFF POWER SET BEFORE BRAKE RELEASE
FLAPS	UP
RUNWAY	PAVED, LEVEL, DRY SURFACE
TAKE-OFF SPEED	IAS AS TABULATED

NOTE: GROUND ROLL IS APPROX. 59% OF TOTAL TAKE-OFF DISTANCE OVER A 50 FT OBSTACLE.

EXAMPLE:

OAT	75°F
PRESSURE ALTITUDE	4000 FT
TAKE-OFF WEIGHT	3200 LBS
HEAD WIND	10 KNOTS
TOTAL TAKE-OFF DISTANCE OVER A 50 FT OBSTACLE	2190 FT
GROUND ROLL (59% OF 2190)	1292 FT
IAS TAKE-OFF SPEED: LIFT-OFF	79 MPH
AT 50 FT	90 MPH

WEIGHT POUNDS	IAS TAKE-OFF SPEED (ASSUMES ZERO INSTR. ERROR)			
	LIFT-OFF MPH	LIFT-OFF KNOTS	50 FEET MPH	50 FEET KNOTS
3400	81	70	92	80
3200	79	69	90	78
3000	76	66	87	76
2800	73	63	84	73
2600	70	61	80	70
2400	67	58	77	67

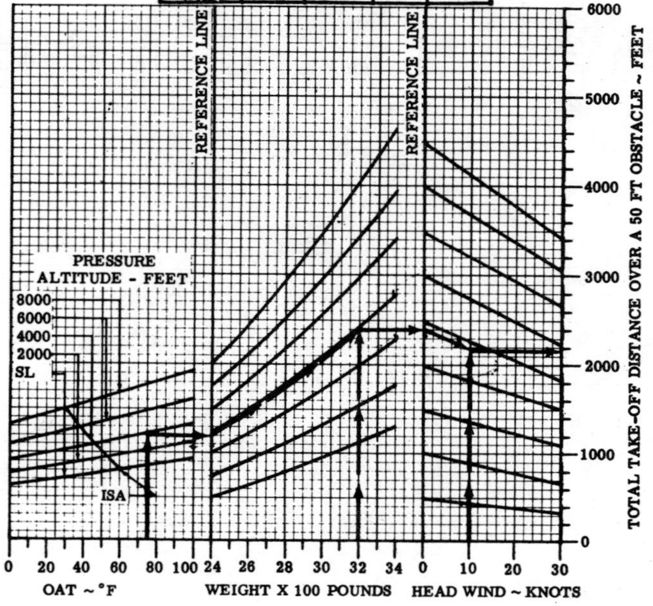

Figure 4–27. *Determining speed for best rate and best angle of climb.*

CHAPTER V—WEATHER

Weather Information for the Pilot Despite the improvements in aircraft design, powerplants, radio aids, and navigation techniques, safety in flight is still subject to conditions of limited visibility, turbulence, and icing.

To avoid hazardous flight conditions, pilots must have knowledge of the atmosphere and of weather behavior.

One may wonder why pilots need more than general information available from the predictions of the "weather man." The answer is well known to the experienced pilot. Meteorologists' predictions are based upon movements of large air masses and upon local conditions at points where weather stations are located. Air masses at times are unpredictable and weather stations in some areas are spaced rather widely apart; therefore, pilots must understand the conditions that could cause unfavorable weather to occur between the stations as well as the conditions that may be different from those indicated by weather reports.

Furthermore, the meteorologist can only predict the weather conditions; the pilot must decide whether the particular flight may be hazardous, considering the type of aircraft being flown, equipment used, flying ability, experience, and physical limitations.

This section is designed to help the pilot acquire a general background of weather knowledge plus the following basic information:

1. Services provided by the National Weather Service and FAA to give the pilot weather information.
2. Sources of weather information available to the pilot.
3. Knowledge required to understand the weather terms commonly used by pilots.
4. Interpretation of weather maps, aviation weather reports and forecasts, and of other data.
5. Conditions of clouds, wind, and weather that are inconvenient or dangerous, and those that the pilot can use to advantage.
6. Suggested methods to use in avoiding dangerous weather conditions.
7. Significance of cloud formations and precipitation areas that may be encountered during flight.

This discussion is intended to give beginning pilots the principles of aviation weather upon which sound judgment can be built as experience is gained and further study is undertaken. There is no substitute for experience in any flight activity, and this is particularly true if good judgment is to be applied to decisions concerning weather.

Services to the Pilot Weather service to aviation is a combined effort of the National Weather Service (NWS), the Federal Aviation Administration (FAA), and other aviation groups and individuals. Because of the increasing need for world-wide weather, countries other than the United States have a vital input into the weather service.

Throughout the conterminous United States a network of approximately 520 airport weather stations, staffed by trained personnel, make weather observations. These observations are transmitted on an hourly basis to central points. Fig. 5-1 shows the development and flow of observations, reports, and forecasts through the various weather processing facilities to the user.

This flow of data takes place through longline communications made up of teletypewriters and facsimiles. Teletypewriter circuits are used to transmit weather reports, forecasts, and warnings, while facsimile is a process of transmitting and reproducing printed matter, such as weather charts.

Each facility that provides weather information to the users has at least one teletypewriter which is connected to an area teletypewriter circuit. This circuit provides complete data within a hundred miles of the facility but only sparse information for more remote areas. Weather reports and forecasts that are not routinely available on the local area circuit are available on a request/reply circuit which will be discussed in more detail later in this chapter. Most facilities are equipped with facsimile service which provides graphic weather analyses and prognostic charts.

Observations Weather observations are measurements and estimates of existing weather at a particular station. These are recorded and transmitted over the teletypewriter system, and at that time the observations become a weather report. The reports are the basis of all weather analyses and forecasts. Observations are made at the surface and also aloft.

A network of airport stations provides routine up-to-date aviation weather reports which include weather elements pertinent to flying. All civilian weather observers are certified by the National Weather Service.

Radar is used to aid in observing weather. Precipitation reflects radar signals which are displayed as echoes on the radar scope. The use of radar is

99

particularly helpful in determining the exact locations of storm areas. Except for some mountainous terrain, radar coverage is complete over the contiguous 48 states.

There are various other observations which have a significant input to the aviation weather service. Upper air observations are taken twice daily at specified stations throughout the United States by releasing weather balloons equipped with a radio that transmits temperature, humidity, pressure, and wind from heights often in excess of 100,000 ft. The station receives this data and compiles a report of the existing atmospheric conditions aloft. Weather satellites scan the earth and provide pictures of cloud cover which are very useful in determining cloud conditions in remote areas. The only means of directly observing turbulence, icing, and height of cloud tops is through pilots reporting weather conditions during flight. Pilot reports are a vital source of weather observations.

Meteorological Centers and Forecast Offices Meteorological centers collect and analyze data and prepare forecasts on a national, hemispheric, or worldwide basis. National Weather Service (NWS) forecast offices prepare forecasts which are generally more detailed.

The National Meteorological Center (NMC) of the NWS is the hub of all weather processing. From worldwide reports NMC prepares forecasts and charts of observed and forecast weather. A few of these charts are prepared manually by forecasters, but many are prepared by computer. NMC prepares weather products that serve many industries other than aviation, but some of their products, such as the wind and temperature forecast, are prepared specifically for aviation.

The National Hurricane Center (NHC) develops hurricane forecasting techniques and issues hurricane forecasts for the Atlantic, Caribbean, Gulf of Mexico, and adjacent land areas.

The National Severe Storms Forecast Center (NSSFC) prepares forecasts of severe convective storms over the contiguous 48 states. It is located at Kansas City, Missouri, which is near the heart of the area most frequented by severe thunderstorms and tornadoes.

The National Environment Satellite Service (NESS) directs the weather satellite program. Through newly-developed radiation measuring techniques, it contributes directly to NMC processing. Satellite cloud photographs are available at a number of weather facilities by facsimile and at some stations by direct picture reception.

The Weather Service Forecast Office (WSFO) issues forecasts, advisories,

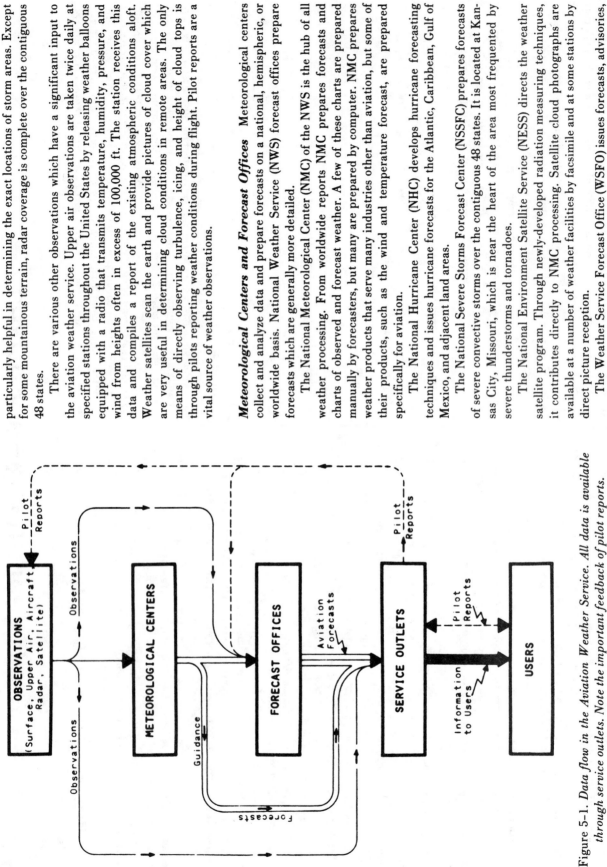

Figure 5–1. *Data flow in the Aviation Weather Service. All data is available through service outlets. Note the important feedback of pilot reports.*

and warnings. This office prepares terminal forecasts for certain airports within its area of responsibility and also the area forecast.

A Weather Service Office (WSO) prepares local forecasts and warnings and provides general weather service. This office has input into the terminal forecast and can adjust these forecasts for a period of two hours or less.

Service Outlets A weather service outlet as used here is any facility, either government or private, that provides aviation weather service to users. Only the FAA and NWS outlets will be discussed in this chapter.

The FAA Flight Service Station (FSS) provides more aviation weather briefing service than any other government service outlet. They provide pre-flight and in-flight briefings, make scheduled and unscheduled weather broadcasts, and furnish weather advisories to flights within the FSS area. Selected Flight Service Stations also disseminate weather information through transcriptions available by telephone or radio. These transcriptions, if heard by the pilot, provide a means of assessing the weather conditions so as to determine any further need for a more detailed briefing. There are two types of transcriptions: (1) Transcribed Weather Broadcast (TWEB), and (2) Pilot's Automatic Telephone Weather Answering Service (PATWAS).

The TWEB is a continuous broadcast on low/medium frequencies (200 to 415 kHz) and selected VORs (108.0 to 117.95 MHz). PATWAS is a recorded telephone briefing service. TWEB and PATWAS transcriptions are on a route concept, i.e., the weather along routes used most regularly is given. A few selected stations also prepare transcriptions for an area within a 50-nautical-mile radius.

The content of the transcriptions is in the following order:

1. Synopsis.
2. Flight Precautions.
3. Route Forecasts.
4. Outlook (Optional).
5. Winds Aloft Forecast.
6. Radar Reports.
7. Surface Weather Reports.
8. Pilot Reports.
9. Notices to Airmen (NOTAMS).

The first five items are forecasts prepared by NWS and will be discussed in detail later in this chapter. The synopsis and route forecasts are prepared especially for TWEB and PATWAS. Flight precautions, outlook, and winds aloft are adapted respectively from in-flight advisories, area forecasts, and the NMC winds aloft forecast. Radar, pilot, and surface reports will also be discussed in detail later in the chapter.

The Airport Facility Directory shows the availability of TWEB at a facility and lists the frequency and shows PATWAS telephone numbers in the section titled, "FSS/CS/T and National Weather Service Telephone Numbers."

FAA Air Traffic Control FAA Air Traffic Control provides a service outlet to disseminate weather. Air Route Traffic Control Centers (ARTCC) advise air traffic under their control of significant weather. The controller may also advise pilots of forecast weather conditions affecting airports that may cause a change in a flight plan. FAA terminal controllers also share in this service by remaining aware of current weather conditions for that terminal and relaying this information to arriving and departing aircraft. The terminal controller also shares the responsibility with NWS for reporting visibility observations at many facilities. At some facilities the terminal controller has the full responsibility for observing, reporting, and classifying aviation weather elements.

The NWS Weather Service Offices (WSO) provide weather briefings in areas not served by FAA Flight Service Stations, and also provide local warnings to aviation. The NWS Weather Service Forecast Offices provide service similar to that of the Weather Service Offices.

Users Many people use the Aviation Weather Service, but the primary users are pilots. As users, pilots should help contribute to the service. This can be done by advising the weather service outlets of weather conditions encountered during flight. This information will help fellow pilots, weather briefers, and forecasters to arrive at a more complete picture of existing weather conditions. In the interest of flight safety, pilots should make a habit of obtaining a complete weather briefing before each flight. This is particularly true of extended flight. If an L/MF radio is available, a preliminary briefing can be obtained by listening to the TWEB at home or office. If a radio is unavailable and PATWAS is, its number should be dialed for a briefing. If the preliminary briefing is such that additional information may be desired, an FSS or WSO should be contacted for a more complete briefing.

When a weather briefing is requested, make it known to the briefer that a pilot is requesting the information. Clear and concise facts about the proposed flight should be given. The following is a list of items that should be given to the briefer:

1. Pilot's name or the aircraft number.
2. Destination, route, and planned altitude.
3. Whether the flight is to be made under Visual Flight Rules or Instrument Flight Rules.

15° C. (59° F.) the year round. As altitude increases, the temperature decreases by about 2° C. (3.5° F.) for every 1,000 ft. (normal lapse rate) until air temperature reaches about −55° C. (−67° F.) at 7 miles above the earth.

For flight purposes, the atmosphere is divided into two layers: the upper layer, where temperature remains practically constant, is the "stratosphere;" the lower layer, where the temperature changes, is the "troposphere" (Fig. 5-2). Although jets routinely fly in the stratosphere, the private pilot usually has no occasion to go that high, but usually remains in the lower layer—the troposphere. It is in this region that all weather occurs and practically all light airplane flying is done. The top of the troposphere lies 5 to 10 miles above the earth's surface.

Obviously, a body of air as deep as the atmosphere has tremendous weight. It is difficult to realize that the normal sea-level pressure upon the body is about 15 pounds per square inch, or about 20 tons on the average person. The body does not collapse because this pressure is equalized by an equal pressure within the body. In fact, if the pressure were suddenly released, the human body would explode. As altitude is gained, the temperature of the air not only decreases (it is usually freezing above 18,000 ft.) but the air becomes thinner; therefore there is less pressure. At first, pressure is rapidly reduced up to 18,000 ft. where the pressure is only half as great as at sea level.

Oxygen and the Human Body

The atmosphere is composed of gases—about four-fifths nitrogen and one-fifth oxygen, with approximately one per cent of various other gases. Oxygen is essential to human life. At 18,000 ft., with only half the normal atmospheric pressure, the body intake of oxygen would be only half the normal amount. Body reactions would be definitely below normal, and unconsciousness might result. In fact, the average person's reactions become affected at 10,000 ft.

To overcome these unfavorable conditions at high altitudes, pilots use oxygen-breathing equipment and wear heavy clothing, often electrically heated; or fly in sealed cabins in which temperature, pressure, and oxygen content of the air can be maintained within proper range.

Significance of Atmospheric Pressure

The average pressure exerted by the atmosphere is approximately 15 pounds per square inch at sea level. This means that a column of air 1 inch square extending from sea level to the top of the atmosphere would weigh about 15 pounds. The actual pressure at a given place and time, however, depends upon several factors—altitude, temperature, and density of the air. These conditions very definitely affect flight.

4. Proposed time of departure.
5. Proposed time of arrival at destination.
6. Any intermediate stops.

With this information, the weather briefer can concentrate on the weather conditions relevant to the proposed flight.

The weather briefing should include:

1. Adverse weather (anything that might cause the pilot to cancel or postpone the flight).
2. Synopsis.
3. Current weather, including PIREPS.
4. En route forecast.
5. Destination forecast.
6. Winds and temperatures aloft.
7. NOTAMS.
8. Request for PIREPS (when appropriate).
9. Closing statement.

A pilot weather briefing is complete when the pilot has a clear picture of the weather to expect on the flight.

A request/reply service mentioned earlier is available at all FSSs, WSOs, and WSFOs. Through this service a pilot may request any weather reports or forecasts not routinely available at a service outlet. Included in the request/reply are route forecasts used in TWEB and recorded briefings used in PATWAS transcriptions.

An alternate plan of action should be available, particularly if the weather is questionable. If during flight weather conditions deteriorate to the extent that safe flight cannot be continued, a turn should be made immediately and in the direction away from the adverse weather. Without preplanning, and knowing where the hazardous weather lies, this turn could be made in the wrong direction and lead to more complex problems. A preplanned diversion to a direction known to be safe is far superior to a wrong decision and panic.

Before discussing the various types of aviation forecasts and reports, the characteristics of weather will be discussed. This discussion will include the nature of the atmosphere, atmospheric pressure, wind, moisture, temperature, condensation, and air masses and fronts.

Nature of the Atmosphere

Life exists at the bottom of an ocean of air called the atmosphere. This ocean extends upward from the earth's surface for many miles, gradually thinning as it nears the top. The exact upper limit has never been determined. Near the surface, the air is relatively warm from contact with the earth. The temperature in the United States averages about

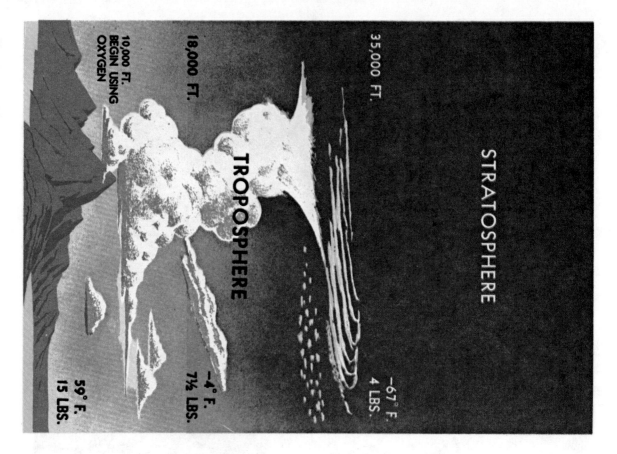

STRATOSPHERE

−67° F.
4 LBS.

TROPOSPHERE

35,000 FT.

18,000 FT.

−4° F.
7½ LBS.

59° F.
15 LBS.

10,000 FT.
BEGIN USING
OXYGEN

Figure 5-2. *The troposphere and stratosphere are the realm of flight.*

Measurement of Atmospheric Pressure A barometer is generally used to measure the height of a column of mercury in a glass tube. It is sealed at one end and calibrated in inches. An increase in pressure forces the mercury higher in the tube; a decrease allows some of the mercury to drain out, reducing the height of the column. In this way, changes of pressure are registered in inches of mercury. The standard sea-level pressure expressed in these terms is 29.92 inches at a standard temperature of 15° C. (59° F.).

The mercury barometer is cumbersome to move and difficult to read. A more compact, more easily read, and more mobile barometer is the aneroid, although it is not so accurate as the mercurial. The aneroid barometer is a partially-evacuated cell sensitive to pressure changes. The cell is linked to an indicator which moves across a scale graduated in pressure units.

If all weather stations were at sea level, the barometer readings would give a correct record of the distribution of atmospheric pressure at a common level. To achieve a common level, each station translates its barometer reading into terms of sea-level pressure. A change of 1,000 ft. of elevation makes a change of about 1 inch on the barometer reading. Thus, if a station located 5,000 ft. above sea level found the mercury to be 25 inches high in the barometer tube, it would translate and report this reading as 30 inches (Fig. 5-3).

Since the rate of decrease in atmospheric pressure is fairly constant in the lower layers of the atmosphere, the approximate altitude can be determined by finding the difference between pressure at sea level and pressure at the given atmospheric level. In fact, the aircraft altimeter is an aneroid barometer with its scale in units of altitude instead of pressure.

Effect of Altitude on Atmospheric Pressure It can be concluded that atmospheric pressure decreases as altitude increases. It can also be stated that pressure at a given point is a measure of the weight of the column of air above that point. As altitude increases, pressure diminishes as the weight of the air column decreases. This decrease in pressure (increase in density altitude) has a pronounced effect on flight.

Effect of Altitude on Flight As previously discussed in this handbook, the most noticeable effect of a decrease in pressure (increase in density altitude), due to an altitude increase, becomes evident in takeoffs, rates of climb, and landings. An airplane that requires a 1,000-foot run for takeoff at a sea-level airport will require a run almost twice as long to take off at an airport which is approximately 5,000 ft. above sea level. The purpose of the takeoff run is to gain enough speed to generate lift from the passage of air over the

103

Figure 5–3. *Barometric pressure at a weather station is expressed as pressure at sea level.*

wings. If the air is thin, more speed is required to obtain enough lift for takeoff—hence, a longer ground run. It is also true that the engine is less efficient in thin air, and the thrust of the propeller is less effective. The rate of climb is also slower at the higher elevation, requiring a greater distance to gain the altitude to clear any obstructions. In landing, the difference is not so noticeable except that the plane has greater groundspeed when it touches the ground (Figs. 5-4 and 5-5).

Effect of Differences in Air Density Differences in air density caused by changes in temperature result in changes in pressure. This, in turn, creates motion in the atmosphere, both vertically and horizontally (currents and winds). This action, when mixed with moisture, produces clouds and precipitation—in fact, these are all the phenomena called "weather."

Pressure Recorded in "Millibars" The mercury barometer reading at the individual weather stations is converted to the equivalent sea-level pressure and then translated from terms of inches of mercury to a measure of pressure called millibars. One inch of mercury is equivalent to approximately 34 millibars; hence, the normal atmospheric pressure at sea level (29.92), expressed in millibars, is 1,013.2 or roughly 1,000 millibars. The usual pressure readings range from 950.0 to 1,040.0. For economy of space, the entry is shortened on some reports by omitting the initial 9 or 10 and the decimal point. On the hourly weather report, a number beginning with 5 or higher presupposes an initial "9," whereas a number beginning with a 4 or lower presupposes an initial "10." For example: 653 = 965.3; 346 = 1034.6; 999 = 999.9; 001 = 1000.1; etc. If the fourth element in the aviation weather report is sea-level pressure coded 132, this is decoded 1013.2 millibars.

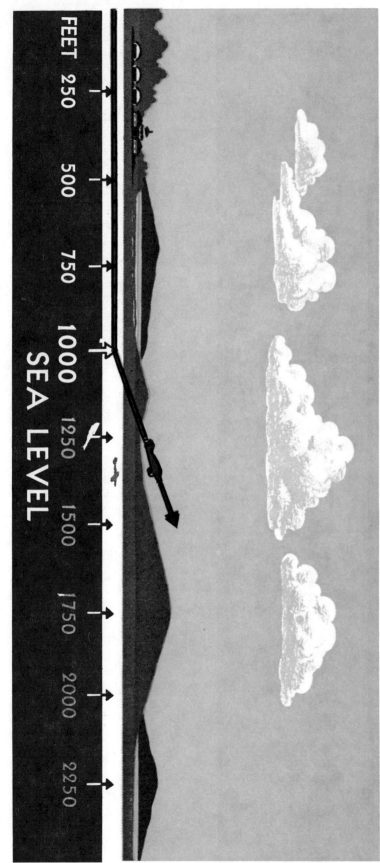

Figure 5-4. *Atmospheric density at sea level enables an airplane to take off in a relatively short distance.*

FEET 250 500 750 1000 1250 1500 1750 2000 2250

SEA LEVEL

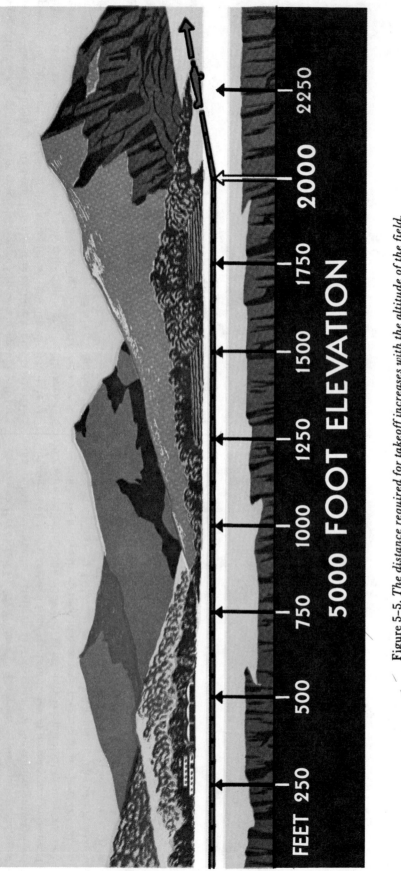

FEET 250 500 750 1000 1250 1500 1750 2000 2250

5000 FOOT ELEVATION

Figure 5–5. *The distance required for takeoff increases with the altitude of the field.*

Individually these pressure readings are of no particular value to the pilot; but when pressures at different stations are compared or when pressures at the same station show changes in successive readings, it is possible to determine many symptoms indicating the trend of weather conditions. In general, a marked fall indicates the approach of bad weather and a marked rise indicates a clearing of the weather.

Wind The pressure and temperature changes discussed in the previous section produce two kinds of motion in the atmosphere—vertical movement of ascending and descending currents, and horizontal flow known as "wind." Both of these motions are of primary interest to the pilot because they affect the flight of aircraft during takeoff, landing, climbing, and cruising flight.

This motion also brings about changes in weather, which may make the difference between a safe flight or a disastrous one.

Conditions of wind and weather occurring at any specific place and time are the result of the general circulation in the atmosphere. This will be discussed briefly in the following pages.

The atmosphere tends to maintain an equal pressure over the entire earth, just as the ocean tends to maintain a constant level. When the equilibrium is disturbed, air begins to flow from areas of higher pressure to areas of lower pressure.

The Cause of Atmospheric Circulation The factor that upsets the normal equilibrium is the uneven heating of the earth. At the equator, the earth receives more heat than in areas to the north and south. This heat is transferred to the atmosphere, warming the air and causing it to expand and become less dense. Colder air to the north and south, being more dense, moves toward the equator forcing the less dense air upward. This air in turn becomes warmer and less dense and is forced upward, thus establishing a constant circulation that might consist of two circular paths; the air rising at the equator, traveling aloft toward the poles, and returning along the earth's surface to the equator, as shown in Fig. 5-6.

This theoretical pattern, however, is greatly modified by many forces, a very important one being the rotation of the earth. In the Northern Hemisphere, this rotation causes air to deflect to the right of its normal path. In the Southern Hemisphere, air is deflected to the left of its normal path. For simplicity this discussion will be confined to the motion of air in the Northern Hemisphere (Fig. 5-7).

As the air rises and moves northward from the equator, it is deflected toward the east, and by the time it has traveled about a third of the distance to the pole, it is no longer moving northward, but eastward. This causes the air to accumulate in a belt at about latitude 30°, creating an area of high

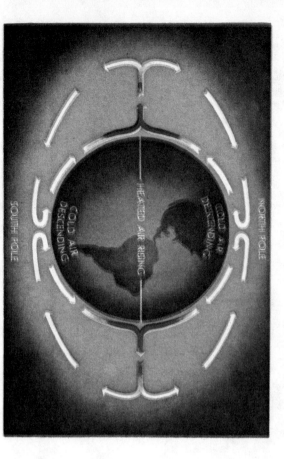

Figure 5-6. *Heat at the equator would cause the air to circulate uniformly, as shown, if the earth did not rotate.*

pressure. Some of this air is then forced down to the earth's surface, where part flows southwestward, returning to the equator, and part flows northeastward along the surface.

A portion of the air aloft continues its journey northward, being cooled en route, and finally settles down near the pole, where it begins a return trip toward the equator. Before it has progressed very far southward, it comes into conflict with the warmer surface air flowing northward from latitude 30°. The warmer air moves up over a wedge of the colder air, and continues northward, producing an accumulation of air in the upper latitudes.

Further complications in the general circulation of the air are brought about by the irregular distribution of oceans and continents, the relative effectiveness of different surfaces in transferring heat to the atmosphere, the daily variation in temperature, the seasonal changes, and many other factors.

Regions of low pressure, called "lows," develop where air lies over land or water surfaces that are warmer than the surrounding areas. In India, for example, a low forms over the hot land during the summer months, but moves out over the warmer ocean when the land cools in winter. Lows of this type are semipermanent, however, and are less significant to the pilot than the "migratory cyclones" or "cyclonic depressions" that form when unlike air masses meet. These lows will be discussed later in this chapter under "Occlusions."

Wind Patterns This is a discussion of wind patterns associated with areas of high and low pressure. As previously stated, air flows from an area of high pressure to an area of low pressure. In the Northern Hemisphere during this flow the air is deflected to the right. Therefore, as the air leaves the high pressure area it is deflected to produce a clockwise circulation. As the air flows toward the low pressure area it is deflected to produce a counterclockwise flow around the low pressure area.

Another important aspect is that air moving out of a high-pressure area depletes the quantity of air. Therefore, highs are areas of descending air. Descending air favors dissipation of cloudiness; hence the association, high pressure—good weather. By similar reasoning, when air converges into a low-pressure area, it cannot go outward against the pressure gradient, nor can it go downward into the ground; it must go upward. Rising air is conducive to cloudiness and precipitation; thus the general association low pressure—bad weather.

A knowledge of these patterns frequently enables a pilot to plan a course to take advantage of favorable winds, particularly during long flights. In flying from east to west, for example, the pilot would find favorable winds to the

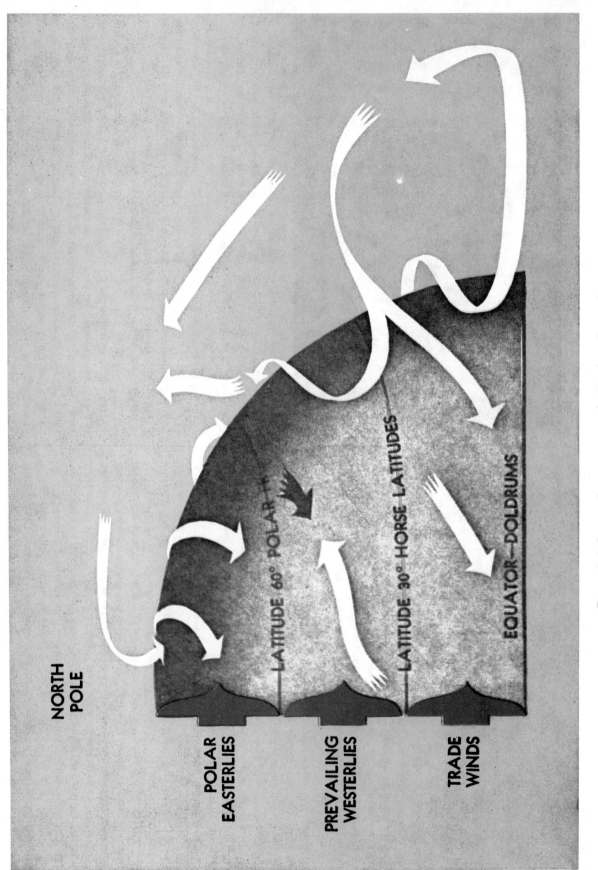

Figure 5–7. *Principal air currents in the Northern Hemisphere.*

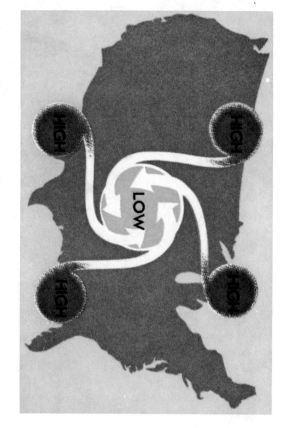

Figure 5-8. *Circulation of wind within a "low."*

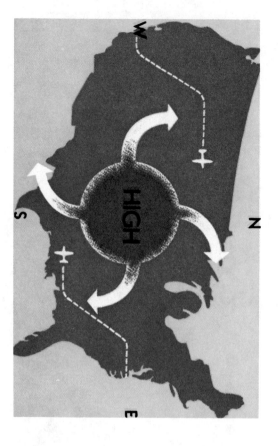

Figure 5-9. *Use of favorable winds in flight.*

south of a high, or to the north of a low (Figs. 5–8 and 5–9). It also gives the pilot a general idea of the type of weather to expect relative to the "highs" and "lows."

The theory of general circulation in the atmosphere, and the wind patterns formed within areas of high pressure and low pressure have been discussed. These concepts account for the large-scale movements of the wind, but do not take into consideration the effects of local conditions that frequently cause drastic modifications in wind direction and speed near the earth's surface.

Convection Currents Certain kinds of surfaces are more effective than others in heating the air directly above them. Plowed ground, sand, rocks, and barren land give off a great deal of heat, whereas water and vegetation tend to absorb and retain heat. The uneven heating of the air causes small local circulations called "convection currents," which are similar to the general circulation just described.

This is particularly noticeable over land adjacent to a body of water. During the day, air over land becomes heated and less dense; colder air over water moves in to replace it forcing the warm air aloft and causing an on-shore wind. At night the land cools, and the water is relatively warmer. The cool air over the land, being heavier, then moves toward the water as an off-shore wind, lifting the warmer air and reversing the circulation (Figs. 5–10 and 5–11).

Convection currents cause the bumpiness experienced by pilots flying at low altitudes in warmer weather. On a low flight over varying surfaces, the pilot will encounter updrafts over pavement or barren places and downdrafts over vegetation or water. Ordinarily, this can be avoided by flight at higher altitudes. When the larger convection currents form cumulus clouds, the pilot will invariably find smooth air above the cloud level (Fig. 5–12).

Convection currents also cause difficulty in making landings, since they affect the rate of descent. For example, a pilot making a constant glide frequently tends to land short of or overshoot the intended landing spot, depending upon the presence and severity of convection currents (Figs. 5–13 and 5–14).

The effects of local convection, however, are less dangerous than the turbulence caused when wind is forced to flow around or over obstructions. The only way for the pilot to avoid this invisible hazard is to be forewarned, and to know where to expect unusual conditions.

Effect of Obstructions on Wind When the wind flows around an obstruction, it breaks into eddies—gusts with sudden changes in speed and

109

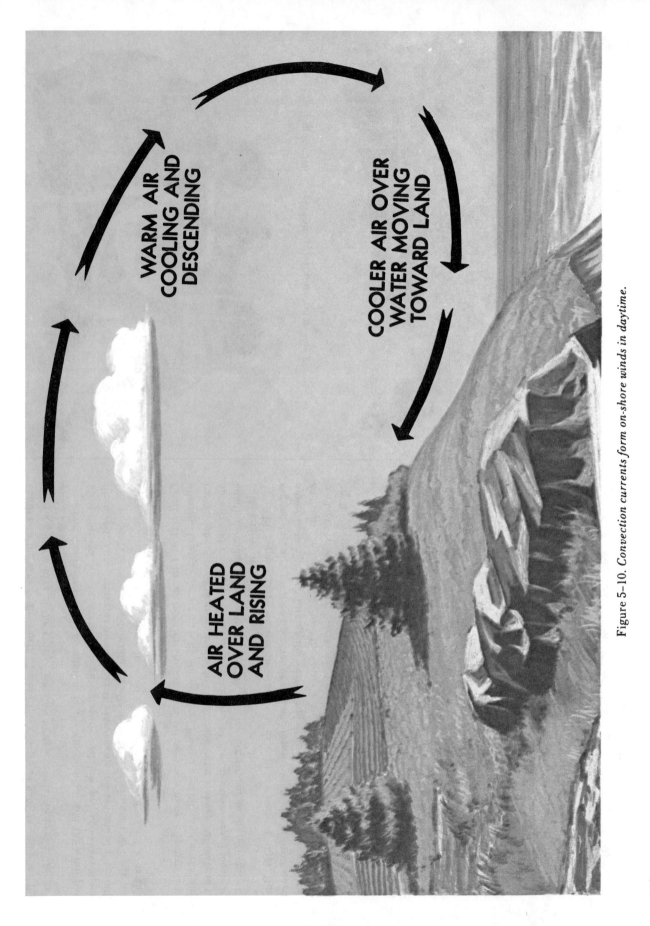

WARM AIR
COOLING AND
DESCENDING

COOLER AIR OVER
WATER MOVING
TOWARD LAND

AIR HEATED
OVER LAND
AND RISING

Figure 5–10. *Convection currents form on-shore winds in daytime.*

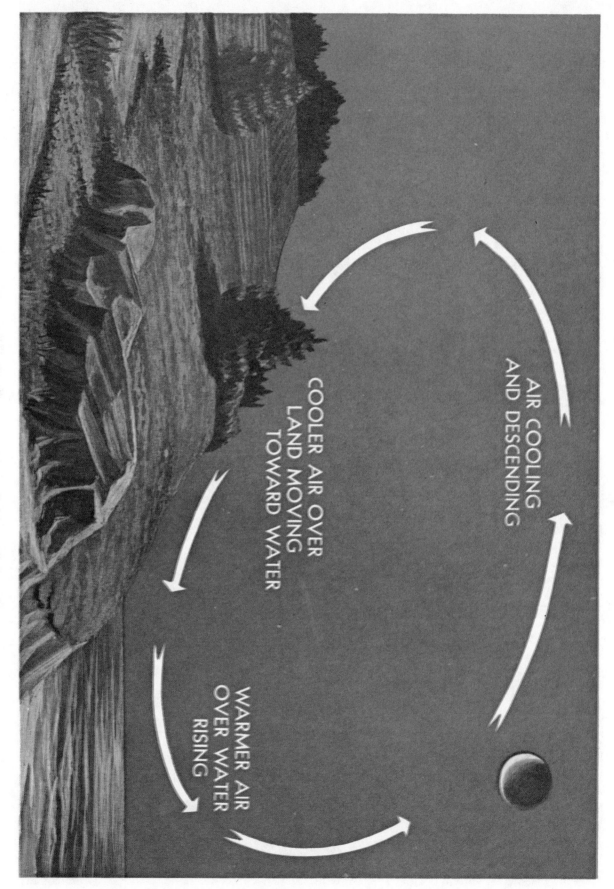

Figure 5-11. *Convection currents form off-shore winds at night.*

AIR COOLING
AND DESCENDING

COOLER AIR OVER
LAND MOVING
TOWARD WATER

WARMER AIR
OVER WATER
RISING

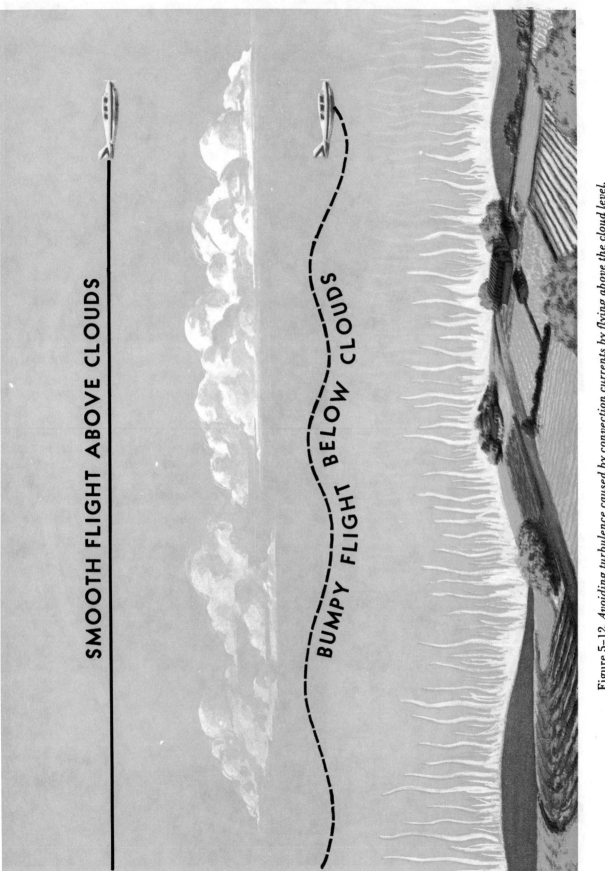

Figure 5–12. *Avoiding turbulence caused by convection currents by flying above the cloud level.*

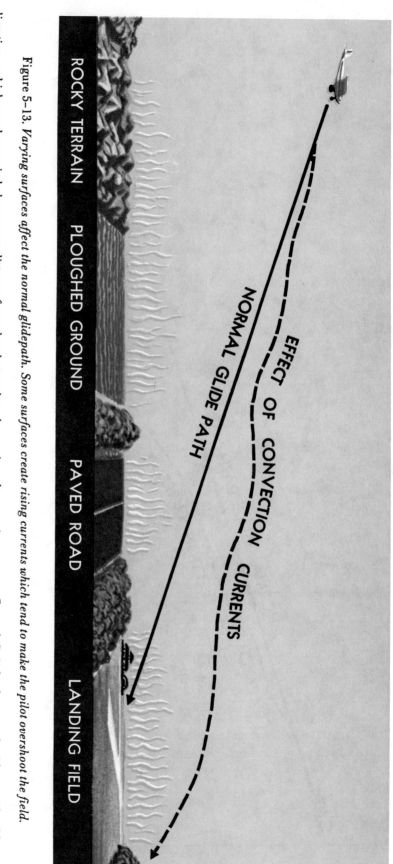

EFFECT OF CONVECTION CURRENTS

NORMAL GLIDE PATH

ROCKY TERRAIN | PLOUGHED GROUND | PAVED ROAD | LANDING FIELD

Figure 5–13. *Varying surfaces affect the normal glidepath. Some surfaces create rising currents which tend to make the pilot overshoot the field.*

direction—which may be carried along some distance from the obstruction. A pilot flying through such turbulence should anticipate the bumpy and un-steady flight that may be encountered. This turbulence—the intensity of which depends upon the size of the obstacle and the velocity of the wind—can present a serious hazard during takeoffs and landings. For example, during landings it can cause a pilot to "drop in"; during takeoffs it could cause the aircraft to fail to gain enough altitude to clear low objects in its path. Any landings or takeoffs attempted under gusty conditions should be made at higher speeds, to maintain adequate control during such conditions (Fig. 5–15).

This same condition is more noticeable where larger obstructions such as bluffs or mountains are involved. As shown in Fig. 5–16, the wind blowing up the slope on the windward side is relatively smooth and its upward current helps to carry the aircraft over the peak. The wind on the leeward side, follow-

ing the terrain contour, flows definitely downward with considerable tur-bulence and would tend to force an aircraft into the mountain side. The stronger the wind, the greater the downward pressure and the accompanying turbulence. Consequently, in approaching a hill or mountain from the leeward side, a pilot should gain enough altitude well in advance. Because of these downdrafts, it is recommended that mountain ridges and peaks be cleared by at least 2,000 ft. If there is any doubt about having adequate clearance, the pilot should turn away at once and gain more altitude. Between hills or moun-tains, where there is a canyon or narrow valley, the wind will generally veer from its normal course and flow through the passage with increased velocity and turbulence. A pilot flying over such terrain needs to be alert for wind shifts and particularly cautious if making a landing.

Low-Level Wind Shear Wind shear is best described as a change in wind

113

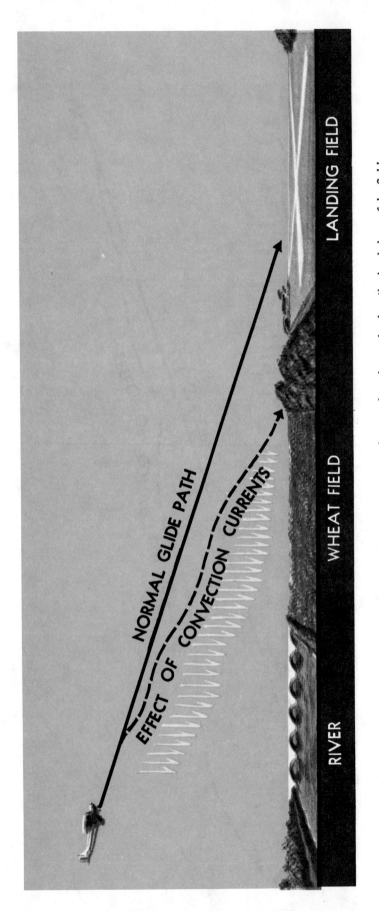

Figure 5–14. *Descending currents prevail above some surfaces and tend to make the pilot land short of the field.*

direction and/or speed within a very short distance in the atmosphere. Under certain conditions, the atmosphere is capable of producing some dramatic shears very close to the ground; for example, wind direction changes of 180° and speed changes of 50 knots or more within 200 ft. of the ground have been observed. This, however, is not something encountered every day. In fact, it is unusual, which makes it more of a problem. It has been thought that wind cannot affect an aircraft once it is flying except for drift and groundspeed. This is true with steady winds or winds that change gradually. It isn't true, however, if the wind changes faster than the aircraft mass can be accelerated or decelerated.

The most prominent meteorological phenomena that cause significant low-level wind shear problems are thunderstorms and certain frontal systems at or near an airport.

Basically, there are two potentially hazardous shear situations. First, a tailwind may shear to either a calm or headwind component. In this instance, initially the airspeed increases, the aircraft pitches up and the altitude increases. Second, a headwind may shear to a calm or tailwind component. In this situation, initially the airspeed decreases, the aircraft pitches down, and the altitude decreases. Aircraft speed, aerodynamic characteristics, power/weight ratio, powerplant response time, and pilot reactions along with other factors have a bearing on wind shear effects. It is important, however, to remember that shear can cause problems for ANY aircraft and ANY pilot.

There are two atmospheric conditions that cause the type of low-level wind shear discussed herein. These are thunderstorms and fronts.

The winds around a thunderstorm are complex. Wind shear can be found on all sides of a cell. The wind shift line or gust front associated with thunder-

Figure 5-15. *Turbulence caused by obstructions.*

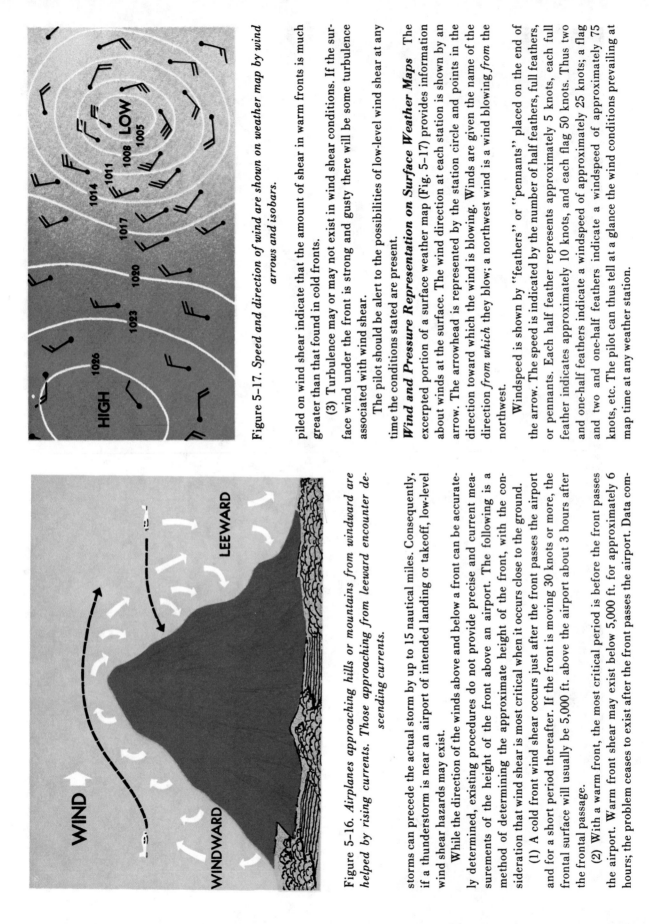

Figure 5–16. *Airplanes approaching hills or mountains from windward are helped by rising currents. Those approaching from leeward encounter descending currents.*

Figure 5-17. *Speed and direction of wind are shown on weather map by wind arrows and isobars.*

storms can precede the actual storm by up to 15 nautical miles. Consequently, if a thunderstorm is near an airport of intended landing or takeoff, low-level wind shear hazards may exist.

While the direction of the winds above and below a front can be accurately determined, existing procedures do not provide precise and current measurements of the height of the front above an airport. The following is a method of determining the approximate height of the front, with the consideration that wind shear is most critical when it occurs close to the ground.

(1) A cold front wind shear occurs just after the front passes the airport and for a short period thereafter. If the front is moving 30 knots or more, the frontal surface will usually be 5,000 ft. above the airport about 3 hours after the frontal passage.

(2) With a warm front, the most critical period is before the front passes the airport. Warm front shear may exist below 5,000 ft. for approximately 6 hours; the problem ceases to exist after the front passes the airport. Data com-

piled on wind shear indicate that the amount of shear in warm fronts is much greater than that found in cold fronts.

(3) Turbulence may or may not exist in wind shear conditions. If the surface wind under the front is strong and gusty there will be some turbulence associated with wind shear.

The pilot should be alert to the possibilities of low-level wind shear at any time the conditions stated are present.

Wind and Pressure Representation on Surface Weather Maps   The excerpted portion of a surface weather map (Fig. 5–17) provides information about winds at the surface. The wind direction at each station is shown by an arrow. The arrowhead is represented by the station circle and points in the direction toward which the wind is blowing. Winds are given the name of the direction *from which* they blow; a northwest wind is a wind blowing *from the* northwest.

Windspeed is shown by "feathers" or "pennants" placed on the end of the arrow. The speed is indicated by the number of half feathers, full feathers, or pennants. Each half feather represents approximately 5 knots, each full feather indicates approximately 10 knots, and each flag 50 knots. Thus two and one-half feathers indicate a windspeed of approximately 25 knots; a flag and two and one-half feathers indicate a windspeed of approximately 75 knots, etc. The pilot can thus tell at a glance the wind conditions prevailing at map time at any weather station.

116

Pilots can obtain this information and forecasts of expected winds from all weather reporting stations.

The pressure at each station is recorded on the weather map, and lines (isobars) are drawn to connect points of equal pressure. Many of the lines make complete circles to surround pressure areas marked **H** (high) or **L** (low).

Isobars are quite similar to the contour lines appearing on aeronautical charts. However, instead of indicating altitude of terrain and steepness of slopes, isobars indicate the amount of pressure and steepness of pressure gradients. If the gradient (slope) is steep, the isobars will be close together, and the wind will be strong. If the gradient is gradual, the isobars will be far apart, and the wind gentle (Fig. 5–18).

Isobars furnish valuable information about winds in the first few thousand feet above the surface. Close to the earth, wind direction is modified by the contours over which it passes, and windspeed is reduced by friction with the surface. At levels 2,000 or 3,000 ft. above the surface, however, the speed is greater and the direction is usually parallel to the isobars. Thus, while wind arrows on the weather map excerpt indicate wind near the surface, isobars indicate winds at slightly higher levels (Fig. 5–17).

In the absence of specific information on upper wind conditions, the pilot can often make a fairly reasonable estimate of the wind conditions in the lower few thousand feet on the basis of the observed surface wind. Generally, it will be found that the wind at an altitude of 2,000 ft. above the surface will veer about 20° to 40° to the right and almost double in speed. The veering will be greatest over rough terrain and least over flat surfaces. Thus, a north wind of 20 knots at the airport would be likely to change to a northeast wind of 40 knots at 2,000 ft. This subject will be reviewed later in this chapter.

Moisture and Temperature The atmosphere always contains a certain amount of foreign matter—smoke, dust, salt particles, and particularly moisture in the form of invisible water vapor. The amount of moisture that can be present in the atmosphere depends upon the temperature of the air. For each increase of 20° F. the capacity of the air to hold moisture is about doubled; conversely, for each decrease of 20° F. the capacity becomes only half as much.

Relative Humidity "Humidity" is commonly referred to as the apparent dampness in the air. A similar term is used by the National Weather Service is *relative humidity*, which is a ratio of the amount of moisture present in any given volume of air to the amount of moisture the air could hold in that volume of air at prevailing temperature and pressure. For instance, "75 per-

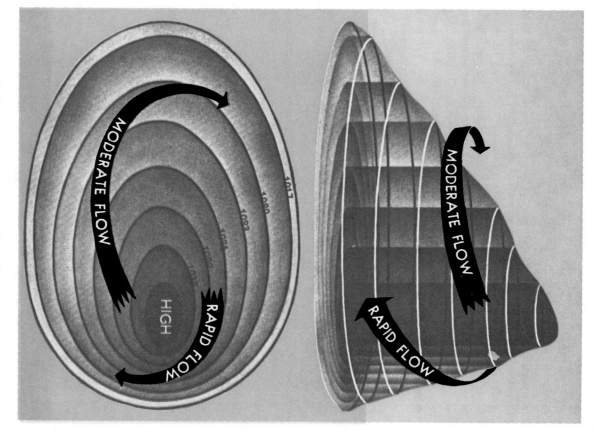

Figure 5–18. *Above: Flow of air around a "high." Below: Isobars on a weather map indicate various degrees of pressure within a high.*

cent relative humidity,'' means that the air contains three-fourths of the water vapor which it is capable of holding at the existing temperature and pressure.

Temperature-Dewpoint Relationship For the pilot, the relationship discussed under relative humidity is expressed in a slightly different way—as ''temperature and dewpoint.'' It is apparent from the foregoing discussion that if a mass of air at 27° C. (80° F.) has a relative humidity of 50 percent and the temperature is reduced 11° C. (20° F.) to 16° C. (60° F.), the air will then be saturated (100 percent relative humidity). In this case, the original relationship will be stated as ''temperature 80° F., dewpoint 60.'' In other words, dewpoint is the temperature to which air must be cooled to become saturated.

Dewpoint is of tremendous significance to the pilot because it represents a critical condition of the air. When temperature reaches the dewpoint, water vapor can no longer remain invisible, but is forced to condense, becoming visible on the ground as dew or frost, appearing in the air as fog or clouds, or falling to the earth as rain, snow, or hail.

NOTE: This is how water can get into the fuel tanks when the tanks are left partially filled overnight. The temperature cools to the dewpoint and the water vapor contained in the fuel tank air space condenses. This condensed moisture then sinks to the bottom of the fuel tank, since water is heavier than gasoline.

Methods by Which Air Reaches the Saturation Point It is interesting to note the various ways by which air can reach the saturation point. As previously discussed, this is brought about by a lowering of temperature such as might occur under the following conditions: when warm air moves over a cold surface; when cold air mixes with warm air; when air is cooled during the night by contact with the cold ground; or when air is forced upward. Only the fourth method needs any special comment.

When air rises, it uses heat energy in expanding. Consequently, the rising air loses heat rapidly. If the air is unsaturated, the loss will be approximately 3.0° C. (5.4° F.) for every 1,000 ft. of altitude.

Warm air can be lifted aloft by three methods; by becoming heated through contact with the earth's surface, resulting in convective currents; by moving up sloping terrain (as wind blowing up a mountainside); and by being forced to flow over another body of air. For example: when air masses of different temperatures and densities meet. Under the last condition, the warmer, lighter air tends to flow over the cooler, denser air. This will be discussed in greater detail in this chapter under ''Fronts.''

Air can also become saturated by precipitation. Whatever the cause,

when temperature and dewpoint at the ground are close together, there is a good possibility for low clouds and fog to form. Temperature and dewpoint are reported in degrees Fahrenheit in the aviation weather report. (Note: The chart in Fig. 5–19 may be used to convert degrees Fahrenheit to degrees Celsius, and vice versa. For example, 0° C. equals 32° F.)

Effect of Temperature on Air Density Atmospheric pressure not only varies with altitude, it also varies with temperature. When air is heated, it expands and therefore has less density. A cubic foot of warm air is less dense

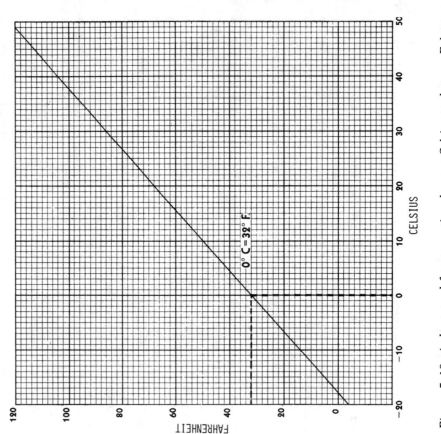

Figure 5-19. *A chart used for converting degrees Celsius to degrees Fahrenheit and vice versa.*

than a cubic foot of cold air. This decrease in air density (increase in altitude), brought about by an increase in temperature, has a pronounced effect on flight.

Effect of Temperature on Flight Since an increase in temperature makes the air less dense (increases density altitude), the takeoff run will be longer, the rate of climb slower, and the landing speed (groundspeed) faster on a hot day than on a cold day. Thus, an increase in temperature has the same effect as an increase in altitude. An airplane which requires a ground run of 1,000 ft. on a winter day when the temperature is −18° C. (0° F.), will require a much longer run on a summer day when the temperature is 38° C. (100° F.). An airplane that requires the greater portion of a short runway for takeoff on a cold winter day may be unable to take off on this runway during a hot summer day.

Effect of High Humidity on Air Density A common misconception is that water vapor weighs more than an equal volume of dry air. *This is not true.* Water vapor weighs approximately five-eighths or 62% of an equal volume of perfectly dry air. When the air contains moisture in the form of water vapor, it is not as heavy as dry air and so is less dense.

Assuming that temperature and pressure remain the same, the air density varies inversely with the humidity—that is, as the humidity increases, the air density decreases, (density altitude *increases*); and, as the humidity decreases, the air density increases (density altitude *decreases*).

The higher the temperature, the greater the moisture-carrying ability of the air. Therefore, air at a temperature of 38° C. (100° F.) and a relative humidity of 80 percent will contain a greater amount of moisture than air at a temperature of 16° C. (60° F.), and a relative humidity of 80 percent.

Effect of High Humidity on Flight Since high humidity makes the air less dense (increases density altitude), the takeoff roll will be longer, rate of climb slower, and landing speed higher.

Combined Effect of High Altitude, High Temperature, and High Humidity on Flight. As indicated earlier in this section, each of the foregoing conditions can seriously affect flight characteristics. When all three conditions are present, the problem is aggravated. Therefore, beware of "high, hot, and humid" conditions (high-density altitudes), and take the necessary precautions, by using performance charts, to assure the runway is long enough for takeoff.

Dew and Frost When the ground cools at night, the temperature of the air immediately adjacent to the ground is frequently lowered to the saturation point, causing condensation. This condensation takes place directly upon objects on the ground as dew if the temperature is above freezing, or as frost if the temperature is below freezing.

Dew is of no importance to aircraft, but frost creates friction which interferes with the smooth flow of air over the wing surfaces, resulting in a tendency to stall during takeoff. FROST SHOULD ALWAYS BE REMOVED BEFORE FLIGHT.

Fog When the air near the ground is four or five degrees above the dew-point, the water vapor condenses and becomes visible as fog. There are many types of fog, varying in degree of intensity and classified according to the particular phenomena which cause them. One type, "ground fog," which frequently forms at night in low places, is limited to a few feet in height, and is usually dissipated by the heat of the sun shortly after sunrise. Other types, which can form any time conditions are favorable, may extend to greater heights and persist for days or even weeks. Along seacoasts fog often forms over the ocean and is blown inland. All fogs produce low visibilities and therefore constitute a serious hazard to aircraft.

Clouds There are two fundamental types of clouds. First, those formed by vertical currents carrying moist air upward to its condensation point are lumpy or billowy and are called "cumulus" (Fig. 5-20), which means an "accumulation" or a "pile." Second, those which develop horizontally and lie in sheets or formless layers like fog are called "stratus" (Fig. 5-21), which means "spread out."

When clouds are near the earth's surface they are generally designated as "cumulus" or "stratus" unless they are producing precipitation, in which case the word "nimbus" (meaning "rain cloud") is added—as "nimbostratus" or "cumulonimbus" (Fig. 5-22).

If the clouds are ragged and broken, the word "fracto" (meaning "broken") is added—as "fractostratus" or "fractocumulus."

The word "alto" (meaning "high") is generally added to designate clouds at intermediate heights, usually appearing at levels of 5,000 to 20,000 ft.—as "altostratus" or "altocumulus."

Clouds formed in the upper levels of the troposphere (commonly between 20,000 and 50,000 ft.) are composed of ice crystals and generally have a delicate, curly appearance, somewhat similar to frost on a windowpane. For these clouds the word "cirro" (meaning "curly") is added—as "cirrocumulus" or "cirrostratus." At these high altitudes there is also a fibrous type of cloud appearing as curly wisps, bearing the single name "cirrus."

CUMULUS

ALTOCUMULUS

CIRROCUMULUS

Figure 5–20. Cumulus clouds as they appear at low, intermediate, and high levels.

Figure 5-21. *Stratus-type clouds at various altitudes.*

STRATOCUMULUS   NIMBOSTRATUS   CUMULONIMBUS   FRACTOCUMULUS

Figure 5–22. *Various types of bad weather clouds.*

Under "Air Masses and Fronts," the relationship will be shown between the various types of clouds and the kind of weather expected. At present the chief concern is with the flying conditions directly associated with the different cloud formations.

The ice-crystal clouds (cirrus group) are well above ordinary flight levels of light aircraft, and normally do not concern the pilots of these aircraft, except as indications of approaching changes in weather.

The clouds in the "alto" group are not normally encountered in flights of smaller planes, but they sometimes contain icing conditions important for commercial and military planes. Altostratus clouds usually indicate that unfavorable flying weather is near.

The low clouds are of great importance to the pilot because they create low ceilings and low visibilities. They change rapidly, and frequently drop to the ground, forming a complete blanket over landmarks and landing fields. In temperatures near freezing, they are a constant threat because of the probability of icing. The pilot should be constantly alert to any changes in conditions, and be prepared to land before visibility lowers to the point where objects are suddenly obscured.

Cumulus clouds vary in size from light "scud" or fluffy powder puffs to towering masses rising thousands of feet in the sky. Usually they are somewhat scattered, and the pilot can fly around them without difficulty. Under some conditions, particularly in the late afternoon, they are likely to multiply, flatten out, and cover the sky. This may leave the pilot with no alternative except to reverse course or find a safe landing field.

Cumulonimbus clouds are very dangerous. When they appear individually or in small groups, they are usually of the type called "air mass thunderstorms" (caused by heating of the air at the earth's surface) or "orographic thunderstorms" (caused by the upslope motion of air in mountainous regions). On the other hand, when these clouds take the form of a continuous or almost continuous line, they are usually caused by a front or squall line. The most common position for a squall line is in advance of a cold front, but one can form in air far removed from a front.

Since cumulonimbus clouds are formed by rising air currents, they are extremely turbulent; moreover, it is possible for an airplane flying nearby to be drawn into the cloud. Once inside, an airplane may encounter updrafts and downdrafts with velocities as great as 3,000 ft. per minute. Airplanes have been torn apart by the violence of these currents. In addition, the clouds frequently contain large hailstones capable of severely damaging aircraft, lightning, and great quantities of water at temperatures conducive to heavy icing.

Many "unexplained" crashes have probably been caused by the disabling effect of cumulonimbus clouds upon airplanes which have been accidentally or intentionally flown into them. The only practical procedure for a pilot caught within a thunderstorm is to reduce airspeed. A recommended safe speed for an airplane flying through turbulence is an airspeed not greater than the maneuvering speed for the particular airplane.

Fig. 5-23 shows the important characteristics of a typical cumulonimbus cloud. The top of the cloud flattens into an anvil shape, which points in the direction the cloud is moving, generally with the prevailing wind. Near the base, however, the winds blow directly toward the cloud and increase in speed, becoming violent updrafts as they reach the low rolls at the forward edge.

Within the cloud and directly beneath it are updrafts and downdrafts; in the rear portion is a strong downdraft which becomes a wind blowing away from the cloud.

The cloud is a storm factory. The updrafts quickly lift the moist air to its saturation point, whereupon it condenses and raindrops begin to fall. Before these have reached the bottom of the cloud, updrafts pick them up and carry them aloft, where they may freeze and again start downward, only to repeat the process many times until they have become heavy enough to break through the updrafts and reach the ground as hail or very large raindrops. As the storm develops, more and more drops fall through the turbulence, until the rain becomes fairly steady. The lightning that accompanies such a storm is probably due to the breakup of raindrops. This produces static electricity that discharges as lightning, thus causing sudden expansion of the air in its path, resulting in thunder.

It is impossible for a small plane to fly over these clouds because they frequently extend to 50,000 ft. and are usually too low to fly under. If they are close together there may be violent turbulence in the clear space between them. If they are isolated thunderstorms, it usually is possible to fly around them safely by remaining a good distance from them. If, however, they are "frontal" or squall line storms, they may extend for hundreds of miles, and the only safe procedure is to land immediately and wait until the cumulonimbus cloud formation has passed.

Ceiling   A ceiling is defined as the height above the surface of the base of the lowest layer of clouds or obscuring phenomena that hide more than half of the sky, and is reported as broken or overcast. A ceiling is also defined as the vertical visibility into a surface based obscuration that hides all of the sky. A layer of clouds or obscuring phenomena classified as thin does not constitute

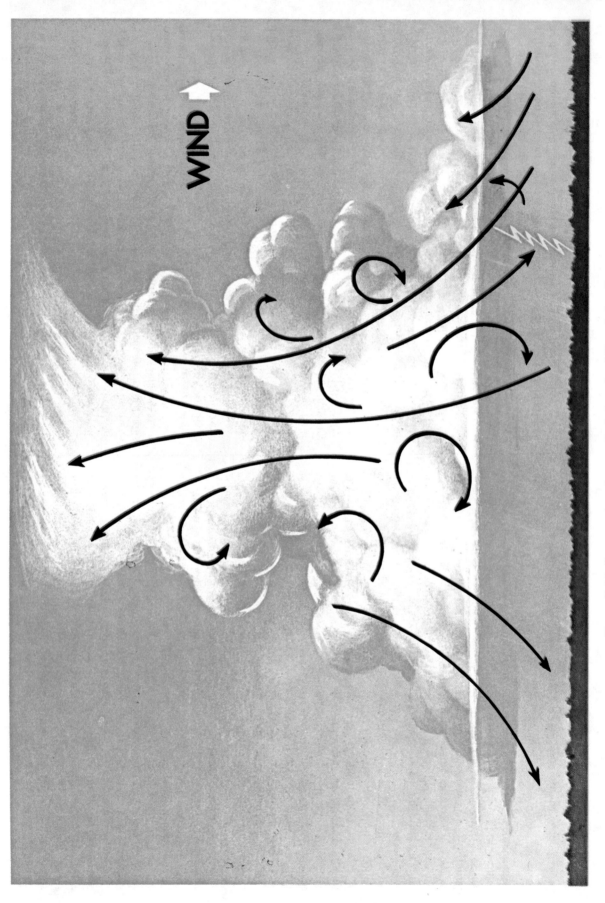

Figure 5–23. Cross-section of a cumulonimbus cloud (thunderhead).

a ceiling. Clouds are reported as broken when they cover six-tenths to nine-tenths of the sky, and as overcast when they cover more than nine-tenths. The ceiling is unlimited if the sky is cloudless or less than six-tenths covered as seen from the ground. The latest information on ceilings can be obtained from the Surface Aviation Weather Reports. Forecasts of expected changes in ceilings and other conditions also are available at weather stations.

Visibility Closely related to ceiling and cloud cover is "visibility"—the greatest horizontal distance at which prominent objects can be distinguished with the naked eye. Visibility, like ceiling, is included in hourly weather reports and in aviation forecasts.

Precipitation In addition to possible damage by hail and the danger of icing, precipitation may be accompanied by low ceilings, and in heavy precipitation visibilities may suddenly be reduced to zero.

It should be obvious that aircraft which may have accumulated snow while on the ground should never be flown until all traces of snow have been removed, including the hard crust that frequently adheres to the surfaces. An aircraft which has been exposed to rain followed by freezing temperatures should be carefully cleared of ice and checked before takeoff to make certain that the controls operate freely.

Air Masses and Fronts Large, high pressure systems frequently stagnate over large areas of land or water with relatively uniform surface conditions. They take on characteristics of these "source regions"—the coldness of polar regions, the heat of the tropics, the moisture of oceans, or the dryness of continents.

As they move away from their source regions and pass over land or sea, the air masses are constantly being modified through heating or cooling from below, lifting or subsiding, absorbing or losing moisture. Actual temperature of the air mass is less important than its temperature in relation to the land or water surface over which it is passing. For example, an air mass moving from polar regions usually is colder than the land and sea surfaces over which it passes. On the other hand, an air mass moving from the Gulf of Mexico in winter usually is warmer than the territory over which it passes.

If the air is colder than the surface, it will be warmed from below and convection currents will be set up, causing turbulence. Dust, smoke, and atmospheric pollution near the ground will be carried upward by these currents and dissipated at higher levels, improving surface visibility. Such air is called "unstable."

Conversely, if the air is warmer than the surface, there is no tendency for convection currents to form, and the air is smooth. Smoke, dust, etc., are concentrated in lower levels with resulting poor visibility. Such air is called "stable."

From the combination of the source characteristics and the temperature relationship just described, air masses can be associated with certain types of weather.

The following are general characteristics of certain air masses but they may vary considerably.

Characteristics of a Cold (Unstable) Air Mass

Type of clouds . . . . generally cumulus and cumulonimbus.

Ceilings . . . . generally unlimited (except during precipitation).

Visibilities . . . . excellent (except during precipitation).

Unstable air . . . . pronounced turbulence in lower levels (because of convection currents).

Type of precipitation . . . occasional local thunderstorms or showers—hail, sleet, snow flurries.

Characteristics of a Warm (Stable) Air Mass

Type of clouds . . . . stratus and stratocumulus (fog, haze).

Ceilings . . . . generally low.

Visibilities . . . . poor (smoke and dust held in lower levels).

Stable air . . . . smooth, with little or no turbulence.

Type of precipitation . . . drizzle.

When two air masses meet, they will not mix readily unless their temperatures, pressures, and relative humidities are very similar. Instead, they set up boundaries called frontal zones, or "fronts," the colder air mass projecting under the warmer air mass in the form of a wedge. This condition is termed a "stationary front" if the boundary is not moving.

Usually, however, the boundary moves along the earth's surface, and as one air mass withdraws from a given area it is replaced by another air mass. This action creates a moving front. If warmer air is replacing colder air, the front is called "warm;" if colder air is replacing warmer air, the front is called "cold."

Warm Front When a warm front moves forward, the warm air slides up over the wedge of colder air lying ahead of it.

Warm air usually has high humidity. As this warm air is lifted, its temperature is lowered. As the lifting process continues, condensation occurs, low nimbostratus and stratus clouds form and drizzle or rain develops. The rain falls through the colder air below, increasing its moisture content so that it also becomes saturated. Any reduction of temperature in the colder air, which might be caused by upslope motion or cooling of the ground after sunset, may result in extensive fog.

As the warm air progresses up the slope, with constantly falling temperature, clouds appear at increasing heights in the form of altostratus and cirrostratus, if the warm air is stable. If the warm air is unstable, cumulonimbus clouds and altocumulus clouds will form and frequently produce thunderstorms. Finally, the air is forced up near the stratosphere, and in the freezing temperatures at that level, the condensation appears as thin wisps of cirrus clouds. The upslope movement is very gradual, rising about 1,000 ft. every 20 miles. Thus, the cirrus clouds, forming at perhaps 25,000 ft. altitude, may appear as far as 500 miles in advance of the point on the ground which marks the position of the front (Fig. 5-24).

Flight Toward an Approaching Warm Front. Although no two fronts are exactly alike, a clearer understanding of the general pattern may be gained if the atmospheric conditions which might exist when a warm front is moving eastward from St. Louis, Mo., is considered. (Refer to Fig. 5-24.)

At St. Louis, the weather would be very unpleasant, with drizzle and probably fog.

At Indianapolis, Ind., 200 miles in advance of the warm front, the sky would be overcast with nimbostratus clouds, and continuous rain.

At Columbus, Ohio, 400 miles in advance, the sky would be overcast with stratus and altostratus clouds predominating. The beginning of a steady rain would be probable.

At Pittsburgh, Pa., 600 miles ahead of the front, there would probably be high cirrus and cirrostratus clouds.

If a flight was made from Pittsburgh to St. Louis, ceiling and visibility would decrease steadily. Starting under bright skies, with unlimited ceilings and visibilities, lowering stratus-type clouds would be noted as Columbus was approached, and soon afterward precipitation would be encountered. After arriving at Indianapolis, the ceilings would be too low for further flight. Precipitation would reduce visibilities to practically zero. Thus, it would be wise to remain in Indianapolis until the warm front had passed, which might require a day or two.

If a return flight to Pittsburgh was made, it would be recommended to wait until the front had passed beyond Pittsburgh, which might require three or four days. Warm fronts generally move at the rate of 10 to 25 miles an hour.

On the trip from Pittsburgh to Indianapolis a gradual increase in tem-

perature would have been noticed, and a much faster increase in dewpoint until the two coincided. Also the atmospheric pressure would be gradually lessening because the warmer air aloft would have less weight than the colder air it was replacing. This condition illustrates the general principle that a falling barometer indicates the approach of stormy weather.

Cold Front When the cold front moves forward, it acts like a snow plow, sliding under the warmer air and forcing it aloft. This causes the warm air to cool suddenly and form cloud types that depend on the stability of the warm air.

Fast-Moving Cold Fronts. In fast-moving cold fronts, friction retards the front near the ground, which brings about a steeper frontal surface. This steep frontal surface results in a narrower band of weather concentrated along the forward edge of the front. If the warm air is stable, an overcast sky may occur for some distance ahead of the front, accompanied by general rain. If the warm air is conditionally unstable, scattered thunderstorms and showers may form in the warm air. At times an almost continuous line of thunderstorms (squall lines) may form along the front or ahead of it. These lines of thunderstorms (squall lines) contain some of the most turbulent weather experienced by pilots.

Behind the fast-moving cold front there is usually rapid clearing, with gusty and turbulent surface winds, and colder temperatures.

Comparison of Cold Fronts With Warm Fronts. The slope of a cold front is much steeper than that of a warm front and the progress is generally more rapid—usually from 20 to 35 miles per hour, although in extreme cases, cold fronts have been known to move at 60 miles per hour. Weather activity is more violent and usually takes place directly at the front instead of in advance of the front. In late afternoon during the warm season, however, squall lines frequently develop as much as 50 to 200 miles in advance of the actual cold front. Whereas warm front dangers are low ceilings and visibilities, cold front dangers are chiefly sudden storms, high and gusty winds, and turbulence.

Unlike the warm front, the cold front rushes in almost unannounced, makes a complete change in the weather within a period of a few hours, and moves on. Altostratus clouds sometimes form slightly ahead of the front, but these are seldom more than 100 miles in advance. After the front has passed, the weather often clears rapidly and cooler, drier air with usually unlimited ceilings and visibilities prevail.

Flight Toward an Approaching Cold Front. If a flight was made from Pittsburgh toward St. Louis (Fig. 5-25) when a cold front was approaching from St. Louis, weather conditions quite different from those associated with a

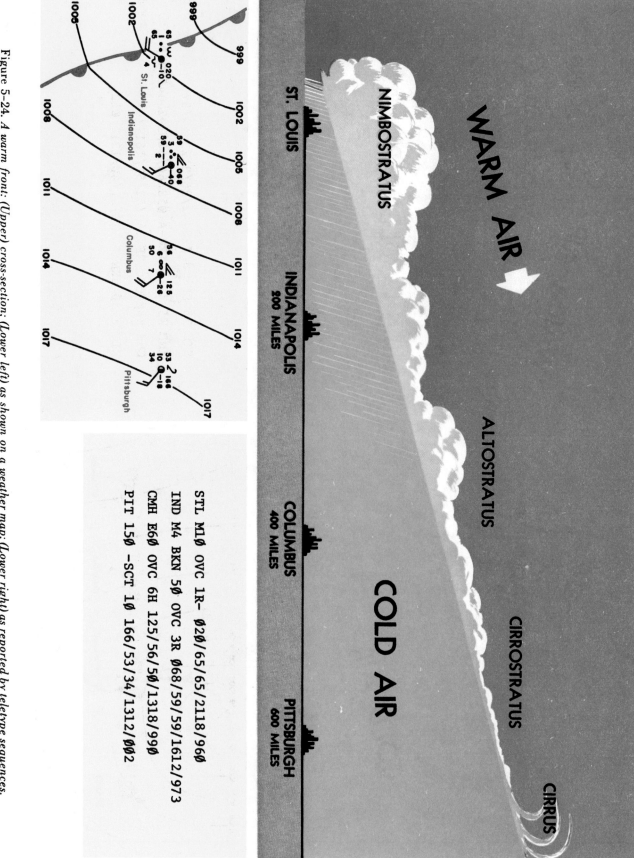

Figure 5-24. *A warm front: (Upper) cross-section; (Lower left) as shown on a weather map; (Lower right) as reported by teletype sequences.*

STL M10 OVC 1R- 020/65/65/2118/960
IND M4 BKN 50 OVC 3R 068/59/59/1612/973
CMH E60 OVC 6H 125/56/50/1318/990
PIT 150 -SCT 10 166/53/34/1312/002

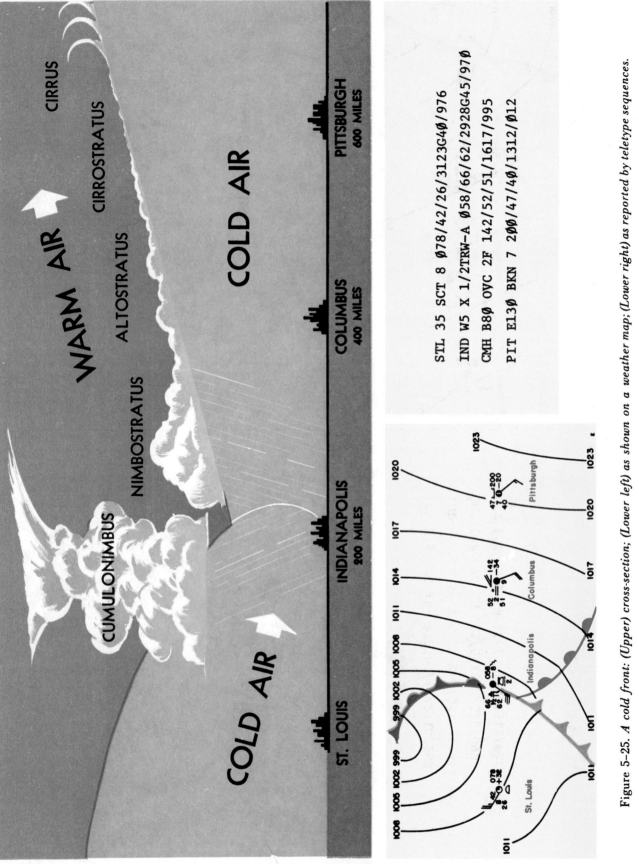

STL 35 SCT 8 Ø78/42/26/3123G4Ø/976
IND W5 X 1/2TRW–A Ø58/66/62/2928G45/97Ø
CMH B8Ø OVC 2F 142/52/51/1617/995
PIT E13Ø BKN 7 2ØØ/47/4Ø/1312/Ø12

Figure 5–25. A cold front: (Upper) cross-section; (Lower left) as shown on a weather map; (Lower right) as reported by teletype sequences.

warm front would be experienced. The sky in Pittsburgh would probably be somewhat overcast with stratocumulus clouds typical of a warm air mass, the air smooth, and the ceilings and visibilities relatively low although suitable for flight.

As the flight proceeded, these conditions would prevail until reaching Indianapolis. At this point, it would be wise to check the position of the cold front by consulting a recent weather map and teletype sequences, or the meteorologist. It would probably be found that the front was now about 75 miles west of Indianapolis. A pilot with sound judgment based on knowledge of frontal conditions, would remain in Indianapolis until the front had passed—a matter of a few hours—and then continue to the destination under near perfect flying conditions.

If, however, through the lack of better judgment the flight was continued toward the approaching cold front, a few altostratus clouds and a dark layer of nimbostratus lying low on the horizon, with perhaps cumulonimbus in the background would be noted. Two courses would now be open: (1) Either to turn around and outdistance the storm, or (2) to make an immediate landing which might be extremely dangerous because of gustiness and sudden wind shifts.

If flight was continued, entrapment in a line of squalls and cumulonimbus clouds could occur. It may be disastrous to fly beneath these clouds; impossible, in a small plane, to fly above them. At low altitudes, there are no safe passages through them. Usually there is no possibility of flying around them because they often extend in a line for 300 to 500 miles.

Wind Shifts. Wind shifts perhaps require further explanation. The wind in a "high" blows in a clockwise spiral. When two highs are adjacent, the winds are in almost direct opposition at the point of contact as illustrated in Fig. 5-26. Since fronts normally lie between two areas of higher pressure, wind shifts occur in all types of fronts, but they usually are more pronounced in cold fronts.

Occluded Front   One other form of front with which the pilot should become familiar is the "exclusion" or "occluded front." This is a condition in which an air mass is trapped between two colder air masses and forced aloft to higher and higher levels until it finally spreads out and loses its identity.

Meteorologists subdivide occlusions into two types, but so far as the pilot is concerned, the weather in any occlusion is a combination of warm front and cold front conditions. As the occlusion approaches, the usual warm front indications prevail—lowering ceilings, lowering visibilities, and precipitation.

Figure 5-26. *Weather map indication of wind shift line (center line leading to low).*

Generally, the warm front weather is then followed almost immediately by the cold front type, with squalls, turbulence, and thunderstorms.

Fig. 5-27 is a vertical cross section of an occlusion. Fig. 5-28 shows the various stages as they might occur during development of a typical occlusion. Usually the development requires three or four days, during which the air mass may progress as indicated on the map.

The first stage (A) represents a boundary between two air masses, the cold and warm air moving in opposite directions along a front. Soon, however, the cooler air, being more aggressive, thrusts a wedge under the warm air, breaking the continuity of the boundary, as shown in (B). Once begun, the process continues rapidly to the complete occlusion as shown in (C). As the warmer air is forced aloft, it cools quickly and its moisture condenses, often causing heavy precipitation. The air becomes extremely turbulent, with sudden changes in pressure and temperature.

Fig. 5-29 shows the development of the occluded front in greater detail. Fig. 5-30 is an enlarged view of (C) in Fig. 5-28, showing the cloud formations and the areas of precipitation.

In Figs. 5-24, 5-25, and 5-27, a panel representing a surface weather map is placed below each cross-sectional view. These panels represent a bird's-eye or plan view, and show how the weather conditions are recorded. A warm front is indicated by a red line, a cold front by a blue line, an occluded

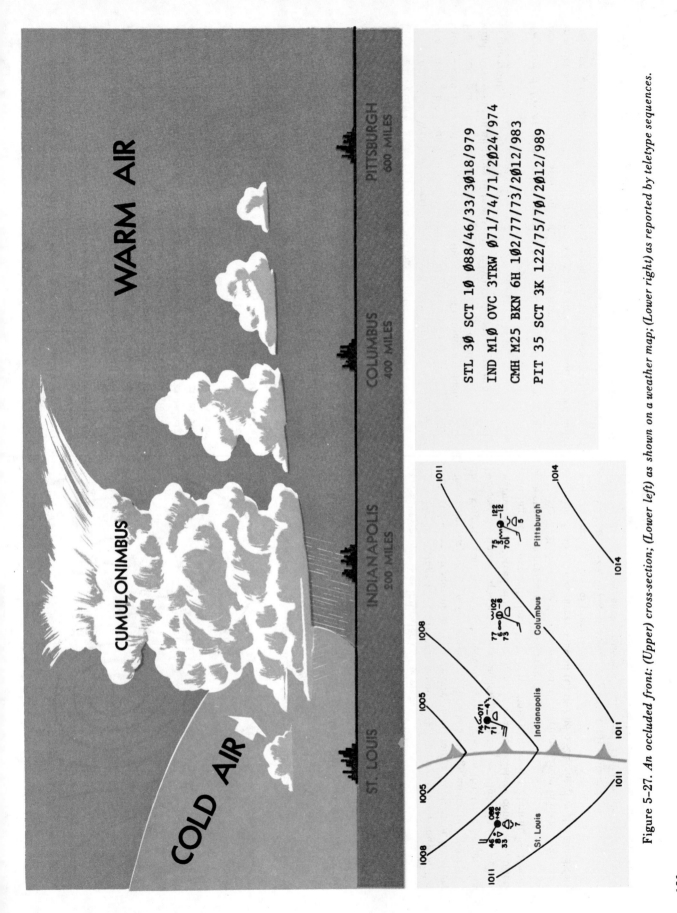

STL 3∅ SCT 1∅ ∅88/46/33/3∅18/979

IND M1∅ OVC 3TRW ∅71/74/71/2∅24/974

CMH M25 BKN 6H 1∅2/77/73/2∅12/983

PIT 35 SCT 3K 122/75/7∅/2∅12/989

Figure 5-27. An occluded front: (Upper) cross-section; (Lower left) as shown on a weather map; (Lower right) as reported by teletype sequences.

130

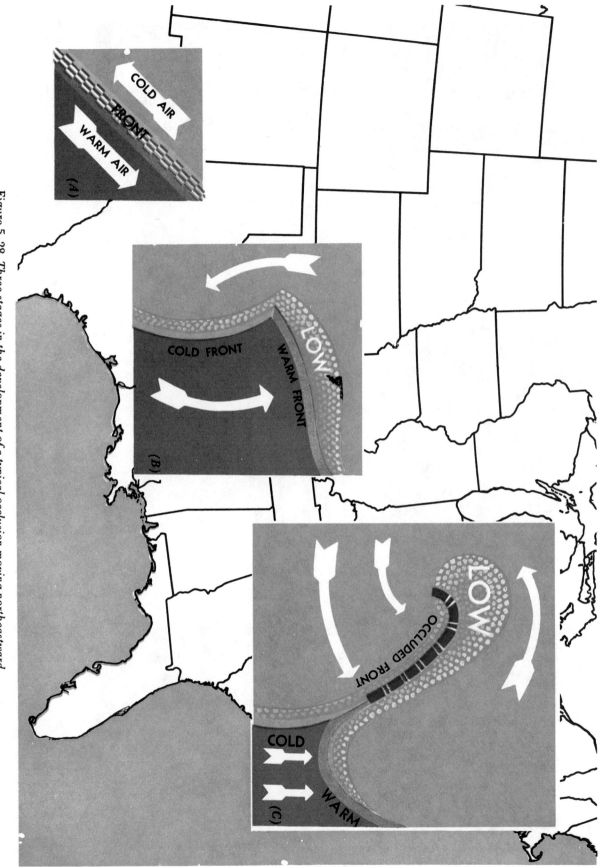

Figure 5-28. Three stages in the development of a typical occlusion moving northeastward.

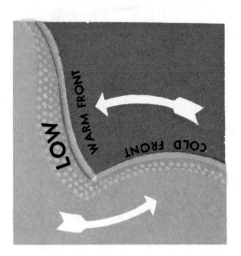

(A) Air flowing along a front in equilibrium.

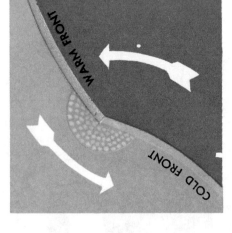

(B) Increased cold-air pressure causes "bend."

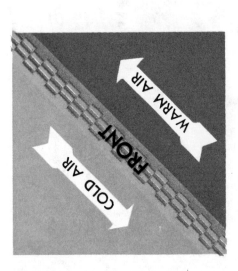

(D) Precipitation becomes heavier.

(C) Cold air begins to surround warm air.

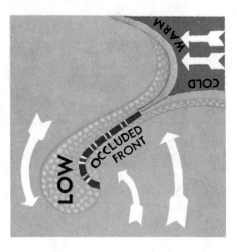

(E) Warm air completely surrounded.

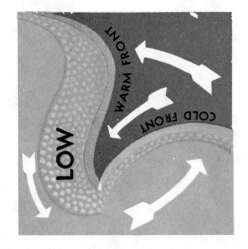

(F) Warm-air sector ends in mild whirl.

Figure 5–29. Development of an occlusion. If warm air were red and cold air were blue, this is how various stages of an occlusion would appear to a person aloft looking toward the earth. (Precipitation is green.)

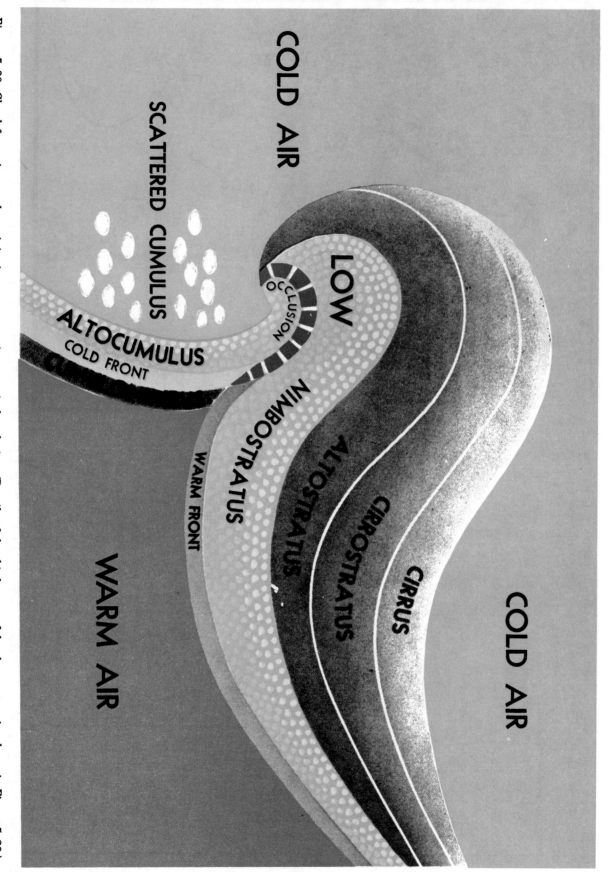

Figure 5–30. *Cloud formations and precipitation accompanying a typical occlusion. (Details of the third stage of development series shown in Figure 5–28.)*

COLD AIR

SCATTERED CUMULUS

ALTOCUMULUS
COLD FRONT

LOW

OCCLUSION

NIMBOSTRATUS

WARM FRONT

ALTOSTRATUS

CIRROSTRATUS

CIRRUS

COLD AIR

WARM AIR

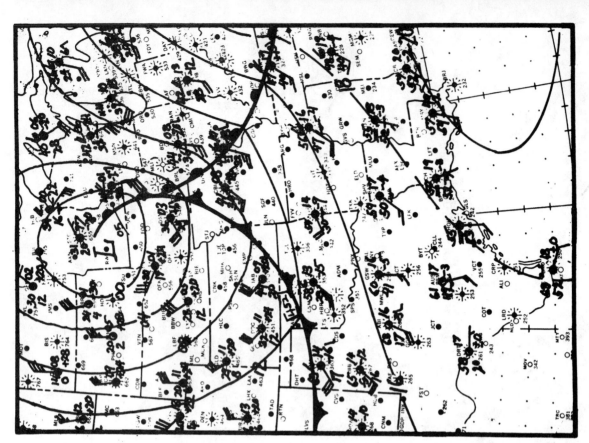

Figure 5-31. *Section of typical weather map showing methods indicating weather facts important to pilots.*

front by a purple line, and a stationary front by alternating red and blue dashes. The rounded and pointed projections are generally omitted from manuscript maps, but are placed on facsimile, printed, or duplicated maps to distinguish the different fronts.

A frontal line on the weather map represents the position of the frontal surface on the earth's surface. A pilot flying west at an altitude of 6,500 ft. would pass through the frontal boundary about 100 miles in advance of the point where the warm front is shown, or about 25 to 50 miles to the rear of the line on the map representing the cold front.

Fig. 5-31 is a section of a surface weather map as transmitted on facsimile. It shows a low pressure center with warm, cold, and occluded fronts.

The preceding discussion categorizes weather with types of fronts. However, weather with a front depends more on the characteristics of the conflicting air masses than on the type of front. A pilot should not attempt to determine expected weather from fronts and pressure centers on the surface chart alone. The pilot must rely heavily on other charts, reports, and forecasts which are discussed in the next section of this chapter.

Aviation Weather Forecasts, Reports, and Weather Charts A few forecasts and reports that are available to pilots, were discussed briefly in previous portions of this chapter. This section will discuss these forecasts and reports in greater detail along with weather charts.

Aviation weather reports and forecasts are displayed at each National Weather Service station. Also, a great number of reports and forecasts are available at the Flight Service Stations (FSS). FSS personnel are trained in weather briefing and assisting pilots with the weather aspects of flights. Pilots are encouraged to take advantage of the aviation weather information available and to request assistance from the weather briefer. There is also every advantage in personally obtaining the weather information rather than doing it by telephone. A personal visit affords the opportunity to see the information first-hand, especially the maps and charts, and therefore the pilot is better able to visualize the general weather situation. However, if a personal visit is not practical, the telephone should be used to obtain the weather information.

Although this section contains a great deal of information about forecasts, reports, and charts, for further information reference should be made to AC 00–45 [latest revision], Aviation Weather Services, which is for sale by the Superintendent of Documents, U.S. Government Printing Office, Washington, D.C. 20402.

Aviation Forecasts Forecasts especially prepared for aviation include Area Forecasts (FA), Terminal Forecasts (FT), In-Flight Weather Advisories (SIGMET and AIRMET), and Winds and Temperatures Aloft Forecasts (FD).

Area Forecasts (FA). An Area Forecast is a prediction of general weather conditions over an area consisting of several states or portions of states. It is used to obtain information about expected en route weather conditions and also to provide an insight to weather conditions that might be expected at airports where weather reports or forecasts are not issued.

Area Forecasts are issued every 12 hours for a total validity period of 30 hours. This validity period includes expected weather for the following 18-hour period with an additional 12-hour categorical outlook. The time used in these forecasts is based on Greenwich Mean Time (GMT) and stated in whole hours with two digits, i.e., 13Z. Wind direction is measured in degrees from true north, and windspeed is given in knots. All distances except visibility are in nautical miles; visibility is in statute miles.

Each Area Forecast is arranged using the same format which consists of the following sections:

1. Heading.
2. Forecast area.
3. Height statement.
4. Flight precaution statement.
5. Synopsis.
6. Clouds and weather, plus categorical outlook.
7. Icing and freezing level.
8. Turbulence.

Contractions are used to facilitate communication in aviation forecasts, reports, and charts. Contractions are a representation of words, titles, or phrases in a shortened form. The contractions usually bear a close similarity to the whole word, so that with a little practice the material can be read and understood without difficulty. A handbook entitled "Contractions," which can be used to decode the more difficult contractions, is available at many weather service outlets.

Fig. 5–32 is an example of an Area Forecast. A plain language interpretation and description of each of the sections are provided following the example.

Plain Language Interpretation:

Heading—

The heading identifies the originating WSFO, states that it is an

area forecast, provides the date and time of issue, and gives the valid periods of the forecast and outlook. For example:

> DFW FA 281240.
> 13Z WED–07Z THU.
> OTLK07Z–19Z THU.

states that this FA was issued by Dallas–Fort Worth (DFW) on the 28th day of the month at 1240Z. The forecast is valid from 1300Z Wednesday until 0700Z Thursday, with a categorical outlook from 0700Z Thursday until 1900Z Thursday.

Forecast Area—

The area is in contraction form identifying states; portions of states; and, where applicable, adjacent waters. For example:

> NMEX OKLA TEX AND CSTL WTRS . . . .

means New Mexico, Oklahoma, Texas, and coastal waters.

DFW FA 281240.
13Z WED–07Z THU.
OTLK 07Z–19Z THU.

NMEX OKLA TEX AND CSTL WTRS . . .

HGTS ASL UNLESS NOTED . . .

FLT PRCTNS RCMDD DUE TO SCT TO NMRS TSTMS

SYNOPSIS . . . WK FNT NR GAG–ROW LN ERY MRNG WL DRFT SLOWLY EWD AND GRDLY DSIPT WITH LO PRES TROF RMNG FM NR PNC–HOB LN BY 07Z. LGT TO MDT SLY FLO OVR OKLA AND TEX SE OF FNT. WK PRES GRAD NW OF FNT.

CLDS AND WX . . .

WRN OKLA AND NW TEX GENLY W OF PNC–30SW SPS LN . . .
SCT TO NMRS TSTMS CNTRD NR 60NE GAG–LTS–AMA LN AT 13Z WL DSIPT
BY LATE MRNG. OUTSIDE TSTMS 50 SCT CIG 100 BKN 160 250 BKN.
16Z GENLY 50 SCT AGL 250 SCT. SCT TSTMS REDVLPG DURG AFTN BCMG
MORE NMRS LATE AFTN. OTLK . . . VFR CIG ABV 10 THSD.

CSTL WTRS . . .
CLR TO 15 SCT WITH A FEW ISOLD SHWRS. OTLK . . . VFR CLR.

ICG AND FRZLVL . . . MDT MXD ICGICIP ABV FRZLVL VCNTY CB. FRZLVL 140–155.

TURBC . . . MDT OR GTR TURB VCNTY TSTMS.

Figure 5–32. *Example of an area forecast.*

into several paragraphs, with each paragraph discussing the expected weather within an area such as a state, well known geographical area, or in reference to location and movement of a pressure system or front.

Obstructions to vision are included when forecast visibility is 6 miles or less. Any expected precipitation or thunderstorms are always included.

All heights are abbreviated by omitting the last two zeros. For example, 10,000 is written as 100 and 1,500 is written as 15.

The following is a list of contractions and their definitions used to denote sky conditions in Area Forecasts. These are listed along with the contractions or designators used in Terminal Forecasts, so that a comparison can be made. Terminal Forecasts will be discussed immediately after the discussion on Area Forecasts.

| FA | FT | |
Contraction	Designator	Definition
CLR	CLR	Sky clear
SCT	SCT	Scattered
BKN	BKN	Broken
OVC	OVC	Overcast
OBSC	X	Obscured, obscure, or obscuring
PTLY OBSC	—X	Partly obscured
THN	—	Thin
VRBL	V	Variable
CIG	C	Ceiling
INDEF	W	Indefinite

The following is a list of adjectives and their meanings as used to describe area coverage of showers and thunderstorms.

Adjectives	Coverage
Isolated	Extremely small number.
Few	15% or less of area or line.
Scattered	16% to 45% of area or line.
Numerous	More than 45% of area or line.

The categorical outlook, identified by the contraction "OTLK," is found at the end of each paragraph in the significant clouds and weather section and describes the outlook (valid for 12 hours) for that particular area.

Both Area and Terminal Forecasts group ceiling and visibilities into categories which are used in the categorical outlook for these forecasts. The categorical outlook extends the Area Forecasts for 12 hours, and the Terminal

Height Statement—

Each area forecast contains the statement

HGTS ASL UNLESS NOTED . . .

to remind the user that most of the heights in area forecasts are measured from "above sea level." The tops and bases of clouds and tops of icing are always stated in ASL.

Heights above ground level (AGL) may be identified in area forecasts in either of two different ways: (1) By definition "ceiling" is measured from above ground level. Therefore, the contractions "CIG" when used means that the base of the clouds was measured from above ground level; (2) the contraction "AGL" when used means above ground level.

Flight Precautions Statement—

Following the height statement each area forecast contains the statement "FLT PRCTNS DUE TO . . . ETC"

The flight precautions statement may include any or all of the following:

—Areas of general thunderstorms.
—Moderate or greater icing.
—Moderate or greater turbulence.
—Widespread IFR conditions.
—Mountain obscurement.
—Low-level wind shear potential.

If none of the above is expected, a negative statement so states.

Synopsis: The synopsis briefly summarizes the location and movements of fronts, pressure systems, and circulation patterns. It may also include moisture and stability conditions. Fig. 5–33 means that a weak front near Gage-Roswell line early morning will drift slowly eastward and gradually dissipate with low-pressure trough remaining from near Ponca City-Hobbs line by 0700Z. Light to moderate southerly flow over Oklahoma and Texas southeast of front. Weak pressure gradient northwest of front.

SYNOPSIS...WK FNT NR GAG-ROW LN ERY MRNG WL DRFT SLOWLY EWD AND GRDLY DSIPT WITH LO PRES TROF RMNG FM NR PNC-HOB LN BY 07Z. LGT TO MDT SLY FLO OVR OKLA AND TEX SE OF FNT. WK PRES GRAD NW OF FNT.

Figure 5–33. *Synopsis.*

Clouds and Weather: This part of the Area Forecast is identified by the contraction CLD AND WX, and forecasts, in broad terms, cloudiness and weather that are considered to be significant to flight. It is usually divided

Forecasts for 6 hours. These outlooks are intended primarily for advanced flight planning.

LIFR (Low IFR)	ceiling less than 500 ft. and/or visibility less than 1 mile.
IFR	ceiling 500 to less than 1,000 ft. and/or visibility 1 to less than 3 miles.
MVFR (Marginal VFR)	ceiling 1,000 to 3,000 ft. and/or visibility 3 to 5 miles inclusive.
VFR	ceiling greater than 3,000 ft. and visibility greater than 5 miles; includes sky clear.

The cause of LIFR, IFR, or MVFR is also given by either ceiling or visibility restrictions or both. The contraction "CIG" and/or weather and obstruction to vision symbols are used. If wind or gusts of 25 knots or greater are forecast for the outlook period, the word "WIND" is also included for all categories, including VFR.

An example of "CLD AND WX" showing two paragraphs is as follows (Fig. 5-34):

```
CLDS AND WX...

WRN OKLA AND NW TEX GENLY W OF PNC-30SW SPS LN...
SCT TO NMRS TSTMS CNTRD NR 60NE GAG-LTS-AMA LN AT 13Z WL DSIPT
BY LATE MRNG. OUTSIDE TSTMS 50 SCT CIG 100 BKN 160 250 BKN.
16Z GENLY 50 SCT AGL 250 SCT. SCT TSTMS REDVLPG DURG AFTN BCMG
MORE NMRS LATE AFTN. OTLK...VFR CIG ABV 10 THSD.

CSTL WTRS...
CLR TO 15 SCT WITH FEW ISOLD SHWRS. OTLK...VFR CLR.
```

Figure 5–34. *Clouds and weather.*

The first paragraph means that the clouds and weather over—Western Oklahoma and northwest Texas generally west of Ponca City 30 nautical miles southwest Wichita Falls line—scattered to numerous thunderstorms center near 60 nautical miles northeast Gage-Altus-Amarillo line at 1300Z will dissipate by late morning. Outside thunderstorms 5,000 ft. Scattered clouds ceilings 10,000-foot broken clouds 16,000-foot 25,000-foot broken clouds. After 1600Z generally 5,000-foot scattered clouds above ground level 25,000-foot scattered. Scattered thunderstorms redeveloping during afternoon becoming more numerous late afternoon. Outlook—Visual flight rules ceilings above 1,000 feet.

The second paragraph means that the clouds and weather over—Coastal waters—Clear to 1,500-foot scattered clouds with a few isolated showers. Outlook—Visual flight rules clear skies.

Icing: The contraction "ICG AND FRZLVL" indentifies the portion of the Area Forecast which deals with icing conditions. It gives the location, type, and extent of expected icing. This section always includes the freezing level in hundreds of feet ASL. It may contain qualifying terms such as "ICG LKLY," icing likely; "MDT MXD ICGICG ABV FRZLVL," moderate mixed icing in clouds above freezing level. For example, Fig. 5–35 means moderate mixed icing in clouds and in precipitation above freezing level vicinity of cumulonimbus clouds. Freezing level 14,000 feet to 15,500 feet.

```
ICG AND FRZLVL...MDT MXD ICGICIP ABV FRZLVL VCNTY CB.  FRZLVL 140-155.
```

Figure 5–35. *Icing and freezing level.*

Turbulence: A "TURBC" section is found at the end of the area forecast and will denote moderate or greater turbulence. This also includes the location, altitude, and cause of the turbulence.

Amended Area Forecasts. Amendments to the Area Forecasts are issued as needed. Only that portion of the FA being revised is transmitted as an amendment. Area Forecasts are also amended and updated by in-flight advisories.

Terminal Forecasts. A Terminal Forecast (FT) differs from an Area Forecast in that the FT is a prediction of weather conditions to be expected for a specific airport rather than a larger area. The size of area covered in a Terminal Forecast is within a 5-mile radius of the center of the runway complex.

Terminal Forecasts are issued in two different codes: (1) domestic U.S. Code (FT) and (2) the ICAO(TAF) code. Only the domestic U.S. Code (FT) will be discussed in this handbook.

Scheduled Terminal Forecasts are issued three times daily by WSFOs, and are valid for a total validity period of 24 hours. This validity period includes expected weather for the following 18 hours with an additional 6-hour categorical outlook.

The issue and valid times of FTs are based on the time zones of the issuing WSFO as follows:

WSFO Location (time zone)	Issue Time	Valid period
Eastern/Central	0940Z	10Z–10Z
	1440Z	15Z–15Z
	2140Z	22Z–22Z
Mountain/Pacific	0940Z	10Z–10Z
	1540Z	16Z–16Z
	2240Z	23Z–23Z

The format of the Terminal Forecast is essentially the same as that of the Surface Aviation Weather Report (SA), which will be discussed later in this chapter. Pilots who can read and interpret the SAs should have no difficulty in reading FTs.

Generally, the Terminal Forecast includes expected ceiling and clouds, visibility, weather and obstructions to vision, and surface wind conditions at each terminal. Also included are remarks that more completely describe expected weather. If a change is expected during the forecast period, this change and expected time of change are included. The last item included on each Terminal Forecast is the 6-hour categorical outlook using the form as discussed for the Area Forecast.

The following are examples of several Terminal Forecasts issued within the state of Texas on the 29th day of the month at 1440Z (Fig. 5-36).

```
TX 291440
ABI 291515 100 SCT 250-BKN 1812. 18Z 50 SCT 100 SCT 250-BKN CHC
C20X 1TRW+ AFT 21Z. 09Z VFR CHC TRW/RW..

ACT 291515 250 SCT 1910. 17Z 50 SCT 1910. 01Z 250 SCT.. 09Z VFR.
ALI 291515 12 SCT SCT V BKN. 17Z 20 SCT 1510 SCT OCNLY BKN. 20Z 35
   SCT 1512 OCNLY C35 BKN. 00Z CLR. 09Z VFR BCMG MVFR CIG 12Z.
AMA 291515 250 SCT 2310. 19Z 40 SCT 2312 SLGT CHC C20 BKN 2TRW
   G30. 02Z 250 SCT 2010. 09Z VFR CLR..
AUS 291515 15 SCT SCT V BKN. 17Z 20 SCT. 19Z 40 SCT 1810. 01Z CLR.
   09Z VFR BCMG 1FR CIG 10Z..
```

Figure 5–36. *Terminal forecasts.*

To aid in the interpretation and understanding, the FT for Abilene (ABI) has been divided into elements and lettered. This is followed by a plain language description of each element (Fig. 5-37):

a. Station identifier. "ABI" identifies the airport for which the forecast applies as Abilene, Texas.

b. Date-Time group. "291515" is the date and valid time for this forecast, beginning on the 29th day of the month at 1500Z and ending at 1500Z on the following day.

c. Sky and ceiling. "100 SCT 250-BKN" means 10,000 scattered, 25,000 thin broken. The broken clouds at 25,000 feet may appear to be a ceiling, but the contraction BKN is preceded by a minus sign which means "thin." Therefore, it does not constitute a ceiling. If the letter "C" precedes the numerical values, i.e., C 250 BKN, it would identify a forecast ceiling layer.

d. Visibility. Absence of the visibility value implies that the visibility is more than 6 miles. When visibility is shown it is stated in statute miles and fractions of statute miles, i.e., 1½ means visibility one and one-half miles.

e. Weather and obstructions to vision. These elements are stated in symbols identical to those used in Surface Aviation Weather Reports (SA), and entered only when expected. An example is "S—BS" which means light snow and blowing snow. This element of FT is missing in the initial forecast period for Abilene, but is shown in the "Expected Changes."

f. Wind "1812" means the expected wind for this forecast period is from 180 degrees at 12 knots. Omission of the wind entry implies that the wind is forecast to be less than 10 knots.

g. Remarks. The remarks are missing in the first portion of the forecast period because there was no significant weather that could be described more completely by using the remarks. An example of a remark, such as "OCNLY C 35 BKN" means occasional ceiling 3,500 feet broken. Note that there is a remarks section in the expected changes section.

h. Expected changes. If weather conditions are expected to change during the forecast period, these expected changes are followed by a period and the time the conditions are expected to change. Expected changes follow the same format and sequence as previously described.

18Z 50 SCT 100 SCT 250-BKN
CHC C20X 1TRW + AFT 21Z

means the expected change at 1800Z will be 5,000-foot scattered 10,000-foot scattered 25,000-foot thin broken with chance ceiling 2,000 sky obscured one mile visibility thunderstorms heavy rain showers after 2100Z.

i. 6-hour categorical outlook. The categorical outlook covers the last 6 hours of the forecast period.

09Z VFR TRW/RW..

means that from 0900Z to 1500Z, which is the end of the forecast period, the weather is expected to be VFR, i.e., ceiling more than 3,000 feet and visibility greater than 5 miles, except there is a chance of thunderstorms with rain showers and rain showers without thunderstorms. The double (..) means the end of the forecast for this specific terminal.

a	b	g	c	d	h	e	f
ABI	291515		1∅∅ SCT 25∅-BKN	18z 5∅ SCT 1∅∅ SCT 25∅-BKN CHC		C2∅X 1TRW+ AFT 21z.	1812.

∅9z VFR CHC TRW/RW..

Figure 5-37. Portions of hourly sequence report.

In-Flight Advisories (WS, WA, WST). In-Flight Advisories are unscheduled forecasts to advise aircraft in flight of the development of potentially hazardous weather. These advisories are available from teletypewriter circuits at weather service outlets and are an excellent source of information for preflight planning and briefing. All heights are stated in ASL unless otherwise noted. Ceilings, of course, are always AGL.

There are three types of In-Flight Advisories. These are SIGMET (WS), AIRMET (WA), and Convective SIGMETS (WST).

The format of these advisories consists of a heading and text. The heading identifies the issuing WSFO, type of advisory, and the valid period.

The text of the advisory contains a message identifier, a flight precautions statement, and further details if necessary.

Message identifier: The WSFO identifies each hazardous area using a phonetic identifier (ALFA, BRAVO, CHARLIE, etc.). Advisories for each hazardous area for that day are numbered in sequence (ALFA 1, ALFA 2, ALFA 3, etc., or BRAVO 1, BRAVO 2, BRAVO 3, etc.).

A new advisory of the same alphabetic series by the same WSFO automatically cancels preceding advisories of the same series; i.e., ALFA 2 cancels ALFA 1, BRAVO 3 cancels BRAVO 2, etc. A new issuance by one WSFO does not cancel an advisory by another WSFO unless specifically stated; i.e., ALFA 2 by Kansas City does not cancel ALFA 1 by Fort Worth.

Flight precautions statement (FLT PRCTN): A flight precautions statement in each advisory gives the location and describes the kind of hazard. It also gives the predicted time of occurrence if the hazard is not already in progress. It is used in the flight precautions statement of the TWEB and PATWAS.

Further details: When necessary, further details are included. If the hazard is expected to continue beyond the valid period shown in the heading

of the WS and WA, this fact is stated by adding the following, "CONDS CONTG BYD ∅2Z," which means conditions continuing beyond ∅2∅∅Z.

The following describe the three types of advisories, and are examples of each.

SIGMET (WS)

A SIGMET is issued to advise pilots of weather considered potentially hazardous to *all* categories of aircraft, and is valid for the period stated in the advisory. SIGMETS are based specifically on forecasts of:

1. Tornadoes.
2. Lines of thunderstorms (squall lines).
3. Embedded thunderstorms.
4. Hail of ¾" or greater in diameter.
5. Severe or extreme turbulence.
6. Severe icing.
7. Widespread sandstorms/duststorms, lowering visibility to less than 3 miles.

Example of SIGMET:

SAT WS 281210
281210–281500Z
SIGMET ALFA 2. FLT PRCTN.
OVR SRN TEX WITHIN AN AREA BNDD BY SAN ANTONIO-PALACIOS 4∅ S OF CORPUS-COTULLA SCTD TO LCLY NMRS SHWRS AND TSTMS FQTLY IN BRKN LNS WITH TOPS TO 45∅. CONDS CONTG BYD 15Z.

Plain language interpretation:

The heading SAT WS 281210
281210–281500Z

means that this SIGMET was issued by SAN ANTONIO WSFO on the 28th day of the month at 121∅Z. Its valid period begins on the 28th day of the month at 1210Z and ends on the same day at 1500Z.

The text contains the message identifier SIGMET ALFA 2, which identifies this advisory as the number 2 SIGMET TO BE ISSUED FOR THE ALFA area.

Flight Precautions: Over southern Texas within an area bounded by San Antonio, Palacios, 40 miles south of Corpus Christi and Cotulla scattered to locally numerous showers and thunderstorms frequently in broken lines with tops to 45,000 feet. Conditions continuing beyond 1500Z.

AIRMET (WA)

An AIRMET is issued to advise pilots of weather that may be hazardous to single-engine and light aircraft and in some cases to all aircraft. The AIRMET is valid for the period stated in the advisory. AIRMETS are based specifically on forecasts of

1. Moderate icing.
2. Moderate turbulence.
3. Sustained winds of 30 knots or greater at or within 2,000 feet from the surface.
4. Possibility of extensive areas of visibility below 3 miles and/or ceilings less than 1,000 feet, including mountain ridges and passes.

Example of AIRMET:

DFW WA 291200
291200-291400
AIRMET BRAVO 1. FLT PRCTN. OVR EXTRM NE OKLA VSBYS BLO 3 MIS FOG. N OF PNC-TUL-FYV LN VSBYS FQTLY BLO 3 MIS FOG. CONDS IPVG RPDLY AFT 13Z. CNL AT 14Z

Plain language interpretation:

The heading DFW WA 291200
291200-291400

means that this AIRMET was issued by DALLAS-FORT WORTH WSFO on the 29th day of the month at 1200Z. Its valid period begins on the 29th day of the month at 1200Z and ends on the same day at 1400Z.

The text contains the message identifier AIRMET BRAVO 1, which identifies this advisory as the number 1 AIRMET issued for the BRAVO area.

Flight precautions: Over extreme northeast Oklahoma visibilities below 3 miles fog. North of Ponca City, Tulsa, Fayetteville line visibilities frequently below 3 miles fog. Conditions improving rapidly after 1300Z. Cancel at 1400Z.

CONVECTIVE SIGMETS (WST)

Three convective SIGMET bulletins (EASTERN, CENTRAL, and WESTERN U.S.) are issued on a scheduled basis each hour at 55 (H+55) minutes past the hour and on an unscheduled basis as specials. Each of the convective SIGMET bulletins are valid for 1 hour and will be removed from the system automatically at 40 (H+40) minutes past the hour. Each bulletin is made up of one or more individually numbered SIGMETS and are designated WST.

The convective SIGMETS are taken from radar observations and use the following criteria as a basis for issuance:

A. Tornadoes.
B. Lines of thunderstorms.
C. Embedded thunderstorms.
D. Isolated thunderstorms greater than or equal to VIP LVL 5.
E. Areas of thunderstorms greater than or equal to VIP LVL 4.
F. Hail greater than or equal to ¾".

Each individual convective SIGMET is made up of two parts: Part A, which contains the location and description of the phenomena in a fixed format, and Part B, which describes the same phenomena in relation to six specified VOR's located throughout the United States. Two of these VOR's are located in the Eastern section, one at Buffalo, New York, and one at Augusta, Georgia; two in the Central section, one at Sioux Falls, South Dakota and one at Dallas/Fort Worth, Texas; and two in the Western section, one at Boise, Idaho, and one at Needles, California.

The format of the CONVECTIVE SIGMET appears as follows:

HEADER

Line 1. MKCx WST DDTTT
 x = E—EAST OF 87 DEGREES LONGITUDE
 = C—CENTRAL, FROM 87 DEGREES WEST TO 107 DEGREES WEST LONGITUDE
 = W—WEST OF 107 DEGREES WEST LONGITUDE
 DD = DAY OF MONTH
 TTT = HOUR OF RADAR OBSERVATION (H+35) GMT
Line 2. CONVECTIVE SIGMET nn
 nn = ISSUANCE NUMBER. INDIVIDUAL SIGMETS WHICH MAKE UP EACH CONVECTIVE SIGMET BULLETIN (MKCE, MKCC, MKCW) WILL BE NUMBERED CONSECUTIVELY (01-99), EACH DAY, BEGINNING AT 0000Z
Line 3. STATES APPLICABLE INCLUDING GREAT LAKES & COASTAL WATERS, e.g., KS, OK, LO, LM, LH, LE, LS, AND LA CSTL WTRS

CONVECTIVE SIGMET MESSAGE

Line 4. PART A BEGINS ON LINE 4 CONTAINS THE LOCATIONS USING VORS AND A DESCRIPTION OF THE PHENOMENA.

PART B, WHICH IS LESS DETAILED, LOCATES THE PHENOMENA BY AZIMUTH (MAGNETIC NORTH) AND DISTANCE, (NAUTICAL MILES) IN RELATION TO THE

FOLLOWING 6 SPECIFIED VORS: BOI, EED, FSD, DFW, BUF, AND AGS, e.g., VOR bbbrrr, VOR bbbrrr, where AZIMUTH=bbb AND DISTANT=rrr

THE FORECASTER'S LAST NAME WILL APPEAR AFTER THE LAST INDIVIDUAL CONVECTIVE SIG-MET IN EACH BULLETIN.

EXAMPLE

ZCZC
MKCC WST 221835
CONVECTIVE SIGMET 19
KS OK
FROM 30E GCK TO 20E GAG.
LN BKN TSTMS 25 WIDE MOVG FROM 2515 WITH AN INTS-LVL5 CELL.
TOPS TO 450 ... HAIL TO 1 IN ... WIND GUSTS TO 55.
LN BKN TSTMS 25 WIDE DFW 340300 DFW 335250
MOVG 2515 TOPS 450
CELL LVL5 DIAM 10 DFW 330280 MOVG 2120 TOPS 450

CONVECTIVE SIGMET 20
ND SD
FROM 90W MOT TO PMB TO 40N MHE TO RAP.
AREA BKN TSTMS MOVG FROM 2530 WITH
A FEW INTS-LVL5 AND EXTRM-LVL6 CELLS
TORNADO RPTD 1820Z VCNTY GFK. MAX TOPS TO 450 ... HAIL TO 1 IN ... WIND GUSTS TO 55. CONDS EXPCD TO INTSFY.

AREA BKN TSTMS FSD 310400 FSD 350270
FSD 310080 FSD 290240 MOVG 2530 TOPS 450
CELL LVL6 DIAM 20 FSD 30010 MOVG 2515 TOPS 420
CELL LVL5 DIAM 10 FSD 330200 MOVG 2515 TOPS 420
PEARSON
NNN

NEGATIVE CONVECTIVE SIGMET

If there are no convective sigmets in effect at the time of issuance the following will be transmitted:

MKCx WST DDTTT
CONVECTIVE SIGMET ...NONE.

SPECIAL CONVECTIVE SIGMETS

A conventive sigmet special will be issued anytime based on the following:

A. SPECIAL RADAR OBSERVATIONS
B. TORNADO REPORTS
C. OTHER INDICATIONS OF RAPIDLY CHANGING CON-DITIONS

EXAMPLE OF SPECIAL CONVECTIVE SIGMET

ZCZC
MKCC WST 131910
CONVECTIVE SIGMET 21
ND
5NE JMS
ISLTD EXTRM-LVL6 TSTM DIAM 20 MOVG FROM 2530.
TORNADO RPRTD 1910Z 5NE JMS ... TOPS 500 ... HAIL TO 2 IN ... WIND GUSTS TO 60.

CELL LVL6 DIAM 20 FSD 340200 MOVG 2530 TOPS 500.
OSTBY
NNNN

NOTES

1. COVERAGE—DENOTED BY THE FOLLOWING CONTRAC-TIONS

   SCATTERED—SCT (1 to 5) TENTHS COVERAGE
   BROKEN —BKN (6 to 9) TENTHS COVERAGE
   SOLID —SLD (10)TENTHS COVERAGE

2. WIDTH OF LINES AND TOPS OF THUNDERSTORMS WILL BE INCLUDED IF AVAILABLE FROM RADAR, SATELLITE PIC-TURES, OR PIREPS.

3. IF LOCATION IS AT ONE OF THE 6 VORS USE 000000, e.g., FSD 000000

Winds and Temperatures Aloft Forecast (FD)

The winds and temperatures aloft are forecast for specific locations in the contiguous United States. These forecasts are also prepared for a network of locations in Alaska

Below are examples and explanations of a Winds and Temperatures Aloft Forecast (FD) giving the heading and five locations.

FD WBC 291745
DATA BASED ON 291200Z
VALID 300600Z   FOR USE 0300–0900Z. TEMPS NEG ABV 24000

FT	3000	6000	9000	12000	18000	24000	30000	34000	39000
ABI		2213+19	2315+14	2313+08	2208−07	9900−19	990034	090644	101155
ABQ		2605+17	9900+10	0710−05	0710−17	9900−18	990033	990043	060655
AMA	2210	2409+16	2406+10	9900−06	9900−18		010534	350543	021055
ATL	2611	2611+17	2612+13	2612+08	2713−06	2712−18	281134	291143	300754
BNA	2414	2617+17	2617+12	2718+07	2819−07	2721−19	272434	272543	262554

In the first line of the heading "FD WBC 291745," the FD identifies this forecast as a Winds and Temperatures Aloft Forecast. "WBC" indicates that the forecast is prepared at the National Meteorological Center through the use of digital computers. In "291745" the first two digits (29) mean the 29th day of the month; 1745 indicates the time of the forecast in Greenwich Mean Time (Z).

The second line "DATA BASED ON 291200Z" indicates that the forecast is based on data collected at 1200Z on the 29th day of the month.

The third line "VALID 300600Z FOR USE 0300–0900Z," means the forecast data are valid at 0600Z on the 30th, and are to be used by pilots between 0300Z and 0900Z. A notation "TEMPS NEG ABV 24,000" is always included. Since temperatures above 24,000 feet are always negative, the minus sign preceding the temperature above this level is omitted.

Forecast levels: The line labeled "FT" shows the 9 standard levels in feet for which the winds and temperatures apply. The levels through 12,000 feet are based on true altitude, and the levels at 18,000 feet and above are based on pressure altitude. The station identifiers denoting the location for which the forecast applies are arranged in alphabetical order in a column along the left side of the data sheet. The coded wind and temperature information in columns under each level and in the line to the right of the station identifier.

Note that at some of the lower levels the wind and temperature information is omitted. The reason for the omission is that winds aloft are not forecast for levels within 1,500 feet of the station elevation. Also, note that no temperatures are forecast for the 3,000-foot level or for a level within 2,500 feet of the station elevation.

Decoding: A 4-digit group shows the wind direction in reference to true north and the windspeed in knots. Refer to the Atlanta (ATL) forecast for the 3,000-foot level. The group "2611" means the wind is forecast to be from 260° true north at a speed of 11 knots. Note that to decode a zero is added to the end of the first two digits giving the direction in increments of 10 degrees, and the second two digits give speed in knots.

A 6-digit group includes the forecast temperature aloft. Refer to the Abilene (ABI) forecast for the 6,000-foot level. The group 2213+19 means the wind is forecast to be from 220° at 13 knots with a temperature of +19° Celsius (C).

If the windspeed is forecast to be 100 to 199 knots the forecaster adds 50 to the direction and subtracts 100 from the speed. To decode, the reverse must be done; i.e., subtract 50 from the direction and add 100 to the speed. For example, if the forecast for the 39,000-foot level appears as "731960," subtract 50 from 73, and add 100 to 19, and the wind would be 230° at 119 knots with a temperature of −60° C.

It is quite simple to recognize when the coded direction has been increased by 50. Coded direction (in tens of degrees) ranges from 01 (010°) to 36 (360°). Thus any coded direction with a numerical value greater than "36" indicates a wind of 100 knots or greater. The coded direction for winds of 100 to 199 knots ranges from 51 through 86.

If the windspeed is forecast to be 200 knots or greater, the wind group is coded as 199 knots, i.e., "7799" is decoded 270° at 199 knots or greater.

When the forecast speed is less than 5 knots, the coded group is "9900" which means "LIGHT AND VARIABLE."

Examples of decoding FD winds and temperatures:

Coded	Decoded
9900+00	Wind light and variable, temperature 0° C.
2707	270° at 7 knots
850552	85−50 = 35; 05+100 = 105
	350° at 105 knots, temperature −52° C.

Aviation Weather Reports Frequently, changes in weather are so rapid that conditions at the time of flight could be quite different from those predicted in forecasts or shown on weather maps or charts previously issued. Therefore, to provide the very latest information on weather conditions, observations are made at numerous stations throughout the country. When these observations are recorded and transmitted, they become weather reports. There are three types of reports available: (1) Surface Aviation Weather Report (SA), sometimes referred to as hourly sequence reports and special surface reports, (2) Pilot Weather Reports (PIREPS), and (3) Radar Weather Reports (RAREPS). These reports will be discussed in detail in this section.

Surface Aviation Weather Reports

These are hourly reports made from on-the-hour observations of existing weather conditions at stations throughout the country. These reports are transmitted over a teletypewriter circuit in a standard abbreviated form, at which time they become available to users.

A pilot must become familiar with a few contractions, abbreviations, and symbols in order to read Surface Aviation Weather Reports. Ease in reading and interpreting these reports can be acquired in a surprisingly short time and the reward is well worth the effort.

Report Identifiers: A heading is found at the beginning of each group of hourly Surface Aviation Weather Reports. The heading identifies the type of report, the circuit number, and the date and time that the observations were taken. For example:

SA21 291400

means surface aviation report (SA); 21 is the circuit number; 29th day of the month; 1400 is the Greenwich Mean Time that the observations were made.

A slightly different heading subdivides each group of SAs. This heading identifies the location of the reporting station. For example:

TX 291702

means that the following SAs pertain to Texas on the 29th day (29); time of observation 1700 GMT; and the relay began at 2 minutes past the hour (02).

Elements of SAs: A Surface Aviation Weather Report contains certain elements that are placed in the same sequence in every report. The following lists these elements:

1. Station designation.
2. Type and time of report.
3. Sky conditions and ceiling.
4. Visibility.
5. Weather and obstructions to vision.
6. Sea level pressure (millibars).
7. Temperature and dewpoint.
8. Wind direction, speed, and character.
9. Altimeter setting (inches of Hg).
10. Remarks.

The following example of a Surface Aviation Weather Report is separated into elements and labeled according to the sequence of elements previously listed.

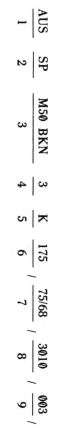

AUS	SP	M50 BKN	3	K	175	75/68	3010	003
1	2	3	4	5	6	7	8	9

UA 5SW AUS 120 OVC

10

The plain language interpretation of this report is: (1) Austin; (2) special report; (3) measured ceiling 5,000 broken; (4) visibility 3 miles; (5) smoke; (6) pressure 1017.5 millibars; (7) temperature 75, dewpoint 68; (8) wind 300° at 10 knots; (9) altimeter setting 30.03 in. Hg; (10) pilot reported 5 southwest of Austin 12,000 overcast.

Those elements not occurring at the time of observation or not pertinent to the observation are omitted from the report. When an element should be included but is unavailable, the letter "M" is transmitted in lieu of the missing element. The elements that are included are transmitted in the sequence described above.

The following describes each of the elements contained in the Surface Aviation Weather Report.

Station Designator: The station designator is a three-letter location identifier assigned to reporting stations which are located at airports. In many cases the identifier, which is derived from the airport name, resembles the name of the airport: For example, John F. Kennedy Airport, New York City (JFK) or Boise, Idaho (BOI). At times the identity must be determined by referring to aeronautical charts, Airport/Facility Directory, or the "Location Identifier" Manual, if available.

Type and Time of Report: There are two basic types of SA reports: (1) reports of observations taken on the hour (hourly), and (2) special reports of observations taken when needed to report significant changes in weather.

The hourly reports are identified in the sequence heading and no further identification is made. In other words, this report can be identified by the omission of the type following the location identifier.

The special report is identified by the letters "SP" which follow the location identifier. Special reports are transmitted either in sequence with other hourly reports or they can be transmitted at any time between the regularly scheduled hourly transmission times. Reports transmitted out of sequence must include the time and type of observation.

Sky Conditions and Ceiling: The sky condition is reported in SAs using the designators shown in the table illustrated in Fig. 5-38.

	Meaning	Spoken
CLR	CLEAR. (Less than 0.1 sky cover.)	CLEAR
SCT	SCATTERED LAYER ALOFT. (0.1 through 0.5 sky cover.)	SCATTERED
BKN*	BROKEN LAYER ALOFT. (0.6 through 0.9 sky cover.)	BROKEN
OVC*	OVERCAST LAYER ALOFT. (More than 0.9, or 1.0 sky cover.)	OVERCAST
−SCT	THIN SCATTERED. At least ½ of the sky cover aloft is transparent at and below the level of the layer aloft.	THIN SCATTERED
−BKN	THIN BROKEN. At least ½ of the sky cover aloft is transparent at and below the level of the layer aloft.	THIN BROKEN
−OVC	THIN OVERCAST. At least ½ of the sky cover aloft is transparent at and below the level of the layer aloft.	THIN OVERCAST
X*	SURFACE BASED OBSTRUCTION. (All of sky is hidden by surface based phenomena.)	SKY OBSCURED
−X	SURFACE BASED PARTIAL OBSCURATION. (0.1 or more, but not all, of sky is hidden by surface based phenomena.)	SKY PARTIALLY OBSCURED

* Sky condition represented by this designator may constitute a ceiling layer.

Figure 5–38. *Summary of sky cover designators.*

A clear sky or a layer of clouds or obscuring phenomena aloft are reported by one of the seven designators shown. A layer is defined as clouds or obscuring phenomena with the base at approximately the same level. The height of the base of a layer precedes the sky cover designator, and is given in hundreds of feet above ground level.

When more than one layer is reported, layers are given in ascending order according to height. For each layer above a lower layer or layers, the sky cover designator for that layer represents the total sky covered by that layer and all lower layers. For example 7 SCT 15 SCT E30BKN, reports three layers, a scattered layer at 700 feet, another scattered layer at 1,500 feet, and a top layer, which in this case it is assumed that the total sky covered by all the layers exceeds 5/10; therefore, the upper layer is reported as Estimated 3,000 broken.

"Transparent" sky cover is clouds or obscuring phenomena aloft through which blue sky or higher sky cover is visible. A scattered, broken, or overcast layer may be reported as "thin." To be classified as thin, a layer must be half or more transparent, and remember that sky cover of a layer includes all sky cover below the layer. For example, if at a station the blue sky had been observed as being visible through half or more of the total sky cover reported by the higher layer, the sky report could appear as

8 SCT350 −SCT.

which reads 800 scattered, 35,000 thin scattered.

Any phenomena based at the surface and hiding all or part of the sky are reported as SKY OBSCURED (X) or SKY PARTIALLY OBSCURED (−X). An obscuration or partial obscuration may be precipitation, fog, dust, blowing snow, etc. No height value precedes the designator for partial obscuration, since vertical visibility is not restricted overhead. A height value precedes the designator for an obscuration and denotes the vertical visibility into the phenomena. For example:

STL −X M40 OVC.

reads as "St. Louis sky partially obscured measured ceiling 4,000 overcast" or

STL W5 X

reads as "St.Louis indefinite ceiling 500 feet sky obscured.

Ceiling is defined as:

1. Height of the lowest layer of clouds or obscuring phenomena aloft that are reported as broken or overcast and not classified as thin;

   or

2. Vertical visibility into a surface-based obscuring phenomenon that hides all the sky.

A ceiling designator always precedes the height of the ceiling layer. The table in Fig. 5–39 lists and explains the ceiling designators that are used.

Coded	Meaning	Spoken
M	MEASURED. Heights determined by ceilometer, ceiling light, cloud detection radar, or by the unobscured portion of a landmark protruding into ceiling layer. (Figure 2–5 illustrates the principle of the ceilometer.)	MEASURED CEILING
E	ESTIMATED. Heights determined from pilot reports, balloons, or other measurements not meeting criteria for measured ceiling.	ESTIMATED CEILING
W	INDEFINITE. Vertical visibility into a surface based obstruction. Regardless of method of determination, vertical visibility is classified as an indefinite ceiling.	INDEFINITE CEILING

Figure 5–39. *Ceiling designators.*

The letter "V" appended to the ceiling height indicates a variable ceiling; the range of the variability is shown in the remarks element of the report. Variable ceiling is reported only when it is critical to terminal operations. For example, M15V OVC and in the remarks CIT 15V18 means "measured ceiling 1,500 variable overcast; ceiling variable between 1,500 and 1,800."

Visibility: Prevailing visibility is the greatest distance objects can be seen and identified through at least 180° of the horizon. Visibility is reported in statute miles and fractions. In the report, prevailing visibility always follows the sky and ceiling element. For example:

means "Chicago Midway measured ceiling 700 overcast visibility one and one-half fog."

When visibility is critical at an airport with a weather observing station and a control tower, both take visibility observations. Of the two observations the one most representative is usually reported as prevailing visibility. If the other is operationally significant it is reported in the remarks element. For example:

MDW M7 OVC 1½F

TWR VSBY ¼

which means "control tower visibility one quarter."

The letter "V" suffixed to prevailing visibility denotes variable visibility; the range of visibility is shown in the remarks. For example

¾V and in the remarks, VSBY ½V1

means "visibility three quarters variable—visibility variable between one-half and one."

Visibility in some directions may vary significantly from prevailing visibility. These differences are reported in the remarks. For example if the prevailing visibility is reported as 1½ miles with a remark VSBY SE2SW1, this would mean visibility to the southeast is 2 miles and to the southwest 1 mile.

Weather and Obstructions to Vision: Weather and obstructions to vision, when occurring at the station during the time of observation, are reported in an element immediately following the visibility. If weather and obstruction to view are observed at a distance from the station, they are reported in the remarks.

The term weather as used for this element refers only to those items listed in Fig. 5-40, rather than to the more general meaning of all atmospheric phenomena. Weather in this sense includes all forms of precipitation plus thunderstorm, tornado, funnel cloud, and waterspout.

Precipitation is reported in one of three intensities, and follows the weather symbol. The intensity symbols and their meanings are as follows:

LIGHT –
MODERATE (No sign)
HEAVY +

No intensity is reported for hail (A) or ice crystals (IC).
A thunderstorm is reported as "T" and a severe thunderstorm, as "T+." A severe thunderstorm is one in which surface wind is 50 knots or greater and/or hail is ¾ inch or more in diameter.

Coded	Spoken
Tornado	TORNADO
Funnel Cloud	FUNNEL CLOUD
Waterspout	WATERSPOUT
T	THUNDERSTORM
T+	SEVERE THUNDERSTORM
R	RAIN
RW	RAIN SHOWER
L	DRIZZLE
ZR	FREEZING RAIN
ZL	FREEZING DRIZZLE
A	HAIL
IP	ICE PELLETS
IPW	ICE PELLET SHOWERS
S	SNOW
SW	SNOW SHOWERS
SP	SNOW PELLETS
SG	SNOW GRAINS
IC	ICE CRYSTALS

Figure 5–40. *Weather symbols and meanings.*

Obstructions to vision include the phenomena listed in the table in Fig. 5–41. No intensities are reported for obstructions to vision.

Coded	Spoken
BD	BLOWING DUST
BN	BLOWING SAND
BS	BLOWING SNOW
BY	BLOWING SPRAY
D	DUST
F	FOG
GF	GROUND FOG
H	HAZE
IF	ICE FOG
K	SMOKE

Figure 5–41. *Obstructions to vision—symbols and meanings.*

The following is an example of a partial SA which includes weather and obstructions to vision:

TUL W4 X 11/2R+F

This means "Tulsa International Airport, indefinite ceiling 400 feet, obscured visibility one and one-half, heavy rain and fog." In this case the weather is heavy rain and the obstruction to vision is fog.

There are two methods used in reporting obscuring phenomena in the remarks element. One pertains to conditions on the surface, and the other to conditions aloft.

When obscuring phenomena are surface based and partially obscure the sky, the remarks element reports tenths of sky hidden. For example:

D3

means 3/10 of the sky is hidden by dust, or

RF2

means 2/10 of the sky is hidden by rain and fog.

A layer of obscuring phenomena aloft is reported in the sky and ceiling portion of the SA in the same manner as a layer of cloud cover. The remark identifies the layer of obscuring phenomena. For example

8-BKN with a remark K8-BKN means a broken layer of smoke with a base at 800 feet above the surface and not concealing the sky (thin).

Sea Level Pressure: Sea level pressure reported in millibars follows the visibility or weather and obstruction to vision in the Surface Aviation Weather Report. It is transmitted in three digits to the nearest tenth millibar, with the decimal point omitted. Sea level pressure normally is greater than 960.0 millibars and less than 1050.0 millibars. The first 9 or 10 is omitted. To decode, prefix a 9 or 10, whichever brings it closer to 1000.0 millibars. The following are a few sea level pressures that are decoded:

As reported	Decoded
980	998.0 millibars
191	1019.1 millibars
752	975.2 millibars
456	1045.6 millibars

Temperature and Dewpoint: The next elements in the SA report are the temperature and dewpoint given in whole degrees Fahrenheit. A slash (/) is used to separate the temperature element from the sea level pressure element. If the sea level pressure is not transmitted, a space is used to separate the temperature from the preceding elements. A slash is also used to separate the temperature and the dewpoint. A minus sign precedes the temperature and dewpoint when these temperatures are below 0° F. The following is an exam-ple of an SA report showing the elements through the temperature and dewpoint:

ELP 250-SCT 80 117/82/59

which means "El Paso, 25,000 thin scattered, visibility eight zero, sea level pressure 1011.7, temperature eight two, dewpoint five nine."

Wind: The surface wind element follows the dewpoint and is separated from it by a slash. The wind is observed for one minute and the average direction and speed are reported in four digits. The *first two digits* are the direction *from* which the wind is blowing. It is in tens of degrees referenced to true NORTH. (Wind direction when verbally broadcast for the local station is referenced to magnetic. This is done so that the pilot can more closely relate the wind direction to the landing runway. Runway alignment is referenced to magnetic north.) When stated, three digits are used, i.e., a wind from 10° is stated as 010°, or 210° is stated as 210°, but to conserve space on the report the last digit (0) is omitted, i.e., 020° would appear as 02 or 220° would appear as 22. The second two digits are the windspeed in knots, except that a calm wind is reported as 0000.

As in the Winds Aloft Forecast, if the windspeed is 100 knots or greater, 50 is added to the direction code and the hundreds digit of the speed is omitted. For example:

8315

can be recognized as a windspeed of 100 knots or more because the first two digits are greater than 36 (36 represents a direction of 360° which is the largest number of degrees in the compass). To decode, 50 is subtracted from 83 (83−50=33) and 100 is added to the last two digits (15+100=115) which results in a wind from 330° at 115 knots.

A gust is a variation in windspeed of at least 10 knots between peak winds and lulls. A squall is a sudden increase in speed of at least 15 knots to a sustained speed of 20 knots or more which lasts for at least one minute. Gusts or squalls are reported by the letter "G" or "Q" respectively, following the average one minute speed. The peak speed in knots follows the letter. For example:

2123G38

means wind 210° at 23 knots with peak speed in gusts to 38 knots.

When any part of the wind report is estimated, the letter "E" precedes the wind group. For example:

E3122Q27

means wind 310° estimated at 22 knots peak speed in squalls estimated at 27 knots.

The following is an example of an SA report showing the elements through the wind element:

HOU SP M8 OVC 10R- 139/74/71/1410

which means, "Houston, special, measured ceiling 800 overcast, visibility 10 miles light rain, pressure 1013.9 millibars, temperature 74, dewpoint 71, wind 140° at 10 knots.

Altimeter Setting: This element follows the wind group in an SA report and is separated by a slash. The normal range for altimeter settings is from 28.00 inches of mercury (Hg) to 31.00 inches. Only the last three digits are transmitted with the decimal point omitted. To decode, prefix the coded value in the report with either a 2 or 3, whichever brings it closer to 30.00 inches. For example:

998 means "altimeter setting 29.98 inches of mercury"

or

025 means "altimeter setting 30.25 inches of mercury."

An estimated altimeter setting is a reading from an instrument that has not been compared to a standard instrument as recently as required. When this reading is used, the altimeter setting element is prefixed with an "E" such as

E992

which means "altimeter setting estimated 29.92 inches of mercury."

Remarks: The remarks element is the last element in a Surface Aviation Weather Report, and is separated from the altimeter setting by a slash. Frequently, remarks are added at the end of the report to cover unusual aspects of the weather, and in many instances contain information that is as important to the pilot as that information found in the main body of the report. Certain remarks are reported routinely, while others may be indicated when considered significant to aviation.

The remarks utilize standard weather symbols, and simple abbreviations or contractions of words. The few special code words used occasionally can be easily memorized.

The following briefly describes the parts of the remarks element. For further information concerning the remarks, reference should be made to AC 00–45, Aviation Weather Services.

The first part of the remarks element, when transmitted, is "runway visibility" or "runway visual range" and is separated from the altimeter setting by a dash. The terms used are defined as follows:

Runway visibility "VV" is the visibility from a particular location along an identified runway, usually determined by a transmissometer instrument. It is reported in miles and fractions. For example:

R22VV1½

means "runway 22, visibility one and one half miles."

Runway visual range is the maximum horizontal distance down a specified runway at which a pilot can see and identify standard high intensity runway lights. It is determined by a transmissometer and reported in hundreds of feet. An example:

R30VR10V20

means "runway 30 visual range variable between 1,000 and 2,000 feet."

The second part of the remarks element pertains to heights of bases and tops of sky cover layers, or obscuring phenomena. These remarks originate from pilots and are prefixed by "UA" which identifies the message as a pilot report. A "U" preceding the "UA" indicates an urgent report. The heights are referenced to mean sea level (MSL). They are separated from the preceding remarks by a slash. An example is:

(U)UA/OV FRR 1745 FL080/TP C310/SK OVC 040/0600VC

and means "over Front Royal VORTAC at 1745Z, flight level 8,000 feet, type aircraft Cessna 310, top of the lower overcast is 4,000 feet, base of the higher overcast is 6,000 feet." NOTE: If the height value appears after the cloud designator, the height value pertains to the top of the layer, i.e., BKN 60 means top of the broken layer is 6,000 feet MSL, if the height value appears before the cloud designator the height value pertains to the base of the layer, i.e., 70 OVC means the base of the overcast is 7,000 feet MSL.

The freezing level data, if applicable, are shown in the remarks of stations equipped with upper air (rawinsonde) observation equipment. This part of the remarks can be identified by the contraction "RADAT." For example:

RADAT 83L026117

"RADAT" means freezing level data; relative humidity at the *lowest* (L) freezing level in percent, 83%. The coded letter "L," "M," or "H" indicates that the relative humidity is for the "lowest," "middle," or "highest" level coded. This letter is omitted when only one level is taken. The digits (026117) mean the height in hundreds of feet above MSL that the 0° isotherm (freezing level) was crossed. In this example two levels were crossed, one at 2,600 feet MSL and the other at 11,700 feet MSL.

When the rawinsonde determines definitely that icing is occurring, this information is transmitted. For example:

RAICG 15MSL

which means, "RAICG" is the contraction used to identify that icing data follows, 15 MSL, is the level in hundreds of feet at which icing occurred. The remarks element also codes "Notices to Airmen." NOTAMS contain pertinent information affecting radio aids, Airports, or other locations.

The following is an example of a "NOTAM" as it appears in the remarks portion of an SA and also the Notam Summary (NOSUM), which usually appears at the end of each group of Surface Aviation Weather Reports. The NOSUM may appear immediately following the report.

→DAL\6/33

This part alerts the reader that a NOTAM is in effect. The horizontal arrow denotes the beginning of the NOTAM for Dallas, using the station identifier "DAL." The second arrow denotes the currency of the NOTAM. The first number 6 in this example, indicates that the NOTAM was issued during the 6th month (June). The second number 33, indicates that this is the 33rd NOTAM to be issued at Dallas during June.

The Notam Summary or (NOSUM) which follows explains, in condensed form, the status of the radio aid facility, airport, or other location. For example:

→DAL 6/33 ADS RWY 15-33 CLSD

The first part of the Notam Summary, "→DAL 6/33" identifies this summary as being pertinent to the NOTAM found in the remarks, which was previously discussed.

"ADS" refers to the Addison Airport at Dallas. There are many airports that are not reporting stations; therefore NOTAMS pertaining to these airports are included in reports from nearby stations.

RWY 15-33 CLSD" means "runway 15-33 closed."

This concludes the discussion on Surface Aviation Weather Reports. Again, for further information concerning these reports reference should be made to AC 00-45, Aviation Weather Services.

Pilot Weather Reports (PIREPS) There is no more timely or helpful weather observation to fill the gap between reporting stations than those observations and reports made by the pilot during flight. Aircraft in flight are the only means of directly observing cloud tops, icing, and turbulence.

A PIREP usually is transmitted by teletypewriter in a prescribed format. First, the letters "UA" which identify the report as a pilot report. Next is the location of the phenomena given in relation to a checkpoint, then in sequence the time in Greenwich, flight level, type of aircraft reporting, sky cover amount with bases, and tops, temperature in Celsius, wind direction and speed, turbulence intensity and type with altitude encountered, icing intensity and type with altitudes encountered, and remarks are given.

Most contractions used in PIREP messages are self explanatory. When possible icing and turbulence are issued, the report states intensity in stan-

dard terminology. If the pilot's description of icing or turbulence cannot readily be translated into standard terminology, the pilot's description is transmitted verbatim.

The following is an example of a pilot report as it would be transmitted over a teletypewriter:

UA/OV CRP 180020 1629 FL050/TP/C182/SK SCT V BKN 040-050

This means that the pilot of a Cessna 182 reported scattered variable to broken clouds with bases 4,000 to 5,000 feet MSL at a point on the 180 radial 20 miles from Corpus Christi, Texas.

Radar Weather Reports (RAREPS) Thunderstorms and general areas of precipitation can be observed by radar. Radar weather reports are routinely transmitted by teletypewriter and some are included in scheduled weather broadcasts by Flight Service Stations.

Most radar stations report each hour with intervening special reports as required. They report location of precipitation along with type, intensity, and trend.

Fig. 5-42 explains symbols denoting intensity and trend. Fig. 5-43 summarizes the order and content of a radar weather report, and Fig. 5-44 explains the contractions used in reporting the operational status of the radar.

Intensity		Intensity Trend	
Symbol	Intensity	Symbol	Trend
−	Light	+	Increasing
(none)	Moderate	−	Decreasing
+	Heavy		
++	Very heavy	NC	No change
X	Intense	NEW	New echo
XX	Extreme		
U	Unknown		

Figure 5-42. *Precipitation intensity and intensity trend.*

When a radar report is transmitted but contains no encoded weather observation, a contraction is sent which indicates operational status of the radar.

OKC 1135 PPINE

Oklahoma City, Oklahoma, radar at 1135 GMT detects no echoes.

Radar weather reports also contain groups of digits, i.e., 00220 00221,

148

OKC 1934 LN 8TRW++/+ 86/40 164/60/ 199/115 15W 2425
MT 570 AT 159/65 2 INCH HAIL RPRTD THIS ECHO

OKC 1934	LN	8	TRW++/+	86/40 164/60 199/115	15W	2425
a.	**b.**	**c.**	**d.**	**e.**	**f.**	**g.**
MT 570 AT 159/65	2 INCH HAIL RPRTD THIS ECHO					
h.	**i.**					

a. Location identifier and time of radar observation (GMT)
b. Echo pattern¹ (line in this example)
c. Coverage in tenths (8/10 of this example)
d. Type, intensity, and trend of weather² (thunderstorm (T), very heavy rainshowers (RW++), increasing in intensity (/+))
e. Azimuth (reference true N) and range in nautical miles (NM) of points defining the echo pattern
f. Dimension of echo pattern³ (15 NM wide)
g. Pattern movement (line moving from 240° at 25 knots); may also show movement of individual storms or "cells"
h. Maximum tops and location (57,000 feet)
i. Remarks; self-explanatory in plain language contractions.

¹ Echo pattern may be a line (LN), fine line (FINE LN), area (AREA), spiral band area (SPIRAL BAND), or single cell (CELL).
² Teletypewriter weather symbols are used.
³ Dimension of an echo pattern is given when azimuth and range define only the center or center line of the pattern.

Figure 5-43. Order and content of a radar weather report.

Contraction	Operational status
PPINE	Equipment normal and operating in PPI (Plan Position Indicator) mode; no echoes observed.
PPIOM	Radar inoperative or out of service for preventative maintenance.
PPINA	Observations omitted or not available for reasons other than PPINE or PPIOM.
ROBEPS	Radar operating below performance standards.
ARNO	"A" scope or azimuth/range indicator inoperative.
RHINO	Radar cannot be operated in RHI (Range-height indicator) mode. Height data not available.

Figure 5-44. Contractions reporting operational status of radar.

etc., which are entered on a line following the RAREP. This manually digitized radar information is omitted from the foregoing examples since it is used primarily by meteorologists and hydrologists for estimating amount of rainfall.

A radar weather report may contain remarks in addition to the coded observation. Certain types of severe storms produce distinctive patterns on the radarscope. For example, a hook-shaped echo may be associated with a tornado; and a spiral band with a hurricane. If hail, strong winds, tornado activity, or other adverse weather is known to be associated with identified echoes on the radarscope, the location and type of phenomena are included as a remark. Examples of remarks are: "TORNADO ON GROUND AT 338/15"; and "HOOK ECHO 243/18."

When using hourly and special radar weather reports in preflight planning, note the location and coverage of echoes, the type of weather reported, the intensity trend, and especially the direction of movement. A word of caution—remember that radar detects only thunderstorms and general areas of precipitation; it is *not* designed to detect en route ceiling and visibility. An area may be blanketed with fog or low stratus, but unless precipitation is also present, the radarscope will be clear of echoes. Use radar reports along with PIREPs and aviation weather reports and forecasts.

RAREPs help pilots to plan ahead to avoid thunderstorm areas. Once airborne, however, the pilot must depend on visual sighting or airborne radar to avoid individual storms.

To assist in interpreting RAREPs, two examples are decoded into plain language:

LIT 1133 AREA 4TRW+/+ 22/100 88/170 196/180 220/115 CELLS 2425 MT 310 AT 16/110

Little Rock, Arkansas, radar weather observation at 1133 GMT

An area of echoes, four-tenths coverage, containing thunderstorms and heavy rain showers, increasing in intensity Area is defined by points (referenced LIT radar) at 22°, 100 NM (nautical miles); 80°, 170 NM; 196°, 180 NM; and 220°, 115 NM. (These points plotted on a map and connected with a line outline the area of echoes.) Individual cells are moving from 240° at 25 knots Maximum tops (MT) are 31,000 feet located at 162° and 110 NM from

LIT HDO 1132 AREA 2TRW ++ 6R–/NC 67/130 308/45 105W CELLS 2240 MT 380 AT 66/54

Hondo, Texas, radar weather report at 1132 GMT

An area of echoes containing two-tenths coverage of thunderstorms, very heavy rain showers, and six-tenths coverage of light rain. No intensity change. (This report suggests thunderstorms embedded in a general area of light rain.) Although the pattern is an "Area," only two points are given followed by "105W". This means the area lies 52½ miles either side of the line defined by the two points—67°, 130 NM and 308°, 45 NM Thunderstorm cells are moving from 220° at 40 knots Maximum tops are 38,00 feet at 66°, 54 NM

Weather Charts The National Weather Service prepares a multitude of

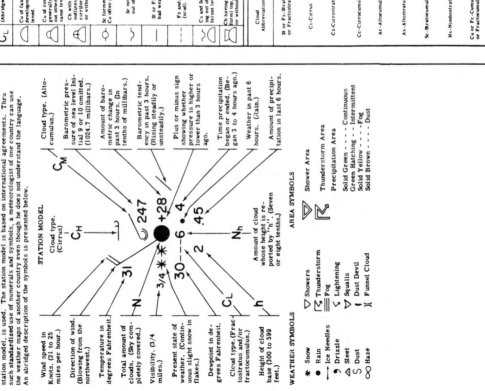

Figure 5-45. Symbols used on the weather map.

weather charts, some of them specifically for aviation. It is easy to learn to interpret and use these charts in determining weather significant to a proposed flight.

Most Weather Service Offices and many Flight Service Stations are equipped with facsimile machines. These machines reproduce entire charts from transmissions received from central locations where the charts are prepared. These charts are used primarily to obtain large-scale weather patterns and weather trends. For local variations, reports and forecasts should be consulted.

This discussion will include the Surface Analysis Chart, Weather Depiction Chart, Radar Summary Chart, and Significant Weather Prognostic Charts.

Surface Analysis Chart: The Surface Analysis Chart, often referred to as a surface weather map, is the basic weather chart. The chart is prepared by the NWS from reports of existing weather conditions and transmitted over the facsimile system to many weather outlets every 3 hours. The valid time of the map corresponds to the time of the plotted observations. A date and time (GMT) group informs the user of when the conditions portrayed on the map were actually occurring. The Surface Analysis Chart displays weather information such as surface wind direction and speed, temperature, dewpoint, and various other weather data. It also includes the position of fronts, and areas of high or low pressure.

Each reporting station is depicted on the chart by a small circle. The weather information pertaining to the station is placed in a standard pattern around this circle, and is called a station model. The standard pattern of a station model, with the explanation of the symbols used, is shown in Fig. 5–45.

Types of fronts are characterized on Surface Analysis Charts according to symbols shown in Fig. 5–46. Some stations color these symbols to facilitate the use of the chart.

The front is positioned on the chart in the same geographic location that it was at the time of the observations. The "pips" on the frontal symbol, as in the stationary front, suggest little or no movement.

A three-digit number, enclosed in brackets, is entered along a frontal symbol to classify the front according to type, intensity, and character. The numbers used can be decoded from the description shown in Fig. 5–47.

Solid lines depicting the pressure pattern are called isobars, and denote lines of equal pressure. These lines can be thought of as being similar to terrain contour lines on many maps. Isobars are usually spaced at 4-millibar intervals, and labeled by a two-digit number which denotes the pressure. For ex-

ample, 35 means 1035.0 MB; 04 means 1004.0 MB; and 93 means 993.0 MB. Isobars usually encircle a high- or low-pressure area.

The letter "H" on the chart marks the center of a high-pressure area and the letter "L" marks the center of a low-pressure area. The actual pressure at each center is indicated by an underlined two-digit number which is interpreted the same as the number along the isobars.

The Surface Analysis Chart gives a pictorial overview of the weather situation, but keep in mind that the weather systems move and this changes the weather. Therefore, this surface map should be used in conjunction with other available weather information. A section of a Surface Weather Map is shown in Fig. 5–48.

Weather Depiction Chart: The Weather Depiction Chart (Fig. 5–49) is prepared from Surface Aviation Weather Reports (SA) and gives a quick picture of the weather conditions as of the valid time stated on the chart. This chart is abbreviated to a certain extent and contains only a portion of the surface weather information. However, areas where clouds and weather may be a factor can be seen at a glance. The chart also shows major fronts and high- and low-pressure centers, and is considered to be a good place to begin a weather briefing for flight planning.

An abbreviated station model is used to plot data consisting of total sky cover, cloud height or ceiling, weather and obstructions to vision, visibility, and an analysis.

Cloud height is the lowest ceiling shown in hundreds of feet; or if there is no ceiling it is the height of the lowest layer. Weather and obstructions to vi-

FRONTAL SYMBOLS

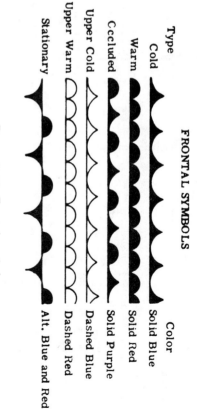

Type		Color
Cold		Solid Blue
Warm		Solid Red
Occluded		Solid Purple
Upper Cold		Dashed Blue
Upper Warm		Dashed Red
Stationary	Alt. Blue and Red	

Figure 5–46. *Frontal symbols.*

Code Figure	Description (Type of front)
0	Quasi-stationary at surface
1	Quasi-stationary above surface
2	Warm front at surface
3	Warm front above surface
4	Cold front at surface
5	Cold front above surface
6	Occlusion
7	Instability line
8	Intertropical front
9	Convergence line

Code Figure	Description (Intensity of front)
0	No specification
1	Weak, decreasing
2	Weak, little or no change
3	Weak, increasing
4	Moderate, decreasing
5	Moderate, little or no change
6	Moderate, increasing
7	Strong, decreasing
8	Strong, little or no change
9	Strong, increasing

Code Figure	Description (Character of front)
0	No specification
1	Frontal area activity decreasing
2	Frontal area activity, little change
3	Frontal area activity increasing
4	Intertropical
5	Forming or existence expected
6	Quasi-stationary
7	With waves
8	Diffuse
9	Position doubtful

Figure 5-47. Numerical classification of fronts.

sion are shown using the same symbol designators as the Surface Analysis. When visibility is less than 7 miles it is entered in miles and fractions of miles.

The chart shows ceilings and visibilities at reporting stations and catergorizes areas as IFR (outlined with scalloped lines), MVFR (outlined with smooth lines), and VFR (not outlined).

Notions used on Weather Depiction Charts are shown in Fig. 5-50.

Radar Summary Chart: Weather radar generally detects precipitation only; it does not ordinarily detect small water droplets such as found in fog and nonprecipitating clouds. The larger the drops the more intense is the return to the radar screen.

Thunderstorms, tornadoes, and hurricanes contain very heavy concentrations of liquid moisture that reflect strong signals to the radarscope. These areas are called "echoes," and are analyzed and plotted on the Radar Summary Chart to give a pictorial view of the storm locations, along with other significant information about this weather. It is for this reason the Radar Summary Chart is an excellent aid in a weather briefing. A sample Radar Summary Chart is shown in Fig. 5-51. Symbols used for echo intensity trend on the radar summary chart are shown in Fig. 5-52.

The information presented on this chart includes echo pattern and coverage, weather associated with echoes, intensity (contours) and trend (+ or −) of precipitation, height of echo bases and tops, and movement of echoes. A chart legend can be found in the lower left corner of the Radar Summary Chart (see Fig. 5-51), and additional information concerning chart symbols and weather symbols are shown in Figs. 5-53 and 5-54.

The arrangement of echoes as seen on the radarscope form a certain pattern which is symbolized on the chart. This pattern of echoes may be a line, area, or an isolated cell. A cell is a concentrated mass of convection that is normally 20 nautical miles or less in diameter. Echo coverage is the amount of space the echoes or cells occupy within an area or line. This should not be confused with sky cover.

The weather associated with echoes is symbolized on the chart. The radar cannot specifically identify the type of precipitation associated with the echoes, but using other sources of weather information the observer can determine precipitation type.

The height of the tops and bases of echoes are shown in hundreds of feet above mean sea level. A horizontal line is used with the heights shown above the line and below the line denoting the top height and base heights respectively. The absence of the number below the line indicates the echo base was not reported. Radar detects tops more readily than bases because as distances increase between the radar antenna and the echo, the earth's curvature is such

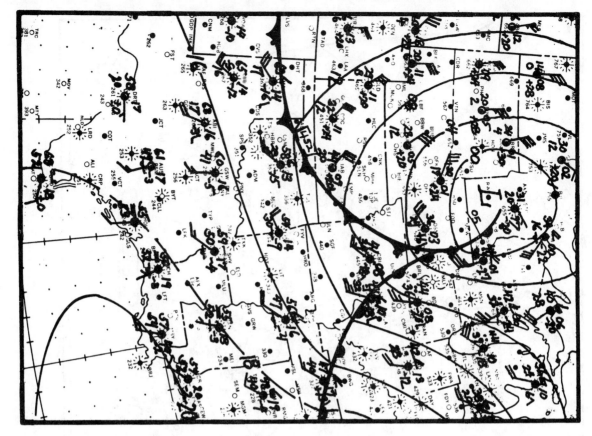

Figure 5-48. Section of a surface weather map as transmitted on facsimile.

that the ground obstructs the signal and cloud bases cannot be detected. Also, radar detects precipitation that reaches the ground and therefore obscures the base of the cloud on the radar screen.

The following are examples of how heights are shown:

$\frac{450}{}$ Average tops 45,000 feet.

$\frac{220}{080}$ Bases 8,000 feet; tops 22,000 feet.

$\frac{330}{650/}$ Top of an individual cell, 33,000 feet. Maximum tops, 65,000 feet.

A350 Tops 35,000 feet reported by aircraft.

The movement of individual storms and also the movement of a line or area is shown on Radar Summary Charts. The movement of the individual storms within a line or area often differs from the movement of the overall storm pattern. The means of depicting these movements are as shown:

35 Individual echo movement to the northeast at 35 knots.

Line or area movement to the east to 20 knots. (Note: A half flag represents 5 knots, and full flag 10 knots.)

Additional information included on these charts, when appropriate, are areas which indicate a severe weather watch in effect. These areas are depicted by dashed line rectangle or square. Also, if reports from a particular radar station do not appear on the chart, a symbol is placed on the chart to explain the reason for no echoes. The following are the symbols used:

Symbol	Meaning
NE	No echo (equipment operating but no echoes observed)
NA	Observation not available
OM	Equipment out for maintenance

Symbols indicating no echoes

Significant Weather Prognostic Charts: Significant Weather Prognostic Charts portray forecast weather to assist in weather briefing. They are issued for both domestic and international flight planning. Only the domestic type will be discussed. The National Weather Service issues the Domestic Prognostic Charts in four panels as illustrated in Fig. 5-55.

The two lower panels are the 12- and 24-hour surface progs. The two upper panels are the 12- and 24-hour progs and portray significant weather from

153

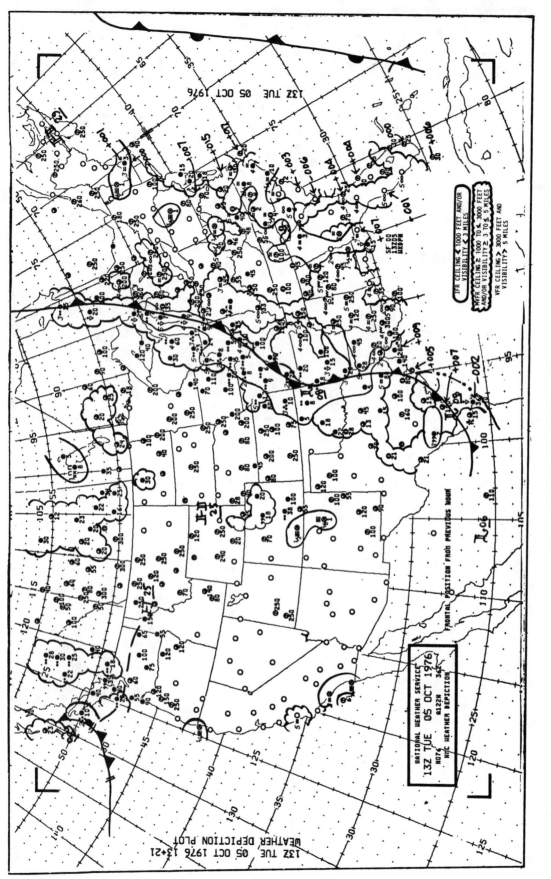

Figure 5-49. *Weather depiction chart.*

154

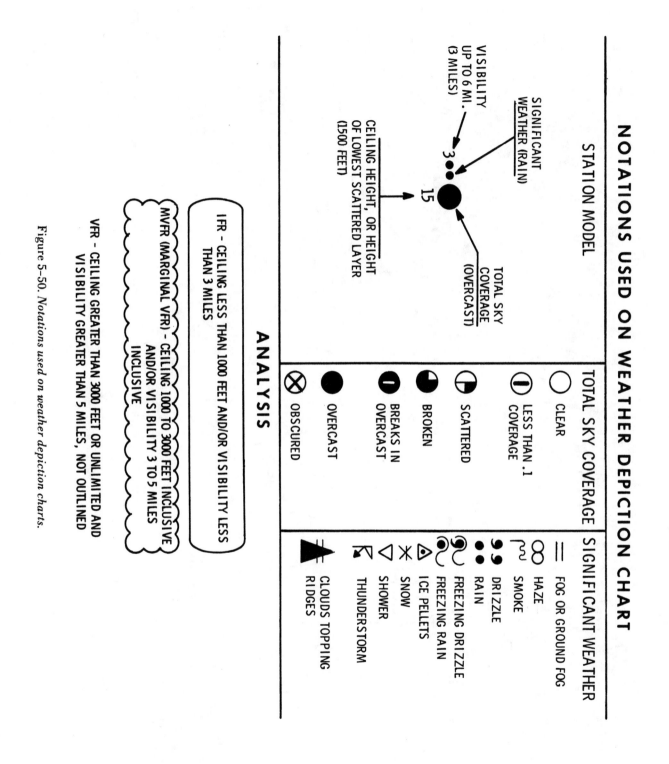

Figure 5-50. *Notations used on weather depiction charts.*

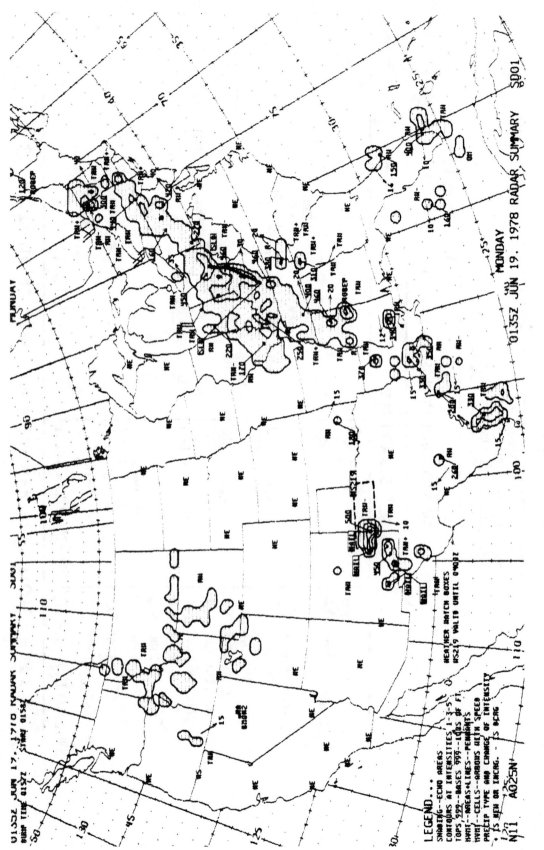

Figure 5-51. *Radar summary chart.*

Symbol	Echo Intensity	Estimated precipitation
—	Weak	Light
(none)	Moderate	Moderate
+	Strong	Heavy
++	Very strong	Very heavy
X	Intense	Intense
XX	Extreme	Extreme
U	Unknown	Unknown

Symbol	Meaning
+	Increasing
–	Decreasing
NC	No change
NEW	New

Figure 5-52. *Symbols used for echo intensity and trend.*

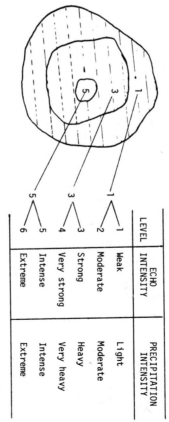

SYMBOL	MEANING	CALLED
(arrow)	A line of echoes	Line
SLD	Over 9/10 coverage	Solid
WS999	Thunderstorm watch area	Thunderstorm watch area
WT999	Tornado watch area	Tornado watch area

LEVEL	ECHO INTENSITY	PRECIPITATION INTENSITY
1	Weak	Light
2	Moderate	Moderate
3	Strong	Heavy
4	Very strong	Very heavy
5	Intense	Intense
6	Extreme	Extreme

NOTE: The numbers representing the intensity level do not appear on the chart. Beginning from the first contour line, bordering the area, the intensity level is 1 - 2; second contour is 3 - 4; and third contour is 5 - 6.

Figure 5-53. *Echo coverage symbols on the radar summary chart.*

the surface up to 400 millibars (24,000 feet). These charts show conditions as they are forecast to be at the valid time of the chart.

The two surface prog panels use the standard symbols to depict fronts and pressure centers. The movement of each pressure center is indicated by an arrow showing the direction and a number indicating speed in knots.

Isobars depicting forecast pressure patterns are included on some 24-hour surface progs.

The surface prog outlines areas in which precipitation and/or thunderstorms are forecast. Smooth lines enclose areas of expected precipitation, either continuous or intermittent; dashed lines enclose areas of expected

157

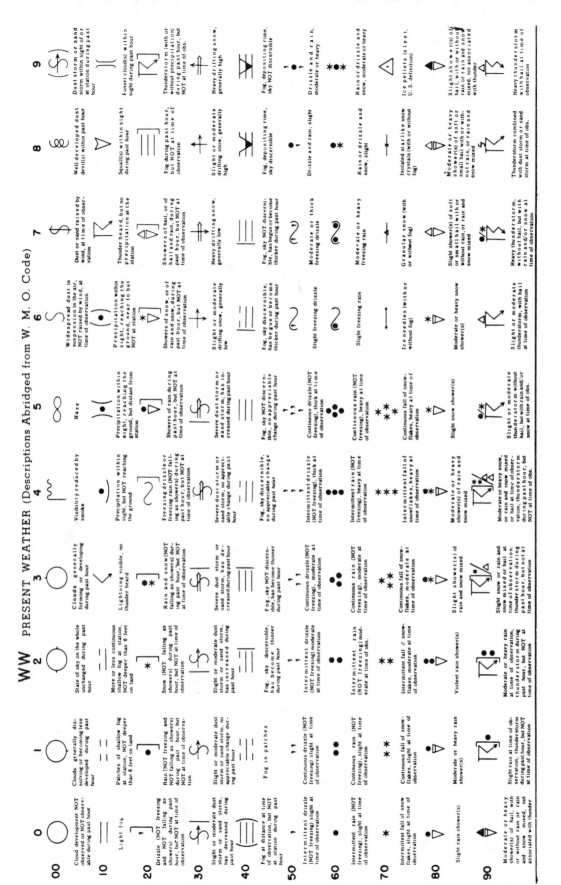

Figure 5-54. *Weather symbols.*

158

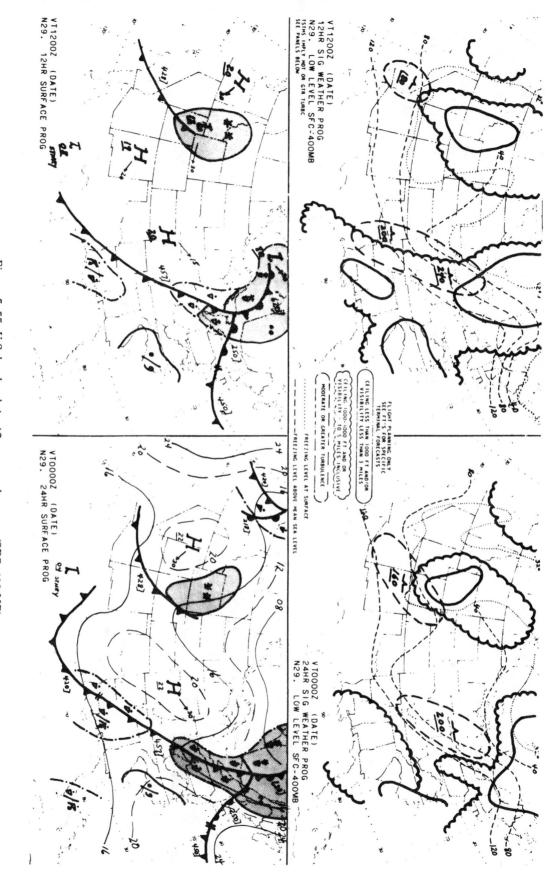

Figure 5-55. *U.S. low-level significant weather prog (SFC–400 MB).*

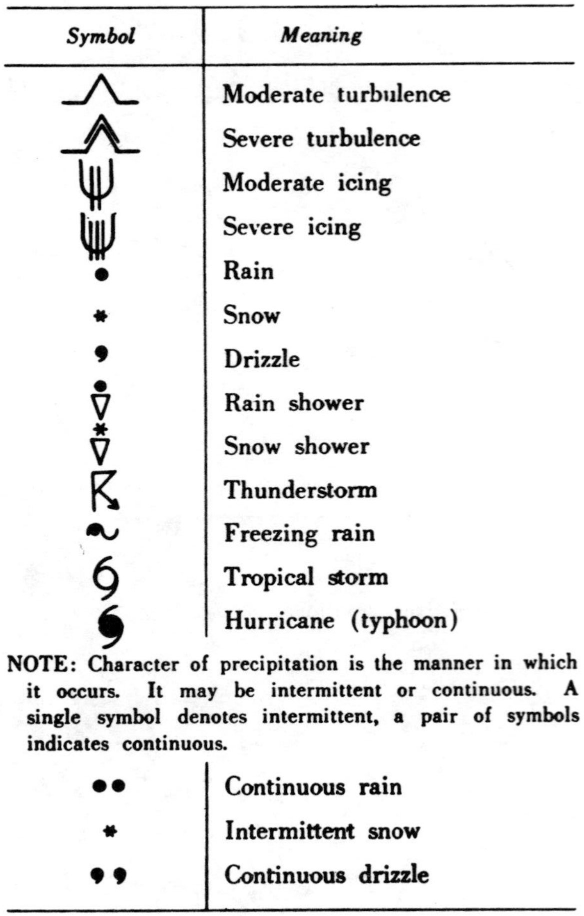

showers or thunderstorms. The symbols indicate the type and character of the precipitation. The area is shaded if the precipitation is expected to cover half or more of the area, and unshaded if less than half.

Fig. 5-56 shows the symbols used on the prog charts and explains their meanings.

The upper panels depict forecasts of significant weather such as ceiling, visibility, turbulence, and freezing level. A legend is placed between the panels which explains methods of depicting weather information on the Significant Weather Prog.

In summarizing this chapter on weather services, pilots are encouraged to use the wealth of information made available for aviation purposes. Although some airports do not have all weather services mentioned in this chapter, a phone call or radio transmission to the nearest facility equipped with weather services will inform the pilot about weather conditions and add much to the safety of the flight.

Symbol	Meaning
	Moderate turbulence
	Severe turbulence
	Moderate icing
	Severe icing
●	Rain
*	Snow
●	Drizzle
▷	Rain shower
▷*	Snow shower
◁	Thunderstorm
	Freezing rain
	Tropical storm
	Hurricane (typhoon)

NOTE: Character of precipitation is the manner in which it occurs. It may be intermittent or continuous. A single symbol denotes intermittent, a pair of symbols indicates continuous.

Symbol	Meaning
●●	Continuous rain
*	Intermittent snow
●●	Continuous drizzle

Figure 5-56. *Some standard weather symbols.*

160

CHAPTER VI—BASIC CALCULATIONS USING NAVIGATIONAL COMPUTERS OR ELECTRONIC CALCULATORS

Almost any type of navigation requires the solution of simple arithmetical problems involving time, speed, distance, and fuel consumption. In addition, the effect of wind on the airplane must be known and included in these calculations. To solve such problems quickly and with reasonable accuracy a variety of navigational computers and electronic calculators are available for use. Most manufacturers furnish instructional booklets explaining the use of their particular computer or calculator.

The following navigational problems are intended primarily for practice. Any navigational computer or electronic calculator can be used to solve these problems by following the procedures outlined in the manufacturers' instructions.

The answers derived from different computers or calculators may vary slightly from answers given in this handbook. The amount of variation will depend upon type of computer or calculator used. However, if the procedure used to solve these problems is correct this variation should be negligible.

Determining En Route Time for a Flight In preflight planning, the pilot should compute the estimated groundspeed based on forecast winds aloft. After computing the groundspeed, along with the distance flown, these values can be used to determine the total time for the flight.

Practice Problems. If a groundspeed of (a) _____ is maintained, how much time will be required to fly a distance of (b) _____? Substitute the following quantities in blanks (a) and (b) and solve:

	(a)		(b)	
	(knots)	(m.p.h.)	(NM)	(miles)
1.	107	123	250	288
2.	123	142	320	370
3.	139	160	205	236
4.	152	175	365	420
5.	157	181	68	78
6.	135	156	43	49

NOTE: Correct answers are shown below:
1. 2 hrs. 20 min.
2. 2 hrs. 36 min.
3. 1 hr. 28 min.
4. 2 hrs. 24 min.
5. 26 min.
6. 19 min.

Determining Groundspeed During Flight During flight a pilot may wish to determine the actual groundspeed. Once on course at cruising altitude, airspeed, and power, the pilot checks the time when passing a known checkpoint. A constant heading is maintained and the time is checked when passing over a second checkpoint. The distance is measured between the checkpoints on the chart and the length of time taken to fly this distance is noted. With these two figures, groundspeed can be determined.

Practice Problems. If an airplane flies (a) _____ miles in (b) _____ minutes, what is its groundspeed? Substitute the following quantities in blanks (a) and (b) and solve:

	(a)		(b)
	(NM)	(miles)	(minutes)
1.	30	34.5	12
2.	10	11.5	5
3.	13	15.0	8
4.	27	31.0	15
5.	32	37.0	16
6.	27	31.0	10.5

NOTE: Correct answers are shown below.
1. 150 knots (172 m.p.h.).
2. 120 knots (138 m.p.h.).
3. 98 knots (112 m.p.h.).
4. 108 knots (124 m.p.h.).
5. 120 knots (138 m.p.h.).
6. 154 knots (177 m.p.h.).

Determining Total Flight Time Available One kind of fuel consumption problem a pilot should be able to solve is determining the total flight time available based on the fuel load.

Practice Problems. If an airplane carries (a) _____ gallons of usable fuel and the rate of fuel consumption is (b) _____ gallons per hour, what is the total flight time available? Substitute the following quantities in blanks (a) and (b) and solve:

	(a) (gals.)	(b) (g.p.h.)
1.	36	9
2.	45	8.5
3.	37	7
4.	55	13
5.	18	6.3

NOTE: *Correct answers are shown below.*
1. 4 hrs.
2. 5 hrs. 18 min.
3. 5 hrs. 17 min.
4. 4 hrs. 14 min.
5. 2 hrs. 51 min.

Determining Total Fuel to be Used on a Flight A pilot should also be able to determine how much fuel will be used during a flight.

Practice Problems. How much fuel will be used during a flight of (a) _____ if the rate of fuel consumption is (b) _____ gallons per hour? Substitute the following quantities in blanks (a) and (b) and solve:

	(a) (time)	(b) (g.p.h.)
1.	3 hrs.	7
2.	3 hrs. 30 min.	11
3.	2 hrs. 20 min.	9.5
4.	4 hrs. 15 min.	10.3
5.	5 hrs. 10 min.	13.7

NOTE: *Correct answers are shown below:*
1. 21 gals.
2. 38.5 gals.
3. 22.2 gals.
4. 43.7 gals.
5. 71 gals.

Determining True Airspeed To compute the groundspeed and heading correctly, a pilot must determine the true airspeed.

Practice Problems. Find the TAS when the following pressure altitudes, temperatures, and IAS are given:

	Altitude (ft.)	Temperature (°C.)	IAS (knots)	IAS (m.p.h.)
1.	5,000	0	120	138
2.	4,000	−10	145	167
3.	4,000	+10	145	167
4.	7,500	+10	145	167
5.	6,500	−15	150	172

NOTE: *Correct answers are shown below.*
1. 129 knots (148 m.p.h.).
2. 149 knots (172 m.p.h.).
3. 155 knots (179 m.p.h.).
4. 162 knots (186 m.p.h.).
5. 158 knots (182 m.p.h.).

Converting Knots to Miles Per Hour Since the winds aloft forecasts give the windspeed in knots, a pilot should be able to convert knots to statute miles per hour, if desired, to determine the correct heading and groundspeed. Since "knots" actually means "nautical miles per hour," the problem is converting nautical miles to statute miles.

Practice Problems. If the following windspeeds are given in knots, find the speed in statute miles per hour.
1. 20 knots.
2. 16 knots.
3. 26 knots.
4. 40 knots.
5. 47 knots.

NOTE: *Correct answers are shown below.*
1. 23 m.p.h.
2. 18.4 m.p.h.
3. 30 m.p.h.
4. 46 m.p.h.
5. 54 m.p.h.

Solution of a Wind Triangle Problem. The problem illustrated is the one the pilot will encounter most often and should solve before taking off on a cross-country flight. This is the problem in which the true course, true airspeed, wind direction, and windspeed are known, and the pilot wants to find the true heading and groundspeed.

Practice Problems. With the TAS, TC, windspeed, and wind direction given find the wind correction angle, true heading, and groundspeed.

TAS	TC	WIND		WCA			
		SPEED FROM		R+	L−	TH	GS
125	010°	35 knots	150°				
122	267°	42 knots	087°				
144	045°	15 knots	315°				
137	140°	36 knots	230°				
135	120°	20 knots	060°				

NOTE: Correct answers are shown below.

	WCA	TH	GS (Knots)
1.	10°R	020°	150
2.	0	267°	164
3.	6°L	039°	143
4.	15°R	155°	134
5.	8°L	112°	123

CHAPTER VII—NAVIGATION

Most pilots take pride in their ability to navigate with precision. To execute a flight which follows a predetermined plan directly to the destination and arrive safely with no loss of time because of poor navigation is a source of real satisfaction.

Lack of navigational skill could lead to unpleasant and sometimes dangerous situations in which adverse weather, approaching darkness, or fuel shortage may force the pilot to attempt a landing under hazardous conditions.

At one time, navigation was considered a difficult art. Recent improvements in instruments, aeronautical charts, pilot techniques, and navigation aids have enabled pilots to plan their flights with confidence and to reach their destinations according to the plan. The prime requirement for success in navigation is a knowledge of a few simple facts and the ability to exercise good judgment based upon these facts.

This discussion of navigation is directed primarily to those pilots who do not hold an instrument rating. As more flight experience is gained, more knowledge of navigation may be desired. This can be obtained from further study using other materials. The primary purpose here it to provide information of practical value when flying under Visual Flight Rules (VFR).

To navigate successfully a pilot should be able to determine the aircraft's position relative to the earth's surface at any time, and navigation or position finding is accomplished by one or more of the following methods:

1. Pilotage (by reference to visible landmarks).
2. Dead reckoning (by computing direction and distance from a known position).
3. Radio navigation (by use of radio aids).
4. Celestial navigation (by reference to the sun, moon, or other celestial bodies).

The basic form of navigation for the beginning pilot is pilotage. This type should be mastered first. An understanding of the principles of dead reckoning, however, will enable the pilot to make necessary calculations of flight time and fuel consumption. The ever-increasing use of radio equipment makes it highly desirable for the pilot to have a thorough knowledge of the use of radio aids to navigation and communications. Celestial navigation is not used to any extent in small aircraft and will not be discussed in this handbook.

Aeronautical Charts The NATIONAL OCEAN SURVEY (NOS) publishes and sells aeronautical charts of the United States and of foreign areas. The type of charts most commonly used by VFR pilots are:

1. Sectional and VFR Terminal Area Charts. The scale of the Sectional Chart is 1:500,000 (1 inch = 6.86 NM), and the scale of a VFR Terminal Area Chart is 1:250,000 (1 inch = 3.43 NM).

These charts are designed for visual navigation of slow/medium speed aircraft. The topographical information featured on these charts consists of the portrayal of relief and a judicious selection of visual checkpoints used for VFR flight. The checkpoints include populated places, drainage, roads, railroads, and other distinctive landmarks.

The aeronautical information on sectional charts includes visual and radio aids to navigation, airports, controlled airspace, restricted areas, obstructions, and related data.

The VFR Terminal Area Charts depict the airspace designated as "Terminal Control Area" which provides for the control or segregation of all aircraft within these areas. The information found on these charts is similar to that found on sectional charts. Both the Sectional and VFR Terminal Area Charts are revised semi-annually.

2. World Aeronautical Charts. The scale of these charts is 1:1,000,000 (1 inch = 13.7 NM). The purpose of these charts is to provide a standard series of aeronautical charts, covering land areas of the world, at a size and scale convenient for navigation by moderate speed aircraft.

The topographical information included on these charts is cities and towns, principal roads, railroads, distinctive landmarks, drainage and relief. The latter is shown by spot elevations, contours, and gradient color tints.

The aeronautical information includes visual and radio aids to navigation, airports, airways, restricted areas, obstructions, and other pertinent data. These charts are revised and updated annually.

3. Aeronautical Planning Charts. The scale of Planning Charts is 1:2,333,232 (1 inch = 32 NM). The purpose of this chart is to fulfill the requirements of preflight planning of long flights. Selected key points may be transferred to more detailed local charts for actual flight use.

A free catalog listing Aeronautical Charts and related publications including prices and instructions for ordering is available upon request from:

Distribution Division (C–44)
National Ocean Survey
Riverdale, Maryland 20840

Fig. 7-1 shows a map of the conterminous United States upon which is superimposed a key identifying the Sectional Charts and indicating the area covered by each.

SECTIONAL AND VFR TERMINAL AREA CHARTS

CONTERMINOUS U.S. AND HAWAIIAN ISLANDS

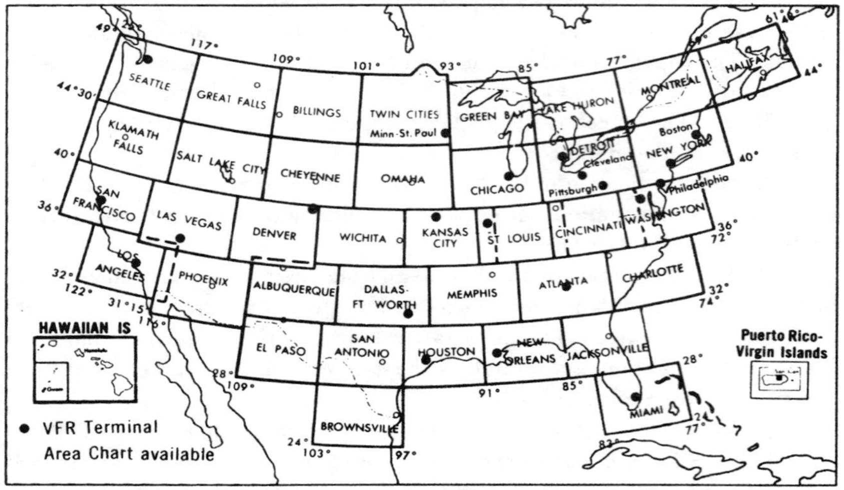

Figure 7-1. Index of sectional and VFR terminal area charts.

It is vitally important that pilots check the publication date on each aeronautical chart to be used. *Obsolete charts should be discarded and replaced by new editions.* This is important because revisions in aeronautical information occur constantly. These revisions include changes in radio frequencies, new obstructions, temporary or permanent closing of certain runways and airports, and other temporary or permanent hazards to flight. To make certain that the sectional aeronautical chart being used is up-to-date, refer to the National Ocean Survey (NOS) Aeronautical Chart Bulletin in the Airport/Facility Directory. This Bulletin provides the VFR pilot with the essential information necessary to update and maintain current charts. It lists the major changes in aeronautical information that have occurred since the last publication date of each chart. Specifically:

1. Changes to controlled airspace and to special use airspace that present hazardous conditions or impose restrictions on the pilot.

2. Major changes to airports and to radio navigational facilities. When a sectional chart is revised and published, the old charts are removed from the Airport/Facility Directory.

To ensure that the latest charts are used, regularly revised lists entitled "Dates of Latest Prints" are published and made available from the same source as the charts.

Sectional Aeronautical Charts While studying this chapter, use the Dallas-Ft. Worth Sectional Chart which is folded inside the back cover of this booklet. The pilot should have little difficulty in reading these aeronautical charts. In many respects, they are similar to automobile road maps. The chart name or title appears on each chart. The chart legend lists various aeronautical symbols as well as information concerning terrain and contour elevations. By referring to the legend, aeronautical, topographical, and obstruction symbols (such as radio and television towers) may be identified. Many landmarks which can be easily recognized from the air, such as stadiums, race tracks, pumping stations, and refineries, are identified by brief descriptions adjacent to small black squares marking their exact locations. Oil fields are shown by small circles, water tanks and oil tanks are shown by small black circles and labeled accordingly. Many items are exaggerated on the chart in order to be more readily seen.

Remember, however, that certain information on aeronautical charts may be obsolete, depending on the date published. Check the information concerning next edition scheduled date of the chart to determine that the latest edition is being used.

Relief The elevation of land surface, *relief,* is shown on the aeronautical charts by brown contour lines drawn at 250-foot intervals. These areas are emphasized by various tints, as indicated in the color legend appearing on each chart.

The manner in which contours express elevation, form, and degree of slope is shown in Fig. 7-2. The sketch in the upper part of the figure represents a river valley between two hills. In the foreground is the sea, and a bay partially enclosed by a hooked sandbar. On each side of the valley is a terrace into which small streams have eroded narrow gullies. The hill to the right

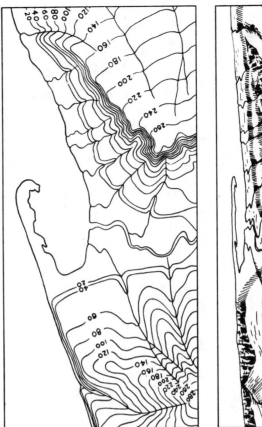
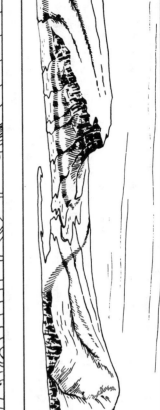

Figure 7-2. Altitude, form, and slope of terrain as indicated by contour lines and numerals.

has a rounded summit and gently sloping spurs separated by ravines. The spurs are cut off sharply at their lower ends by a sea cliff. The hill at the left terminates abruptly at the valley in a steep and almost vertical bluff, from which it slopes gradually away and forms an inclined tableland traversed by a few shallow gullies. Each of these features is represented directly beneath its position in the sketch by contour lines. In Fig. 7-2 the contours represent successive differences in elevation of 20 feet—that is, the contour interval is 20 feet. A small interval was used to better illustrate the terrain features that may be visualized through contours.

Aeronautical Data The aeronautical information on the sectional charts is for the most part self-explanatory. Information concerning very high frequency (VHF) radio facilities such as tower frequencies, omnidirectional radio ranges (VOR), and other VHF communications frequencies is shown in blue. A narrow band of blue tint is also used to indicate the center lines of Victor Airways (VOR civil airways between omnirange stations). Low frequency—medium frequency (LF/MF) radio facilities are shown in magenta (purplish shade of red).

In most instances, FAA navigational aids can be identified by call signs broadcast in International Morse Code. VOR stations and Nondirectional Radiobeacons use three-letter identifiers which are printed on the chart near the symbol representing the radio facility.

Runway patterns are shown for all airports having permanent hard-surfaced runways. These patterns provide for positive identification as landmarks. All recognizable runways, including those that may be closed, are shown to aid in visual identification. Airports and information pertaining to airports having an airport traffic area (operating control tower) are shown in blue. All other airports and information pertaining to these airports are shown in magenta adjacent to the airport symbol which is also in magenta.

The symbol for obstructions is another important feature. The elevation of the top of obstructions above sea level is given in blue figures (without parentheses) adjacent to the obstruction symbol. Immediately below this set of figures is another set of lighter blue figures enclosed in parentheses which represents the height of the top of the obstruction above ground level. Obstructions which extend less than 1,000 feet above the terrain are shown by one type of symbol and those obstructions that extend 1,000 feet or higher above ground level are indicated by a different symbol (see sectional chart). Specific elevations of certain high points in terrain are shown on charts by dots accompanied by small black figures indicating the number of feet above sea level.

The chart also contains larger bold face blue numbers which denote Maximum Elevation Figures (MEF). These figures are shown in quadrangles bounded by ticked lines of latitude and longitude, and are represented in THOUSANDS and HUNDREDS of feet above mean sea level. The MEF is based on information available concerning the highest known feature in each quadrangle, including terrain and obstructions (trees, towers, antennas, Etc.).

An explanation for most symbols used on aeronautical charts appears in the margin of the chart. Additional information appears at the bottom of the chart.

Airport and Air Navigation Lighting and Marking Aids On sectional charts, lighting aids are shown by a blue star or dot along with certain other symbols and coded information. For the most part, the lighting aids represented on sectional charts pertain to rotating or flashing lights. The color or color combination displayed by a particular beacon and/or its auxiliary lights indicates whether the beacon identifies a landing place, landmark, point of the Federal Airways, or hazard. Coded flashes of the auxiliary lights, if employed, further identify the beacon site. A detailed description of airport and air navigation lighting aids can be found in the Airport/Facility Directory.

Meridians and Parallels   The equator is an imaginary circle equidistant from the poles of the earth. Circles parallel to the equator (lines running east and west) are *parallels of latitude.* They are used to measure degrees of

latitude north or south of the equator. The angular distance from the equator to the pole is one-fourth of a circle or 90°. The 48 conterminous states of the United States are located between 25° and 49° N. latitude. The arrows in Fig. 7-3 labeled "LATITUDE" point to lines of latitude.

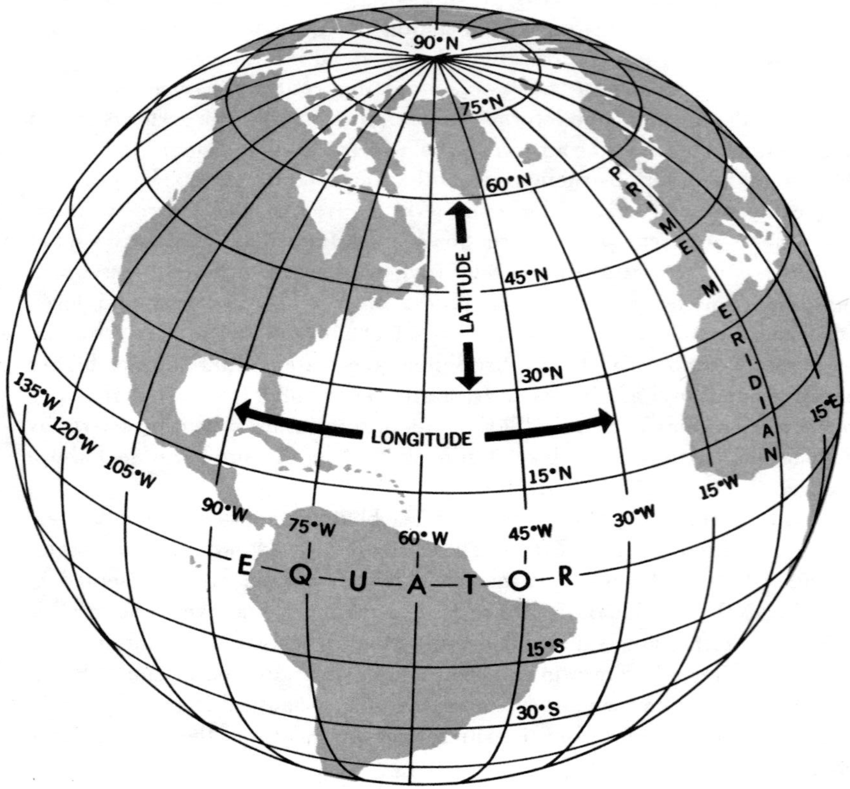

Meridians of longitude are drawn from the North Pole to the South Pole and are at right angles to the equator. The "Prime Meridian" which passes through Greenwich, England, is used as the zero line from which measurements are made in degrees east and west to 180°. The 48 conterminous states of the United States are between 67° and 125° W. longitude. The arrows in Fig. 7-3 labeled "LONGITUDE" point to lines of longitude.

Any specific geographical point can thus be located by reference to its longitude and latitude. Washington, D.C., for example, is approximately 39° N. latitude, 77° W. longitude. Chicago is approximately 42° N. latitude, 88° W. longitude (Fig. 7-3).

The meridians are also useful for designating time belts. A day is defined as the time required for the earth to make one complete revolution of 360°. Since the day is divided into 24 hours, the earth revolves at the rate of 15° an hour. Noon is the time when the sun is directly above a meridian; to the west of that meridian is forenoon, to the east is afternoon.

The standard practice is to establish a time belt for each 15° of longitude. This makes a difference of exactly 1 hour between each belt. In the United States there are four time belts—Eastern (75°), Central (90°), Mountain (105°), and Pacific (120°). The dividing lines are somewhat irregular because communities near the boundaries often find it more convenient to use time designations of neighboring communities or trade centers.

Fig. 7-4 shows the time zones in the United States. When the sun is directly above the 90th meridian, it is noon Central Standard Time. At the same time it will be 1 p.m. Eastern Standard time, 11 a.m. Mountain Standard Time, and 10 a.m. Pacific Standard Time. When "daylight saving" time is in effect, generally between the last Sunday in April and the last Sunday in October, the sun is directly above the 75th meridian at noon, Central Daylight Time.

These time zone differences must be taken into account during long flights eastward—especially if the flight must be completed before dark. Remember, an hour is lost when flying eastward from one time zone to another, or perhaps even when flying from the western edge to the eastern edge of the same time zone. Determine the time of sunset at the destination by consulting the FSS or National Weather Service Station and take this into account when planning an eastbound flight.

Figure 7-3. *Meridians and parallels—the basis of measuring time, distance, and direction.*

Figure 7-4. *When the sun is directly above the meridian, the time at points on that meridian is noon. This is the basis on which time zones are established.*

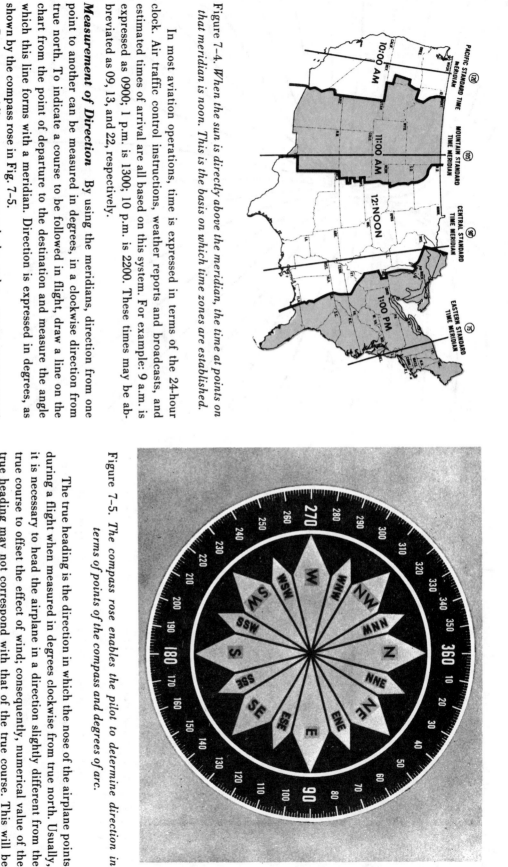

In most aviation operations, time is expressed in terms of the 24-hour clock. Air traffic control instructions, weather reports and broadcasts, and estimated times of arrival are all based on this system. For example: 9 a.m. is expressed as 0900; 1 p.m. is 1300; 10 p.m. is 2200. These times may be abbreviated as 09, 13, and 22, respectively.

Measurement of Direction. By using the meridians, direction from one point to another can be measured in degrees, in a clockwise direction from true north. To indicate a course to be followed in flight, draw a line on the chart from the point of departure to the destination and measure the angle which this line forms with a meridian. Direction is expressed in degrees, as shown by the compass rose in Fig. 7-5.

Because meridians converge toward the poles, course measurement should be taken at a *meridian near the midpoint* of the course rather than at the point of departure. The course measured on the chart is known as the *true course*. This is the direction measured by reference to a meridian or true north. It is the direction of *intended* flight as measured in degrees clockwise from true north. As shown in Fig. 7-6, the direction from A to B would be a true course of 065°, whereas the return trip (sometimes called the reciprocal) would be a true course of 245°.

Figure 7-5. *The compass rose enables the pilot to determine direction in terms of points of the compass and degrees of arc.*

The true heading is the direction in which the nose of the airplane points during a flight when measured in degrees clockwise from true north. Usually, it is necessary to head the airplane in a direction slightly different from the true course to offset the effect of wind; consequently, numerical value of the true heading may not correspond with that of the true course. This will be discussed more fully in subsequent sections in this chapter. For the purpose of this discussion, assume a no-wind condition exists under which heading and course would coincide. Thus, for a true course of 065°, the true heading would be 065°. To use the compass accurately, however, corrections must be made for magnetic variation and compass deviation.

Variation. Variation is the angle between true north and magnetic north. It is expressed as *east variation* or *west variation* depending upon whether

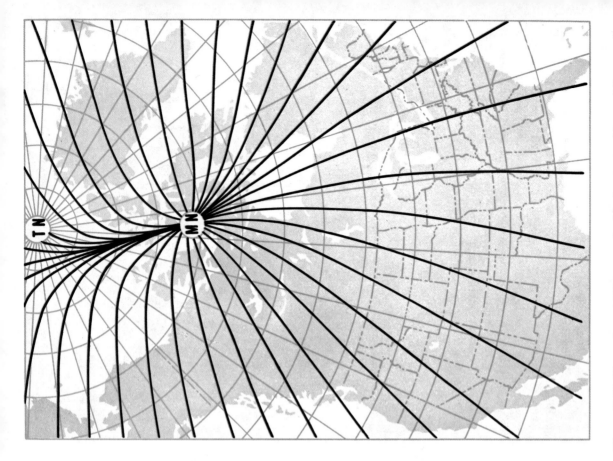

Figure 7-7. *Magnetic meridians are in black, geographic meridians and parallels are in blue. Variation is the angle between a magnetic and geographic meridian.*

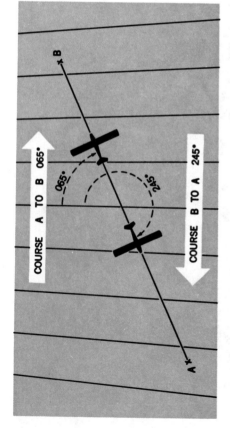

Figure 7-6. *Courses are determined by reference to meridians on aeronautical charts.*

magnetic north (MN) is to the east or west of true north (TN), respectively.

The north magnetic pole is located close to latitude 71° N., longitude 96° W.—about 1,300 miles from the geographic or true north pole, as indicated in Fig. 7-7. If the earth were uniformly magnetized, the compass needle would point toward the magnetic pole, in which case the variation between true north (as shown by the geographical meridians) and magnetic north (as shown by the magnetic meridians) could be measured at any intersection of the meridians.

Actually, the earth is not uniformly magnetized. In the United States the needle usually points in the general direction of the magnetic pole but it may vary in certain geographical localities by many degrees. Consequently, the exact amount of variation at thousands of selected locations in the United States has been carefully determined by the National Ocean Survey. The amount and the direction of variation, which change slightly from time to time, are shown on most aeronautical charts as broken red lines, *called isogonic lines*, which connect points of equal magnetic variation. (The line connecting points at which there is no variation between true north and magnetic north is the *agonic line*.) An isogonic chart is shown in Fig. 7-8. Minor bends and turns in the isogonic and agonic lines probably are caused by unusual geological conditions affecting magnetic forces in these areas.

On the west coast of the United States, the compass needle points to the east of true north; on the east coast the compass needle points to the west of true north. Zero degree variation exists on the agonic line which runs roughly through Lake Michigan, the Appalachian Mountains, and off the coast of Florida, where magnetic north and true north coincide. (Compare Figs. 7-8 and 7-9.)

Because courses are measured in reference to geographical meridians which point toward true north, and these courses are maintained by reference to the compass which points along a magnetic meridian in the general direction of magnetic north, the true direction must be converted into magnetic direction for the purpose of flight. This conversion is made by adding or subtracting the *variation* which is indicated by the nearest isogonic line on the chart. The true heading, when corrected for variation, is known as *magnetic heading*.

At Providence, R.I., the variation is shown as "14° W." This means that magnetic north is 14° west of true north. If a true heading of 360° is to be flown, 14° must be added to 360°, which results in a magnetic heading of 014°. The same correction for variation must be applied to the true heading to obtain any magnetic heading at Providence, or at any point close to the isogonic line "14° W." Thus, to fly east, a magnetic heading of 104° (090° + 14°) would be flown. To fly south, the magnetic heading would be 194° (180° + 14°). To fly west, it would be 284° (270 + 14°). To fly a true heading of 060°, a magnetic heading of 074° (060 + 14°) would be flown (Fig. 7-10).

Suppose a flight is to take place near Denver, Colo. The isogonic line shows the variation to be "14° E." This means that magnetic north is 14° to the east of true north. Therefore, to fly a true heading of 360°, 14° must be subtracted from 360°, and a magnetic heading of 346° would be flown. Again the 14° would be subtracted from the appropriate true heading to obtain the magnetic heading at any point close to the isogonic line "14° E." Thus, to fly 090° a magnetic heading of 076° (090° − 14°) would be used. To fly 180°, the magnetic heading would be 166° (180° − 14°). To fly west it would be 256° (270° − 14°). To fly a true heading of 060°, a magnetic heading of 046° (060° − 14°) would be used (Fig. 7-10).

To summarize: To convert TRUE course or heading to MAGNETIC course or heading, note the variation shown by the nearest isogonic line. If variation is west, add; if east, subtract.

Some method should be devised for remembering whether to add or subtract variation. The following may be helpful: *West is best* (add); *East is least* (subtract).

Deviation   Determining the magnetic heading is an intermediate step necessary to obtain the correct compass reading for the flight. To determine compass heading a correction for deviation must be made. Because of magnetic influences within the airplane such as electrical circuits, radio, lights, tools, engine, magnetized metal parts, etc., the compass needle is frequently deflected from its normal reading. This deflection is *deviation*. The deviation is different for each airplane; it also may vary for different headings in the same airplane. For instance, if magnetism in the engine attracts the north end of the compass, there would be no effect when the plane is on a heading of magnetic north. On easterly or westerly headings, however, the compass indications would be in error, as shown in Fig. 7-11. Magnetic attraction can come from many other parts of the airplane; the assumption of attraction in the engine is merely used for the purpose of illustration.

Some adjustment of the compass, referred to as *compensation*, can be made to reduce this error, but the remaining correction must be applied by the pilot.

Proper compensation of the compass is best performed by a competent technician. Since the magnetic forces within the airplane change, because of landing shocks, vibration, mechanical work, or changes in equipment, the

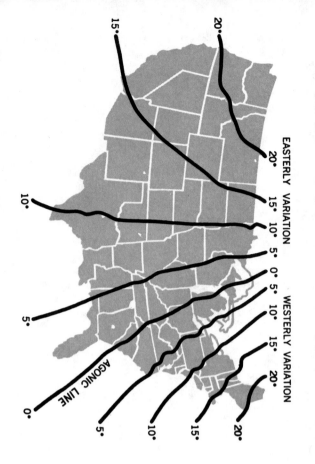

EASTERLY VARIATION   WESTERLY VARIATION   AGONIC LINE

20° 15° 10° 5° 0° 5° 10° 15° 20°

Figure 7-8. *A typical isogonic chart. The black lines are isogonic lines which connect geographic points with identical magnetic variation.*

171

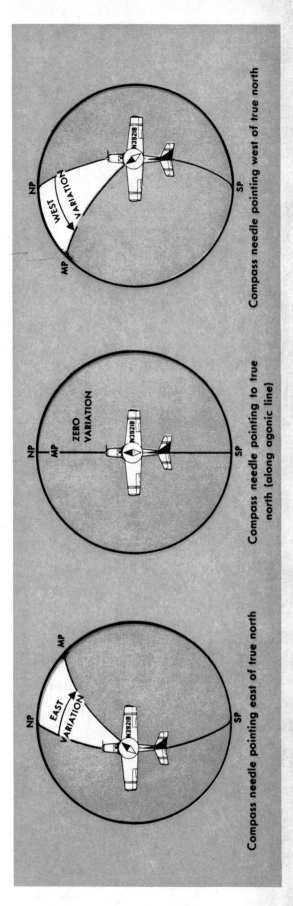

Figure 7-9. In an area of west variation a compass needle points west of true north. In an area of zero variation it points to true north. In an area of east variation it points east of true north.

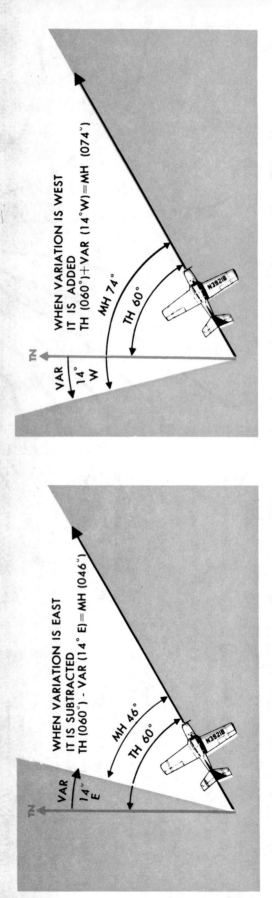

Figure 7-10. The relationship between true heading, magnetic heading, and variation in areas of east and west variation.

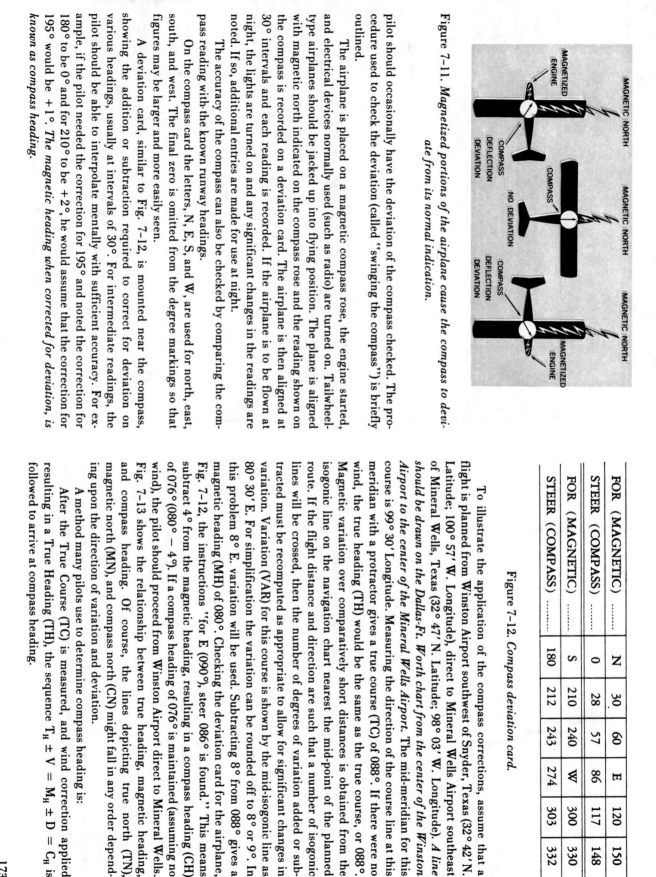

Figure 7-11. *Magnetized portions of the airplane cause the compass to deviate from its normal indication.*

pilot should occasionally have the deviation of the compass checked. The procedure used to check the deviation (called "swinging the compass") is briefly outlined.

The airplane is placed on a magnetic compass rose, the engine started, and electrical devices normally used (such as radio) are turned on. Tailwheel-type airplanes should be jacked up into flying position. The plane is aligned with magnetic north indicated on the compass rose and the reading shown on the compass is recorded on a deviation card. The airplane is then aligned at 30° intervals and each reading is recorded. If the airplane is to be flown at night, the lights are turned on and any significant changes in the readings are noted. If so, additional entries are made for use at night.

The accuracy of the compass can also be checked by comparing the compass reading with the known runway headings.

On the compass card the letters, N, E, S, and W, are used for north, east, south, and west. The final zero is omitted from the degree markings so that figures may be larger and more easily seen.

A deviation card, similar to Fig. 7-12, is mounted near the compass, showing the addition or subtraction required to correct for deviation on various headings, usually at intervals of 30°. For intermediate readings, the pilot should be able to interpolate mentally with sufficient accuracy. For example, if the pilot needed the correction for 195° and noted the correction for 180° to be 0° and for 210° to be +2°, he would assume that the correction for 195° would be +1°. *The magnetic heading when corrected for deviation, is known as compass heading.*

To illustrate the application of the compass corrections, assume that a flight is planned from Winston Airport southwest of Snyder, Texas (32° 42' N. Latitude; 100° 57' W. Longitude), direct to Mineral Wells Airport southeast of Mineral Wells, Texas (32° 47' N. Latitude; 98° 03' W. Longitude). *A line should be drawn on the Dallas-Ft. Worth chart from the center of the Winston Airport to the center of the Mineral Wells Airport.* The mid-meridian for this course is 99° 30' Longitude. Measuring the direction of the course line at this meridian with a protractor gives a true course (TC) of 088°. If there were no wind, the true heading (TH) would be the same as the true course, or 088°.

Magnetic variation over comparatively short distances is obtained from the isogonic line on the navigation chart nearest the mid-point of the planned route. If the flight distance and direction are such that a number of isogonic lines will be crossed, then the number of degrees of variation added or subtracted must be recomputed as appropriate to allow for significant changes in variation. Variation (VAR) for this course is shown by the mid-isogonic line as 80° 30' E. For simplification the variation can be rounded off to 8° or 9°. In this problem 8° E. variation will be used. Subtracting 8° from 088° gives a magnetic heading (MH) of 080°. Checking the deviation card for the airplane, Fig. 7-12, the instructions "for E (090°), steer 086° is found." This means subtract 4° from the magnetic heading, resulting in a compass heading (CH) of 076° (080° − 4°). If a compass heading of 076° is maintained (assuming no wind), the pilot should proceed from Winston Airport direct to Mineral Wells.

Fig. 7-13 shows the relationship between true heading, magnetic heading, and compass heading. Of course, the lines depicting true north (TN), magnetic north (MN), and compass north (CN) might fall in any order depending upon the direction of variation and deviation.

A method many pilots use to determine compass heading is:

After the True Course (TC) is measured, and wind correction applied resulting in a True Heading (TH), the sequence $T_H \pm V = M_H \pm D = C_H$ is followed to arrive at compass heading.

Figure 7-12. *Compass deviation card.*

FOR (MAGNETIC)	N	30	60	E	120	150
STEER (COMPASS)	0	28	57	86	117	148
FOR (MAGNETIC)	S	210	240	W	300	330
STEER (COMPASS)	180	212	243	274	303	332

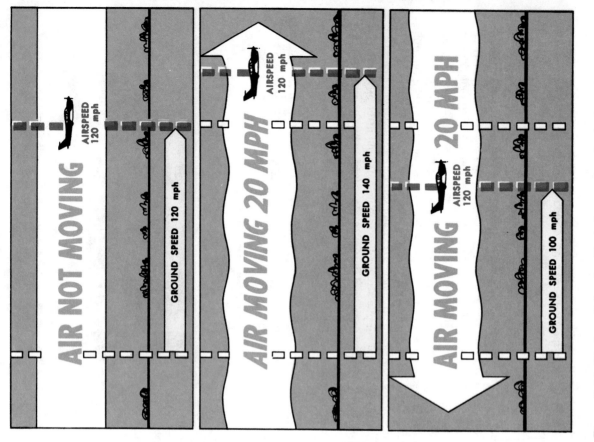

Figure 7-14. *Motion of the air affects the speed with which airplanes move over the earth's surface. Airspeed, the rate at which an airplane moves through the air, is not affected by air motion.*

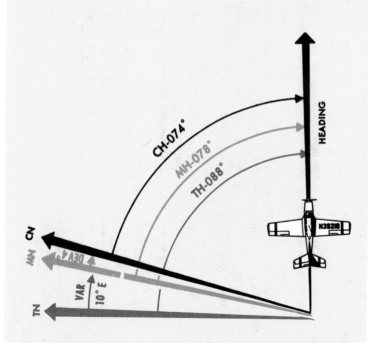

Figure 7-13. *Relationship between true, magnetic, and compass headings for a particular instance.*

Basic Calculations   The preceding discussion explained how to measure a true course on the aeronautical chart and how to make corrections for variation and deviation, but one important factor has not been considered—wind effect. While most of the discussion and calculations in this section use miles and miles-per-hour, it should be pointed out that there is a definite trend toward using nautical miles and knots in calculations.

Effect of Wind   As discussed in the study of the atmosphere, wind is a mass of air moving over the surface of the earth in a definite direction. When the wind is blowing from the north at 25 knots, it simply means that air is moving southward over the earth's surface at the rate of 25 nautical miles in 1 hour.

Under these conditions, any inert object free from contact with the earth will be carried 25 nautical miles southward in 1 hour. This effect becomes apparent when clouds, dust, toy balloons, etc., are observed being blown along by the wind. Obviously, an airplane flying within the moving mass of air will be similarly affected. Even though the airplane does not float freely with the wind, it moves through the air at the same time the air is moving over the ground, thus is affected by wind. Consequently, at the end of 1 hour of flight, the airplane will be in a position which results from a combination of these two motions: the movement of the air mass in reference to the ground, and the forward movement of the airplane through the air mass.

Actually, these two motions are independent. So far as the airplane's flight through the air is concerned, it makes no difference whether the mass of air through which the airplane is flying is moving or is stationary. A pilot flying in a 70-knot gale would be totally unaware of any wind (except for possible turbulence) unless the ground were observed. In reference to the ground, however, the airplane would appear to fly faster with a tailwind or slower with a headwind, or to drift right or left with a crosswind.

As shown in Fig. 7-14, an airplane flying eastward at an airspeed of 120 MPH in calm wind, will have a groundspeed exactly the same—120 MPH. If the mass of air is moving eastward at 20 MPH, the airspeed of the airplane will not be affected, but the progress of the airplane over the ground will be 120 plus 20, or a groundspeed of 140 MPH. On the other hand, if the mass of air is moving westward at 20 MPH, the airspeed of the airplane still remains the same, but groundspeed becomes 120 minus 20 or 100 MPH.

Assuming no correction is made for wind effect, if the airplane is heading eastward at 120 MPH, and the air mass moving southward at 20 MPH, the airplane at the end of 1 hour will be 120 miles east of its point of departure because of its progress through the air and 20 miles south because of the motion of the air (Fig. 7-15). Under these circumstances the airspeed remains 120 MPH, but the groundspeed is determined by combining the movement of the airplane with that of the air mass. Groundspeed can be measured as the distance from the point of departure to the position of the airplane at the end of 1 hour. The groundspeed can be computed by the time required to fly between two points a known distance apart. It also can be determined before flight by constructing a wind triangle, which will be explained later in this chapter.

The direction in which the plane is pointing as it flies is *heading*. Its actual path over the ground, which is a combination of the motion of the airplane and the motion of the air, is *track*. The angle between the heading and the track is *drift angle*. If the airplane's heading coincides with the true course and the wind is blowing from the left, the track will not coincide with the true course. The wind will drift the airplane to the right, so the track will fall to the right of the desired course or true course (Fig. 7-16).

By determining the amount of drift, the pilot can counteract the effect of the wind and make the track of the airplane coincide with the desired course. If the wind is moving across the course from the left, the airplane will drift to the right, and a correction must be make by heading the airplane sufficiently to the left to offset this drift. To state it another way, if the wind is from

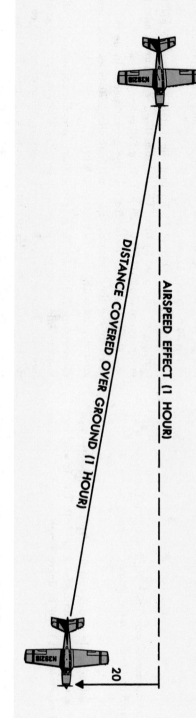

AIRSPEED EFFECT (1 HOUR)

DISTANCE COVERED OVER GROUND (1 HOUR)

20

Figure 7-15. *Airplane flightpath resulting from its airspeed and direction, and the windspeed and direction.*

pressed as the angle measured from a specific reference datum clockwise from 0° through 360° to the line.

HEADING is the direction in which the nose of the airplane points during flight.

TRACK is the actual path made over the ground in flight. (If proper correction has been made for the wind, track and course will be identical.)

DRIFT ANGLE is the angle between heading and track.

WIND CORRECTION ANGLE is correction applied to the course to establish a heading so that track will coincide with course.

AIRSPEED is the rate of the airplane's progress through the air.

GROUNDSPEED is the rate of the airplane's in-flight progress over the ground.

Calculating Time, Speed, Distance, and Fuel Consumption Before a cross-country flight, a pilot should make common calculations for time, speed, and distance, and the amount of fuel required. These calculations should present no difficulty.

Converting Minutes to Equivalent Hours Because speed is sometimes expressed in miles per hour, it frequently is necessary to convert minutes into equivalent hours when solving speed, time, and distance problems. To convert minutes to hours, divide by 60 (60 minutes = 1 hour). Thus, 30 minutes equals $\frac{30}{60} = 0.5$ hour. To convert hours to minutes, multiply by 60. Thus, 0.75 hour equals $0.75 \times 60 = 45$ minutes.

Time $T = \frac{D}{GS}$. To find the time (T) in flight, divide the distance (D) by the groundspeed (GS). The time to fly 210 miles at a groundspeed of 140 MPH is 210 divided by 140, or 1.5 hours. (The 0.5 hour multiplied by 60 minutes equals 30 minutes.) Answer: 1:30.

Distance $D = GS \times T$. To find the distance flown in a given time, multiply groundspeed by time. The distance flown in 1 hour 45 minutes at a groundspeed of 120 MPH is 120×1.75, or 210 miles.

Groundspeed $GS = \frac{D}{T}$. To find the groundspeed, divide the distance flown by the time required. If an airplane flies 270 miles in 3 hours, the groundspeed is 270 divided by 3 = 90 MPH.

Converting Knots to Miles Per Hour Another conversion is changing knots to miles per hour. The aviation industry is beginning to use knots more frequently than miles per hour, but it might be well to discuss the conversion for those who do use miles per hour when working speed problems. The National Weather Service reports both surface winds and winds aloft in knots. However, airspeed indicators in some personal-type airplanes are calibrated

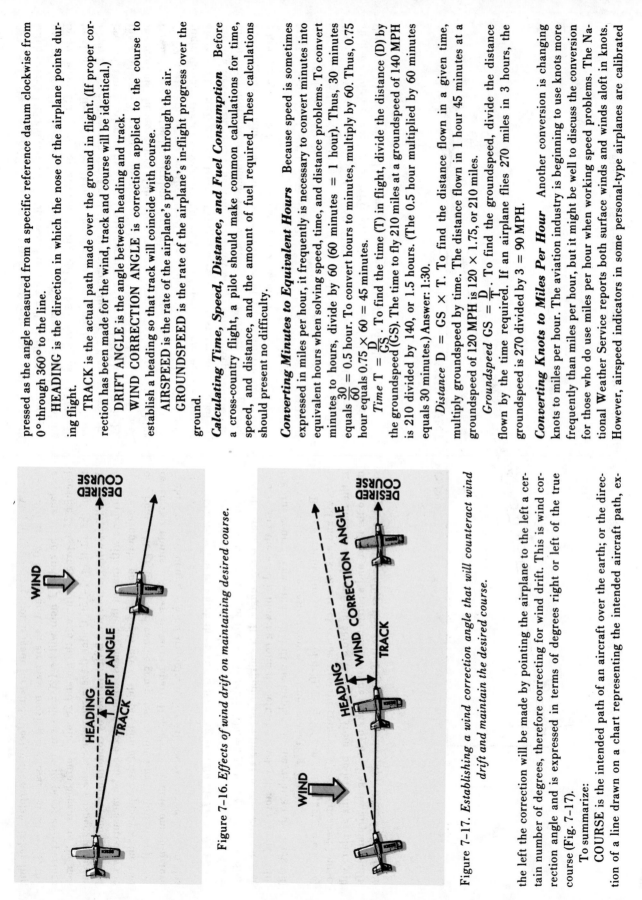

Figure 7-16. *Effects of wind drift on maintaining desired course.*

Figure 7-17. *Establishing a wind correction angle that will counteract wind drift and maintain the desired course.*

the left the correction will be made by pointing the airplane to the left a certain number of degrees, therefore correcting for wind drift. This is wind correction angle and is expressed in terms of degrees right or left of the true course (Fig. 7-17).

To summarize:

COURSE is the intended path of an aircraft over the earth; or the direction of a line drawn on a chart representing the intended aircraft path, exin miles per hour (although many are now calibrated in both miles per hour and knots). Pilots, therefore, should learn to convert windspeeds in knots to miles per hour.

A knot is 1 nautical mile per hour. Because there are 6,076.1 feet in a nautical mile and 5,280 feet in a statute mile, the conversion factor is 1.15. To convert knots to miles per hour, multiply knots by 1.15. For example: a wind-speed of 20 knots is equivalent to 23 MPH.

Most navigational computers or electronic calculators have a means of making this conversion simply by reading the scale. Another quick method of conversion is to use the scales of nautical miles and statute miles at the bottom of aeronautical charts.

Fuel Consumption   Airplane fuel consumption rate is computed in gallons per hour. Consequently, to determine the fuel required for a given flight, the time required for the flight must be known. Time in flight multiplied by rate of consumption gives the quantity of fuel required. For example, a flight of 400 miles at a groundspeed of 100 MPH requires 4 hours. If the plane consumes 5 gallons an hour, the total consumption will be 4 × 5, or 20 gallons.

The rate of fuel consumption depends on many factors: condition of the engine, propeller pitch, propeller r.p.m., richness of the mixture, and particularly the percentage of horsepower used for flight at cruising speed. The pilot should know the approximate consumption rate from cruise performance charts, or from experience. In addition to the amount of fuel required for the flight, there should be sufficient fuel for an adequate reserve.

The Wind Triangle   The wind triangle is a graphic explanation of the effect of wind upon flight. Groundspeed, heading, and time for any flight can be determined by using the wind triangle. It can be applied to the simplest kind of cross-country flight as well as the most complicated instrument flight. The experienced pilot becomes so familiar with the fundamental principles that estimates can be made which are adequate for visual flight without actually drawing the diagrams. *The beginning student, however, needs to develop skill in constructing these diagrams as an aid to the complete understanding of wind effect.* Either consciously or unconsciously, every good pilot thinks of the flight in terms of wind triangle.

If a flight is to be made on a course to the east, with a wind blowing from northeast, the airplane must be headed somewhat to the north of east to counteract drift. This can be represented by a diagram as shown in Fig. 7-18. Each line represents direction and speed. The long dotted line shows the direction the plane is heading, and its length represents the airspeed for 1

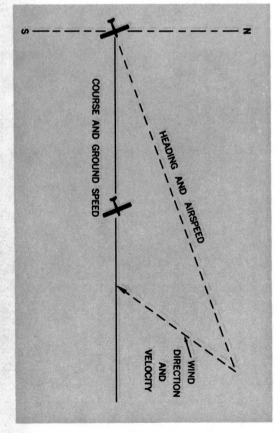

Figure 7-18. *Principle of the wind triangle.*

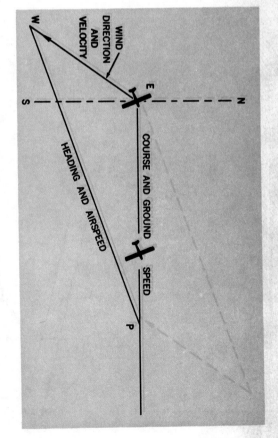

Figure 7-19. *The wind triangle as it is drawn in navigation practice. Blue lines show the triangle as drawn in Figure 7-18.*

hour. The short dotted line at the right shows the wind direction, and its length represents the wind velocity for 1 hour. The solid line shows the direction of the track, or the path of the airplane as measured over the earth, and its length represents the distance traveled in 1 hour, or the groundspeed.

In actual practice, the triangle illustrated in Fig. 7-18 is not drawn; instead, construct a similar triangle as shown by the black lines in Fig. 7-19, which is explained in the following example.

Suppose a flight is to be flown from E to P. Draw a line on the aeronautical chart connecting these two points, measure its direction with a protractor, or plotter, in reference to a meridian. This is the true course which in this example is assumed to be 090° (east). From the National Weather Ser-

vice it is learned that the wind at the altitude of the intended flight is 40 knots from the northeast (045°). Since the National Weather Service reports the windspeed in knots, if the true airspeed of the airplane is 120 knots there is no need to convert speeds from knots to MPH or vice versa.

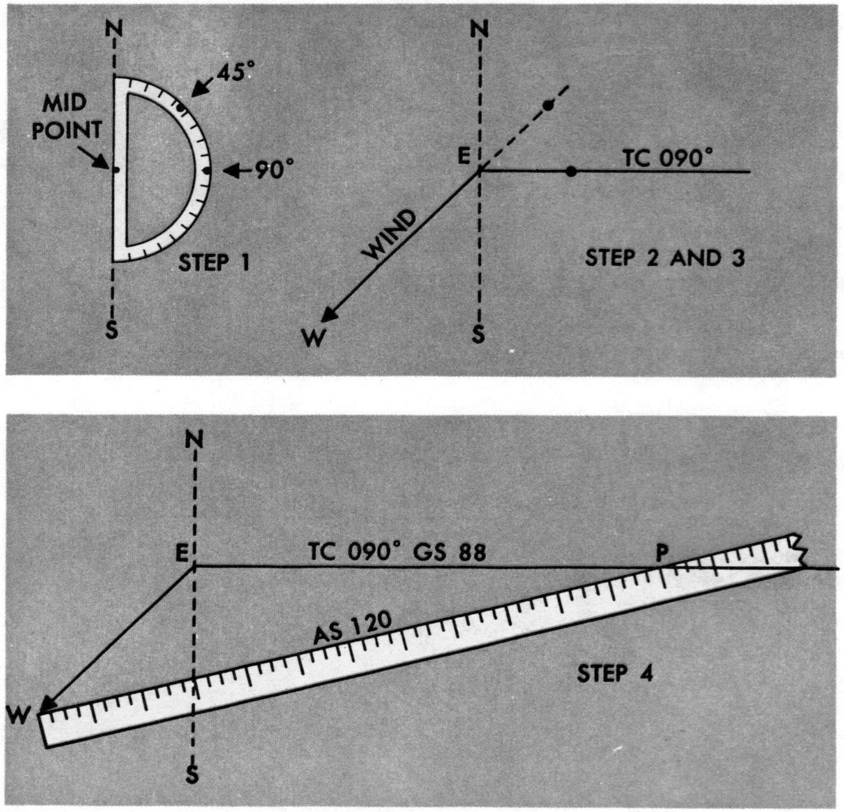

Now on a plain sheet of paper draw a vertical line representing north and south. (The various steps are shown in Fig. 7-20.)

Place the protractor with the base resting on the vertical line and the curved edge facing east. At the center point of the base, make a dot labeled "E" (point of departure), and at the curved edge, make a dot at 90° (indicating the direction of the true course) and another at 45° (indicating wind direction).

With the ruler draw the true course line from E, extending it somewhat beyond the dot at 90°, and labeling it "TC 090°."

Next, align the ruler with E and the dot at 45°, and draw the wind arrow from E, not toward 045°, but downwind in the direction the wind is blowing, making it 40 units long, to correspond with the wind velocity of 40 knots. Identify this line as the wind line by placing the letter "W" at the end to show the wind direction. Finally, measure 120 units on the ruler to represent the airspeed, making a dot on the ruler at this point. The units used may be of any convenient scale or value (such as ¼" = 10 knots), but once selected, the same scale must be used for each of the linear measurements involved. Then place the ruler so that the end is on the arrowhead (W) and the 120 knot dot intercepts the true course line. Draw the line and label it "AS 120." The point "P" placed at the intersection, represents the position of the airplane at the end of 1 hour. The diagram is now complete.

The distance flown in 1 hour (groundspeed) is measured as the number of units on the true course line (88 nautical miles per hour or 88 knots).

The true heading necessary to offset drift is indicated by the direction of the airspeed line which can be determined in one of two ways:

1. By placing the straight side of the protractor along the north-south line, with its center point at the intersection of the airspeed line and north-south line, read the true heading directly in degrees (076°) (Fig. 7-21).

2. By placing the straight side of the protractor along the true course line, with its center at P, read the angle between the true course and the airspeed line. This is the wind correction angle (WCA) which must be applied to the true course to obtain the true heading. If the wind blows from the right of true course, the angle will be added; if from the left, it will be subtracted. In the example given, the WCA is 14° and the wind is from left; therefore, subtract 14° from true course of 090°, making the true heading 076° (Fig. 7-22).

Figure 7-20. *Steps in drawing the wind triangle.*

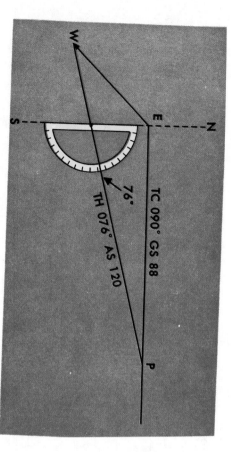

Figure 7-21. *Finding true heading by direct measurement.*

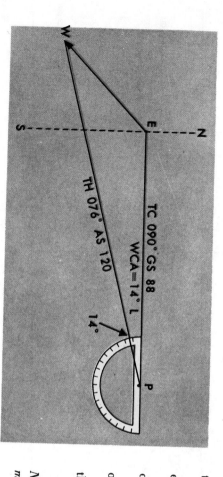

Figure 7-22. *Finding true heading by the wind correction angle.*

distance to the destination by measuring the length of the course line drawn on the aeronautical chart (*using the appropriate scale at the bottom of the chart*). If the distance measures 220 miles, divide by the groundspeed of 88 MPH, which gives 2.5 hours or (2:30) as the time required. If fuel consumption is 8 gallons an hour, 8 × 2.5 or about 20 gallons will be used. Briefly summarized, the steps in obtaining flight information are as follows:

TRUE COURSE.—Direction of the line connecting two desired points, drawn on the chart and measured clockwise in degrees from true north on the mid-meridian.

WIND CORRECTION ANGLE.—Determined from wind triangle. (Added to TC if the wind is from the right; subtracted if wind is from the left.)

TRUE HEADING.—The direction, measured in degrees clockwise from true north, in which the nose of the plane should point to make good the desired course.

VARIATION.—Obtained from the isogonic line on the chart. (Added to TH if west; subtracted if east.)

MAGNETIC HEADING.—An intermediate step in the conversion. (Obtained by applying variation to true heading.)

DEVIATION.—Obtained from the deviation card on the airplane. (Added to MH or subtracted from, as indicated.)

COMPASS HEADING.—The reading on the compass (found by applying deviation to MH) which will be followed to make good the desired course.

TOTAL DISTANCE.—Obtained by measuring the length of the TC line on the chart (using the scale at the bottom of the chart).

GROUNDSPEED.—Obtained by measuring the length of the TC line on the wind triangle (using the scale employed for drawing the diagram).

TIME FOR FLIGHT.—Total distance divided by groundspeed.

FUEL RATE.—Predetermined gallons per hour used at cruising speed.

NOTE: Additional fuel for adequate reserve should be added as a safety measure.

A useful combination Planning Sheet and Flight Log form is shown in Fig. 7-23.

Data for Return Trip. The true course for the return trip will be the reciprocal of the outbound course. This can be measured on the chart, or found more easily by adding 180° to the outbound true course (090° + 180° = 270°), if the outbound course is less than 180°. If the outbound course is greater than 180°, the 180° should be subtracted instead of added. For example, if the outbound course is 200°, the reciprocal will be 200° − 180° = 020°. The wind correction angle will be the same number of degrees as for the

After obtaining the true heading, apply the correction for magnetic variation to obtain magnetic heading, and the correction for compass deviation to obtain a compass heading. The compass heading can be used to fly to the destination by dead reckoning.

To determine the time and fuel required for the flight, first find the

179

PILOT'S PLANNING SHEET

PLANE IDENTIFICATION   DATE

CRUISING AIRSPEED	TC	WIND MPH	WIND FROM	WCA R+ L−	TH	VAR W+ E−	MH	DEV	CH	TOTAL MILES	GS	TOTAL TIME	FUEL RATE	TOTAL FUEL
From:														
To:														
From:														
To:														

VISUAL FLIGHT LOG

TIME OF DEPARTURE	RADIO FREQUENCIES TOWER / RANGE	DISTANCE POINT TO POINT / CUMULATIVE	ELAPSED TIME ESTIMATED / ACTUAL	CLOCK TIME ESTIMATED / ACTUAL	GS ESTIMATED / ACTUAL	CH ESTIMATED / ACTUAL	REMARKS BRACKETS, WEATHER, ETC.
POINT OF DEPARTURE							
CHECKPOINTS							
1.							
2.							
3.							
4.							
5.							
DESTINATION							
6.							

The Pilot's Planning Sheet provides space for entering dead-reckoning data.

The Visual Flight Log may be prepared in advance by entering the selected checkpoints, together with the following data: Distance between checkpoints, and cumulative distance; estimated time between checkpoints; clock or cumulative time; groundspeed and Compass Heading.

As the flight progresses, the actual time, groundspeed and Compass Heading should be filled in, thus completing the log.

Figure 7–23. *Pilot's planning sheet and visual flight log.*

outbound course, but since the wind will be on the opposite side (right) of the airplane, the correction will have to be added to the true course instead of subtracted (270° + 14° = 284°). Thus, the true heading for the return trip will be 284°.

To find the groundspeed, construct a new wind triangle. Instead of drawing another complete diagram, however, consider point E on the previous diagram as the starting point for the return trip and extend the true course line in the direction opposite to the outbound course. The wind line is then in the proper relationship and does not need to be redrawn. The airspeed line (120 units long) can be drawn from the point of the wind arrow (W) to intersect the return-trip true course line as indicated in Fig. 7-24. The distance measured on this course line from the north-south line to the intersection gives the groundspeed for the return trip (147 knots).

Fig. 7-25 shows the various steps for constructing the wind triangle and measuring the true heading and the wind correction angle for the problem in which the true course is 110°, the wind is 20 knots from the southwest (225°), and the airspeed is 100 knots. Notice that the true heading line has to be extended to intersect the north-south line to measure the true heading directly.

Before attempting to use a computer or electronic calculator, the relationships involved by constructing other wind triangles for various airspeeds, winds, and true courses should be understood.

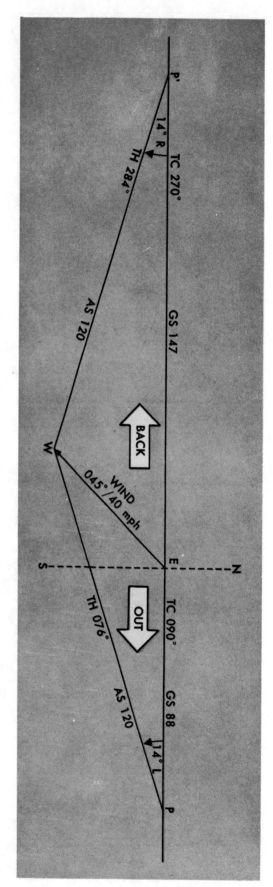

Figure 7-24. *Computations for a round-trip flight.*

Practice Problems. By constructing a wind triangle, find the wind correction angle (WCA), true heading (TH), and groundspeed (GS) for each of the following conditions:

	WIND Direction (degrees)	WIND Speed (MPH)	TRUE Course (degrees)	TRUE Airspeed (MPH)
1.	135	30	240	120
2.	215	20	260	130
3.	050	33	260	150
4.	330	45	350	150
5.	300	45	100	150
6.	220	30	130	150

NOTE: *Correct answers are shown below.*

	WCA	TH	GS (MPH)
1.	14° L	226°	124
2.	6° L	254°	115
3.	6° R	266°	178
4.	6° L	344°	107
5.	6° L	94°	191
6.	12° R	142°	147

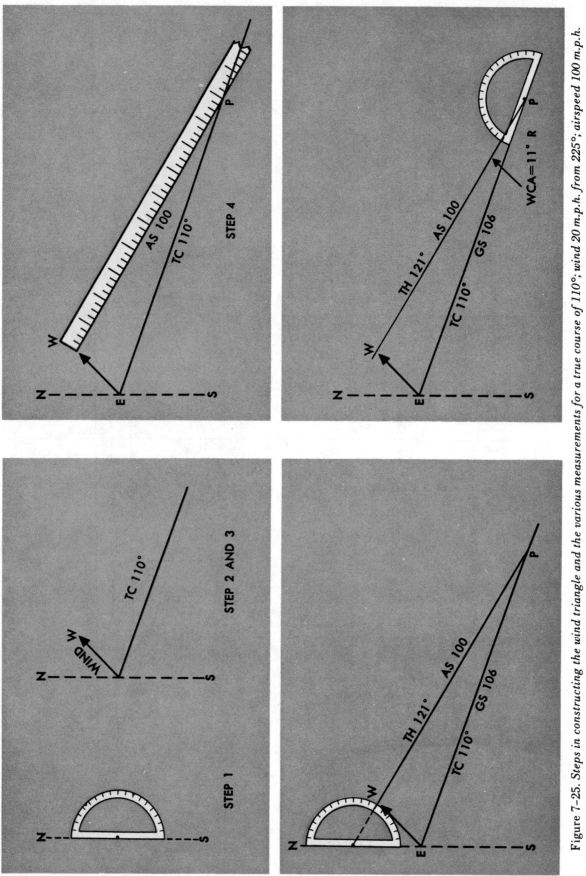

Figure 7–25. Steps in constructing the wind triangle and the various measurements for a true course of 110°; wind 20 m.p.h. from 225°; airspeed 100 m.p.h.

182

Most airplanes flown in today's environment are equipped with radios that provide a means of navigation and communication with ground stations.

Advances in navigational radio receivers installed in airplanes, the development of aeronautical charts which show the exact location of ground transmitting stations and their frequencies, along with refined cockpit instrumentation make it possible for pilots to navigate with precision to almost any point desired. Although precision in navigation is obtainable through the proper use of this equipment, beginning pilots should use this equipment to supplement navigation by visual reference to the ground (pilotage). If this is done, it provides the pilot with an effective safeguard against disorientation in the event of radio malfunction.

There is a variety of radio navigation systems available for use, but only the systems most commonly used for VFR navigation will be discussed in this handbook. These are the VHF Omnidirectional Range (VOR) and the Non-directional Radiobeacon (NDB).

VHF Omnidirectional Range (VOR) The word "omni" means *all*, and an omnidirectional range is a very high frequency (VHF) radio transmitting ground station that projects straight line courses (radials) from the station in all directions. From a top view it can be visualized as being similar to the spokes from the hub of a wheel. The distance VOR radials are projected depends upon the power output of the transmitter.

The courses or radials projected from the station are referenced to magnetic north. Therefore, a radial is defined as "a line of magnetic bearing extending outward from the VOR station." Radials are identified by numbers beginning with 001, which is one degree east of magnetic north, and progress in sequence through all the degrees of a circle until reaching 360. To aid in orientation, a compass rose reference to magnetic north is superimposed on aeronautical charts at the station location.

VOR ground stations transmit within a VHF frequency band of 108.0 — 117.95 MHz. Because the equipment is VHF, the signals transmitted are subjected to line-of-sight restrictions (Fig. 7-26). Therefore, its range varies in direct proportion to the altitude of the receiving equipment. Generally, the reception range of the signals at an altitude of 1,000 feet above ground level is about 40 to 45 miles. This distance increases with altitude.

For the purpose of this discussion, the term VOR will be used to include both VOR and VORTAC. Briefly a VORTAC station provides, in addition to azimuth information, range information. If the airplane is equipped with

distance measuring equipment (DME) the distance from the station in nautical miles is displayed on the instrument.

VORs and VORTACs are classed according to operational use. There are three classes:

T (Terminal).
L (Low altitude).
H (High altitude).

The normal useful range for the various classes is shown in the following table:

VOR/VORTAC NAVAIDS

Normal Usable Altitudes and Radius Distances

Class	Altitudes	Distance (Miles)
T	12,000' and below	25
L	Below 18,000'	40
H	Below 18,000'	40
H	Within the conterminous 48 states only, between 14,500' and 17,999'	100
H	18,000' – FL 450	130
H	Above FL 450	100

The useful range of certain facilities may be less than 50 miles. For further information concerning these restrictions, refer to the Comm/NAVAID Remarks in the Airport/Facility Directory.

The accuracy of course alignment of VOR radials is considered to be excellent. It is generally within plus or minus 1°. However, certain parts of the VOR receiver equipment deteriorate, and this affects its accuracy. This is particularly true at greater distances from the VOR station. The best assurance of maintaining an accurate VOR receiver is periodic checks and calibrations. VOR accuracy checks are not a regulatory requirement for VFR flight; however, to assure accuracy of the equipment these checks should be accomplished quite frequently along with a complete calibration each year. The following means are provided for pilots to check VOR accuracy: (1) FAA VOR Test Facility (VOT), (2) certified airborne checkpoints, and (3) certified ground checkpoints located on airport surfaces. A list of these checkpoints is published in the Airport/Facility Directory.

Basically these checks consist of verifying that the VOR radials the airplane equipment receives are aligned with the radials the station transmits. There are not specific tolerances in VOR checks required for VFR flight, but as a guide to assure acceptable accuracy the required IFR tolerances can be

from the VOR or the airplane is too low and therefore, is out of the line-of-sight of the station's transmitting signals.

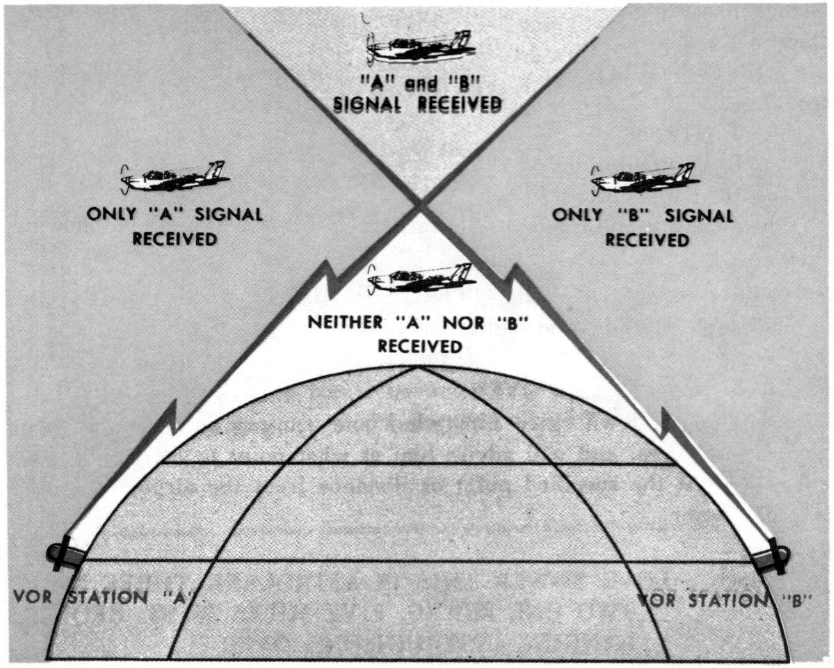

Using the VOR Using the VOR is quite simple once the basic concept is understood. The following information coupled with practice in actually using this equipment should erase all the mysteries and also provide a real sense of security in navigating with the VOR.

In review, for VOR radio navigation there are two components required: the ground transmitter, and the aircraft receiving equipment. The ground transmitter is located at specific positions on the ground and transmits on an assigned frequency. The aircraft equipment includes a receiver with a tuning device and a VOR or omninavigation instrument. The navigation instrument consists of (1) an omnibearing selector (OBS) sometimes referred to as the Course Selector, (2) a course deviation indicator needle (Left-Right Needle), and (3) a TO-FROM indicator.

Figure 7-27. *Omnihead.*

Figure 7-26. *VHF transmission follows a line-of-sight course.*

used which are ± 4° for ground checks and ± 6° for airborne checks. These checks can be performed by the pilot.

The VOR transmitting station can be positively identified by its Morse code identification or by a recorded voice identification which states the name of the station followed by the word "VOR." Many Flight Service Stations transmit voice messages on the same frequency that the VOR operates. Voice transmissions should not be relied upon to identify stations, because many FSS's remotely transmit over several omniranges which have different names than the transmitting FSS. If the VOR is out of service for maintenance the coded identification is removed and not transmitted. This serves to alert pilots that this station should not be used for navigation. VOR receivers are designed with an alarm flag to indicate when signal strength is inadequate to operate the navigational equipment. This happens if the airplane is too far

184

The course selector is an azimuth dial that can be rotated to select a desired radial or to determine the radial over which the aircraft is flying. In addition, the magnetic course "TO" or "FROM" the station can be determined.

When the course selector is rotated it moves the course deviation indicator or needle to indicate the position of the radial relative to the aircraft. If the course selector is rotated until the deviation needle is centered the radial (magnetic course from the station) or its reciprocal (magnetic course to the station) can be determined. The course deviation needle will also move to the left or right if the aircraft is flown or drifting away from the radial which is set in the course selector.

By centering the needle the course selector will indicate either the course "FROM" the station or the course "TO" the station. If the flag displays a "TO", the course shown on the course selector must be flown to the station. If "FROM" is displayed and the course shown is followed, the aircraft will be flown away from the station (Fig. 7-27).

Tracking With Omni. The following describes a step by step procedure to use when tracking to and from a VOR station. Fig. 7-28 illustrates the discussion:

1. First, tune the VOR receiver to the frequency of the selected VOR station. For example: 115.0 to receive Bravo VOR. *Next*, check the identifiers to verify that the desired VOR is being received. As soon as the VOR is properly tuned, the course deviation needle will deflect either left or right; then rotate the azimuth dial of the course selector until the course deviation needle centers and the TO-FROM indicates "TO." If the needle centers with a FROM indication, the azimuth dial should be rotated 180 degrees because in this case it is desired to fly "TO" the station. Now, turn the aircraft to the heading indicated on the omni azimuth dial or course selector. In this example 350°.

2. If a heading of 350° is maintained with a wind from the right as shown, the airplane will drift to the left of the intended track. As the airplane drifts off course, the VOR course deviation needle will gradually move to the right of center or indicate the direction of the desired radial or track.

3. To return to the desired radial, the aircraft heading must be altered approximately 30° to the right. As the aircraft returns to the desired track, the deviation needle will slowly return to center. When centered, the aircraft will be on the desired radial and a left turn must be made toward, but not to the original heading of 350° because a wind drift correction must be established. The amount of correction depends upon the strength of the wind. If the wind

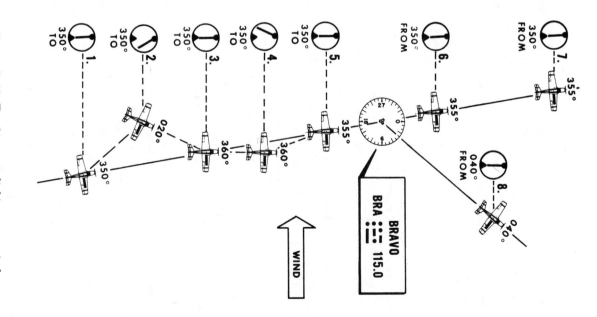

Figure 7-28. *Tracking a radial in a crosswind.*

velocity is unknown, a trial and error method can be used to find the correct heading. Assume, for this example a 10° correction or a heading of 360° is maintained.

4. While maintaining a heading of 360°, assume that the course deviation begins to move to the left. This means that the wind correction of 10° is too great and the airplane is flying to the right of course. A slight turn to the left should be made to permit the airplane to return to the desired radial.

5. When the deviation needle centers, a smaller wind drift correction of 5° or a heading correction of 355° should be flown. If this correction is adequate, the airplane will remain on the radial. If not, small variation in heading should be made to keep the needle centered, and consequently keep the airplane on the radial.

6. As the VOR station is passed, the course deviation needle will fluctuate then settle down, and the "TO" indication will change to "FROM." If the aircraft passes to one side of the station, the needle will deflect in the direction of the station as the indicator changes to "FROM."

7. Generally, the same techniques apply when tracking outbound as those used for tracking inbound. If the intent is to fly over the station and track outbound on the reciprocal of the inbound radial, the course selector should not be changed. Corrections are made in the same manner to keep the needle centered. The only difference is that the omni will indicate "FROM."

8. If tracking outbound on a course other than the reciprocal of the inbound radial, this new course or radial must be set in the course selector and a turn made to intercept this course. After this course is reached, tracking procedures are the same as previously discussed.

Tips on Using the VOR 1. Positively identify the station by its code or voice identification.

2. Keep in mind that VOR signals are "line-of-sight." A weak signal or no signal at all will be received if the aircraft is too low or too far from the station.

3. When navigating to a station, determine the inbound radial and use this radial. If the aircraft drifts, do not reset the course selector, but correct for drift and fly a heading that will compensate for wind drift.

4. If minor needle fluctuations occur, avoid changing headings immediately. Wait momentarily to see if the needle recenters; if it doesn't, then correct.

5. When flying "TO" a station, always fly the selected course with a "TO" indication. When flying "FROM" a station, always fly the selected course with a "FROM" indication. If this is not done, the action of the course

deviation needle will be reversed. To further explain this reverse action, if the aircraft is flown toward a station with a "FROM" indication or away from a station with a "TO" indication, the course deviation needle will indicate in an opposite direction to that which it should. For example, if the aircraft drifts to the right of a radial being flown, the needle will move to the right or point away from the radial. If the aircraft drifts to the left of the radial being flown, the needle will move left or in the opposite direction of the radial.

Automatic Direction Finder (ADF) Many general aviation-type airplanes are equipped with ADF radio receiving equipment. To navigate using the ADF the pilot tunes the receiving equipment to a ground station known as a NONDIRECTIONAL RADIOBEACON (NDB). The NDB stations normally operate in a low or medium frequency band of 200 to 415 kHz. The frequencies are readily available on aeronautical charts or in the Airport/Facility Directory.

All radiobeacons except compass locators transmit a continuous three-letter identification in code except during voice transmissions. A compass locator, which is associated with an Instrument Landing System, transmits a two-letter identification.

Standard broadcast stations can also be used in conjunction with ADF. Positive identification of all radio stations is extremely important and this is particularly true when using standard broadcast stations for navigation.

Nondirectional Radiobeacons have one advantage over the VOR. This advantage is that low or medium frequencies are not affected by line-of-sight. The signals follow the curvature of the earth, therefore if the aircraft is within the power range of the station, the signals can be received regardless of altitude.

The following table gives the class of NDB stations, their power, and usable range:

NONDIRECTIONAL RADIOBEACON (NDB)
(Usable Radius Distances for All Altitudes)

Class	Power (Watts)	Distance (Miles)
Compass Locator	Under 25	15
MH	Under 50	25
H	50–1999	*50
HH	2000 or more	75

*Service range of individual facilities may be less than 50 miles.

One of the disadvantages that should be considered when using low frequency for navigation is that low-frequency signals are very susceptible to electrical disturbances, such as lightning. These disturbances create excessive static, needle deviations, and signal fades. There may be interference from distant stations. Pilots should know the conditions under which these disturbances can occur so they can be more alert to possible interference when using the ADF.

Basically, the ADF aircraft equipment consists of a tuner, which is used to set the desired station frequency, and the navigational display.

The navigational display consists of a dial upon which the azimuth is printed, and a needle which rotates around the dial and points to the station to which the receiver is tuned.

Some of the ADF dials can be rotated so as to align the azimuth with the aircraft heading, others are fixed with the 0°–180° points on the azimuth aligned with the longitudinal axis of the aircraft. Thus, the zero degree position on the azimuth represents the nose of the aircraft. Only the fixed azimuth dial will be discussed in this handbook (Fig. 7-29).

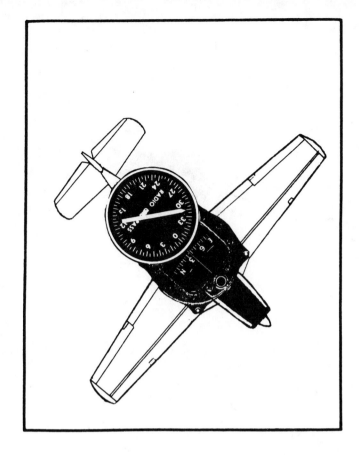

Figure 7-29. ADF with fixed azimuth and magnetic compass.

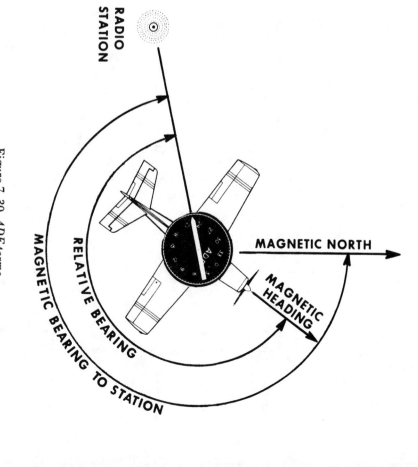

RADIO STATION

MAGNETIC NORTH

RELATIVE BEARING

MAGNETIC HEADING

MAGNETIC BEARING TO STATION

Figure 7-30. ADF terms.

Fig. 7-30 illustrates the following terms that are used with the ADF and should be understood by the pilot.

Relative Bearing is the value to which the indicator (needle) points on the azimuth dial. When using a fixed dial, this number is relative to the nose of the aircraft and is the angle measured clockwise from the nose of the aircraft to a line drawn from the aircraft to the station.

Magnetic Bearing to the station is the angle formed by a line drawn from the aircraft to the station and a line drawn from the aircraft to magnetic north. The magnetic bearing to the station can be determined by adding the relative bearing to the magnetic heading of the aircraft. For example, if the relative bearing is 060° and the magnetic heading is 130°, the magnetic bearing to the station is 060° plus 130° or 190°. This means that in still air a

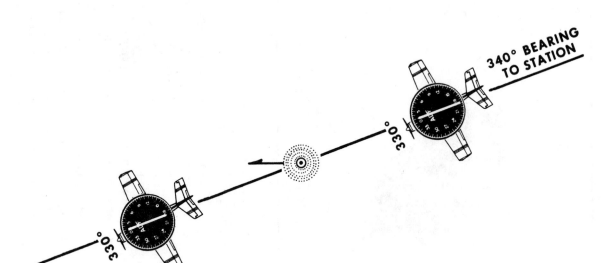

Figure 7–31. *ADF tracking.*

magnetic heading of approximately 190° would be flown to the station. If the total is greater than 360°, subtract 360° from the total to obtain the magnetic bearing to the station. For example, if the relative bearing is 270° and magnetic heading is 300°, 360° is subtracted from the total, or 570° − 360° = 210°, which is the magnetic bearing *to* the station.

To determine the magnetic bearing *from* the station 180° is added to or subtracted from the magnetic bearing to the station. This is the reciprocal bearing and is used when plotting position fixes.

Keep in mind that the needle of fixed azimuth points to the station in relation to the nose of the aircraft. If the needle is deflected 30° to the left or a relative bearing of 330°, this means that the station is located 30° left. If the aircraft is turned left 30°, the needle will move to the right 30° and indicate a relative bearing of 0° or the aircraft will be pointing toward the station. If the pilot continues flight toward the station keeping the needle on 0°, the procedure is called homing to the station. If a crosswind exists, the ADF needle will continue to drift away from zero. To keep the needle on zero the aircraft must be turned slightly resulting in a curved flightpath to the station. Homing to the station is a common procedure, but results in drifting downwind, thus lengthening the distance to the station.

Tracking to the station requires correcting for wind drift and results in maintaining flight along a straight track or bearing to the station. When the wind drift correction is established, the ADF needle will indicate the amount of correction to the right or left. For instance, if the magnetic bearing to the station is 340°, but to correct for a left crosswind the magnetic heading is 330°, the ADF needle would indicate 10° to the right or a relative bearing of 010° (Fig. 7-31).

When tracking away from the station, wind corrections are made similar to tracking to the station but the ADF needle points toward the tail of the aircraft or the 180° position on the azimuth dial. Attempting to keep the ADF needle on the 180° position during winds results in the aircraft flying a curved flight leading further and further from the desired track.

Although the ADF is not so popular as the VOR for radio navigation, with proper precautions and intelligent use the ADF can be a valuable aid to navigation.

Flight Planning FAA regulations state, in part, that before beginning a flight, the pilot in command of an aircraft shall become familiar with all available information concerning that flight. For flights not in the vicinity of an airport, this must include information on available current weather reports and forecasts, fuel requirements, alternatives available if the planned flight cannot be completed, and any known traffic delays reported by ATC.

Careful preflight planning is extremely important. With adequate planning the pilot can complete the flight with greater confidence, ease, and safety. Without it the pilot may become a statistic—figures show inadequate preflight planning is a significant cause of fatal accidents.

Assembling Necessary Materials The pilot should collect the necessary material well before the flight to be sure nothing is missing. An appropriate current sectional chart and charts of areas adjoining the flight route should be among this material if the route of flight is near the border of a chart. By having this information, the pilot will be prepared to circumnavigate weather or minimize the possibility of becoming lost. To determine the charts that cover surrounding areas, check the small replica of a map of the U.S. which appears on each of the new sectional charts. This map gives the coverage of each chart and identifies them by name. For example, the charts surrounding the Dallas-Ft. Worth chart are: Wichita, Kansas City, Memphis, Houston, San Antonio, and Albuquerque.

The latest Airport/Facility Directory should be among the material. Subscriptions to this publication are available from:

National Ocean Survey (NOS)
Distribution Division C-44
Riverdale, MD 20840

Additional equipment should include a computer or electronic calculator, plotter, and any other item appropriate to the particular flight—for example, if a night flight is to be undertaken, carry a flashlight; if a flight is over desert country, carry a supply of water and other necessities.

Weather Check It may be wise to check the weather before continuing with other aspects of flight planning to see, first of all, if the flight is feasible and, if it is, which route is best. A visit should be made to the local National Weather Service airport station or, if available, to the nearest FAA Flight Service Station. A personal visit is best because it provides access to the latest weather maps and charts, area forecasts, terminal forecasts, SIGMETS and AIRMETS, hourly sequence reports, PIREPS, and winds-aloft forecasts, and a weather briefer is available to aid in interpreting the weather.

If a visit is impractical, telephone calls are welcomed. Some National Weather Service Stations have "restricted" (unlisted) telephone numbers on which *only* aviation weather information is given. These numbers, along with other National Weather Service and FSS numbers, are listed in the Airport/Facility Directory. When telephoning for aviation weather information, identify yourself as a pilot, state the intended route, destination, intended time of takeoff, approximate time en route, and advise if the flight is intended to be only VFR.

Unfortunately, there are still many general aviation pilots who are inclined to scoff at or ignore aviation weather forecasts and briefings, since the weather information they received has not always proved entirely accurate. The adequacy or accuracy of the forecasts concerned in more than 1,000 cases was examined in considerable detail and it was determined that in 80% to 85% of these cases, the forecasts adequately depicted the weather conditions with which the pilots would have been faced if they had received these forecasts. Accordingly, it is considered that, with those odds, a pilot simply cannot afford to initiate a flight without full knowledge of the available weather data, including the forecasts. Of course, the weatherwise pilot looks upon weather briefings and forecasts as help and advice and not fixed absolutes. It is almost as bad for a pilot to have blind faith in weather information and forecasts as it is to have none at all. The pilot who understands not only the weather information as given, but appreciates its limitations as well, is the pilot who will be able to make the most effective use of all the weather service available, and will always be wary of the marginal weather situation.

Much of the following has been discussed previously in this handbook in one way or another, but it bears repeating. It is presented here in step-by-step sequence and should be carefully reviewed.

Weather briefings are provided in the interest of assuring that the pilot receive the best available weather information for effective flight planning. Adequate briefings, properly understood and applied, are the best bases for determining whether the flight should be executed as planned, postponed, altered, or cancelled. Safety demands careful consideration of current and forecast weather before departing on any flight.

In-Flight Visibility and the VFR Pilot Basic information relative to the atmosphere and weather behavior has been presented in this handbook. In this section, emphasis will be placed on the necessity for proper planning and use of the weather information available, based on a clear understanding of what it can and cannot do. Also included are specific hazards and how to cope with them, but in the final analysis, it is the pilot's judgment that is the critical factor. A pilot should establish weather limitations based on a realistic assessment of personal limitations and that of the equipment. In nearly every instance, if weather conditions are marginal, or if there is any suspicion of worsening weather, the safest rule is—do not go!

Statistics indicate that on a national basis, over 25 percent of all fatal accidents and over 31 percent of all fatalities result from taking off or continuing into adverse weather with subsequent loss of aircraft control. In other words, cold hard facts indicate that weather-involved accidents continue to account for an unnecessarily high number of fatalities for VFR pilots. A VFR pilot is one who does not have an instrument rating or an instrument rated pilot who is not proficient. It is also a fact that the average pilot who has had no training in instrument flight will lose control of the aircraft in a matter of seconds without outside references are lost.

The inescapable conclusion to the preceding is that VFR pilots must stay out of weather and to do this limitations, as well as the capabilities of present day meteorology, must be understood. The impossible must not be expected, nor the attainable neglected. Recent studies of aviation forecasts indicate the following:

1. For up to 12 hours and even beyond, a forecast of good weather (ceiling 3,000 ft. or more and visibility 3 miles or greater) is much more likely to be correct than is a forecast of conditions below 1,000 ft. or below 1 mile.

2. However, for 3 or 4 hours in advance, the probability that below VFR conditions will occur is more than 80 percent accurate, if below VFR is forecast.

3. Forecasts of single reportable values of ceiling or visibility instead of a range of values imply an accuracy that the present forecasting system does not possess beyond the first 2 or 3 hours of the forecast period.

4. Forecasts of poor flying conditions during the first few hours of the forecast period are most reliable when there is a distinct weather system, such as a front, a trough, precipitation, etc., which can be tracked and forecast.

5. The weather associated with fast-moving cold fronts and squall lines is the most difficult to forecast accurately.

6. Forecasting the specific time of occurrence of bad weather is less accurate than forecasting whether or not the weather will occur within a span of time.

7. Surface visibility is more difficult to forecast than ceiling height, and snow increases the visibility forecasting problem.

Available evidence shows that forecasters CAN predict the following at least 75 percent of the time:

1. The passage of fast-moving cold fronts or squall lines within plus or minus 2 hours, as much as 10 hours in advance.

2. The passage of warm fronts or slow-moving cold fronts within plus or minus 5 hours, up to 12 hours in advance.

3. The rapid lowering of ceiling below 1,000 ft. in prewarm front conditions within plus or minus 200 ft. and within plus or minus 4 hours.

4. The onset of a thunderstorm 1 or 2 hours in advance if radar is available.

5. The time rain or snow will begin within plus or minus 5 hours.

6. Rapid deepening of a low pressure center.

Although recent improvements in forecasting have been made, there are still limitations in predicting the following with an accuracy which completely satisfies present aviation operational requirements:

7. The time freezing rain will begin.

8. The location and occurrence of severe or extreme turbulence.

9. The location and occurrence of heavy icing.

10. The location or the occurrence of a tornado.

11. Ceilings of 100 ft. or zero before they exist.

12. The potential development of a thunderstorm.

13. The position of a hurricane center to nearer than 100 miles for more than 12 hours in advance.

14. The occurrence of ice/fog.

These indications of what can and cannot be predicted will vary, depending on the climatology and general weather conditions of the area. In general, rare events are more difficult to predict than common events. Weather conditions which have a pronounced daily variation, such as the occurrence of nighttime radiation fog or of afternoon convective clouds, can be forecast more reliably than conditions which have small daily variation.

Similarly, weather conditions which depend on interaction of wind flow with mountain ranges, coastal areas, or large bodies of water are more reliably forecast than similar weather conditions which are associated with cyclonic storms moving slowly over flat, uniform terrain. In either instance, however, the pilot who plans ahead and keeps informed of what the weather is doing and what it is forecast to do, is exercising good judgment.

The pilot who predicates safety for several hours of flying on one forecast or briefing may well be gambling with life.

Visibility vs. Time   In order to stay out of clouds, VFR pilots are often forced to low altitudes. Even when they are able to remain clear of the clouds, visibility, more often than not, is marginal, and it is here that visibility in a very real sense relates to time as much as to distance. That is, how many seconds ahead can a pilot see with 1 mile visibility? How many seconds does the pilot have to perceive, interpret, act and obtain aircraft reaction?

Cruising at 95 knots, an aircraft travels 160 feet per second. Thus, related to 1 mile visibility, the pilot can see 33 seconds ahead. Matters get worse as speed increases and/or visibility decreases! The pilot can see ahead only 20 seconds if cruising at 154 knots with 1 mile visibility. IT MAY BE THOUGHT that this is plenty of time to turn before reading zero-zero conditions or to miss an obstruction, but is it?

Reaction Time. It takes time, very precious time under marginal conditions of visibility, for the eye to see something, for the brain to send a message for the muscles to react, and finally for the airplane to respond to control usage. On the average, this all takes from 4 to 5 seconds. If this time is subtracted from the 10 to 20 seconds that a pilot can see ahead, it can be determined that at certain speeds and/or angles of bank it would be impossible to miss the zero-zero conditions or an obstruction. At 160 knots with a reaction time of 5 seconds, the aircraft will travel 1,400 ft. before anything even begins to happen in the way of evasive action.

Radius of Turn. Suppose a pilot must turn away from a range of hills or a mountain, or a low-lying cloud bank. If flying at 154 knots, the aircraft will have moved in the direction of this obstruction approximately 3,900 ft. by the time 90° of turn is completed (1,400 ft. for reaction time, plus 2,457 ft. for radius of turn). If visibility is less than three-quarters of a mile, it would be impossible to miss the obstruction without the help of a good headwind. If a tailwind prevails and only 1 mile visibility, it would be even more hazardous. Obviously, the safe thing to do if caught in such restricted visibility, low-altitude conditions is to fly at a reduced speed.

The Cockpit Cut-Off Angle and In-Flight Visibility. All too often, adequate visibility at the surface becomes marginal, or even below minimums at altitude, yet the VFR pilot may continue simply because surface visibilities are reported at values comfortably above minimums. Some method of determining in-flight visibility with reasonable accuracy is, therefore, important. A rule of thumb (Fig. 7-32) which will not be equally accurate for all airplanes, but which is usually better than guessing is as follows:

The approximate visibility in miles will equal the number of thousands of feet above the surface when the surface is just visible over the nose of the airplane. In other words, at that point where the surface first appears over the nose of the airplane, the slant-range visibility will be approximately 2 miles if the flight is at 2,000 ft. above the surface. This rule of thumb is based on the cockpit cut-off angle. All airplanes do not have the same cut-off angle, therefore the rule of thumb will not be equally accurate for all airplanes. As will be subsequently explained, the cockpit cut-off angle for any airplane can

RULE OF THUMB

when surface is just visible over nose of aircraft the forward visibility will be approximately 1 mile for each 1000 feet altitude.

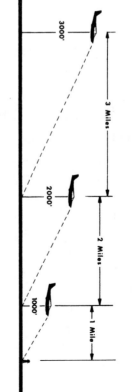

Figure 7-32. *Rule of thumb.*

be determined rather easily. Once it is determined for a given airplane, it will remain constant as long as the eye level of the pilot is not changed. The steps in determining this cut-off angle on the ground are as follows (see Fig. 7-33):

(1) Adjust the aircraft's ground attitude to correspond as closely as possible to its normal cruise pitch attitude.

(2) While normally seated, adjust the pilot's seat to the same position used in flight.

(3) Measure the vertical distance from eye level and the ground (six feet in the example).

(4) Look straight out over the nose of the airplane (cockpit cut-off angle) and determine the spot where the surface is first visible.

(5) Measure the distance from directly under the eye position to the spot established in step 4 (30 ft. in the example).

(6) At this point the visibility can be determined either by establishing a simple proportion, or by solving for the tangent value using the information in the following table. In either case, the result represents the least slant-range visibility that could be obtained when flying at 1,000 ft. above the ground.

COMPUTING CUTOFF ANGLE

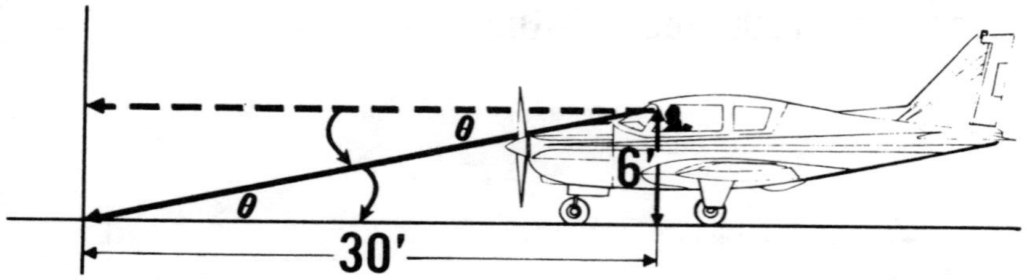

all measurements are from pilots eye level

Figure 7–33. Cockpit cut-off angle.

Tangent Value	Angle (°)	Approximate Visibility at 1,000' AGL
0.052	3	19,000 ft. (5,280 ft. = 1 statute mile)
.070	4	14,280
.087	5	11,500
.105	6	9,530
.123	7	8,130
.141	8	7,090
.158	9	6,330
.176	10	5,750
.194	11	5,150
.213	12	4,710
.231	13	4,320
.249	14	4,010
.268	15	3,730
.287	16	3,480
.306	17	3,270
.325	18	3,070
.344	19	2,910
.364	20	2,750

NOTE: At 500 ft. above the ground, visibility in ft. would be approximately half of the 1,000 ft. value.

Tangent Value Method: The tangent value (tan Θ) is equal to 6 ft. divided by 30 ft. or,

$$\tan \Theta = \frac{6}{30}$$

$$\tan \Theta = .20$$

Referring to the information in the table and locating the value closest to .2, it is apparent that the cut-off angle is somewhere between 11° and 12°. Accurate interpolation reveals that with this cut-off angle the visibility is 5,000 ft.

Proportion Method: Six ft. is to 30 ft. as 1,000 ft. is to "X" ft.

$$\frac{6 \text{ ft.}}{30 \text{ ft.}} = \frac{1,000 \text{ ft.}}{X \text{ ft.}} \text{ or}$$

$$6 X = 30,000 \text{ ft.}$$

$$X = 5,000 \text{ ft.}$$

It must be understood that in either case the visibility thus obtained would be the very least that could be obtained from the cockpit to the ground straight ahead. It can, of course, be more than this in respect to slant range, but if the ground can be seen at this altitude, it cannot be less. Horizontally or laterally, visibility may be more or significantly less, depending on in-flight weather conditions. However, the visibility to the ground ahead (a primary VFR reference for aircraft control) would, in the example cited, be at least 5,000 ft. If the pilot must descend in order to see the ground over the nose of the airplane, the slant-range visibility is proportionally less. (See Note in table.)

How to Get a Briefing

1. Person to person—Visit the nearest National Weather Service airport station or FAA Flight Service Station.

2. By telephone—Call the nearest National Weather Service airport station or FAA Flight Service Station, or call PATWAS (Pilot Automatic Telephone Answering Service). The telephone numbers of these facilities may be found in the Airport/Facility Directory.

3. By radio—Tune to any L/MF (Low/Medium Frequency) "H" Radio Beacon for continuous transcribed weather broadcasts (TWEBS). Call the nearest FAA Flight Service Station radio facility.

Information for the Briefer

1. Name, type of pilot certificate held; e.g., student, private, commercial, and whether instrument rated.

2. Type of aircraft and aircraft number.
3. Point of departure and destination.
4. Proposed route and flight altitude.
5. Estimated time of departure and arrival plus time needed to reach alternate if required.
6. Whether the flight is IFR or VFR.

Items the Weather Briefing Should Contain

1. Weather synopsis (positions and movement of pressure systems, fronts, precipitation areas, etc.).
2. Current weather (at point of departure, en route, including pilot reports, terminal, and alternate if weather is marginal).
3. Forecast weather (at point of departure, en route, terminal, and alternate if required).
4. Alternate routes.
5. Hazardous weather (tornadoes, tropical storms, thunderstorms, hail, turbulence, icing, duststorms, or sandstorms).
6. Forecast winds aloft.
7. A request for pilot reports (help the briefer and fellow pilots by reporting via radio immediately any adverse weather, particularly that which is significantly different from that forecast).

Recommendations

1. If possible, obtain a complete weather briefing to determine if the flight can be conducted safely. Consider personal skill and experience and equipment limitations. Again, if there is any doubt, don't go. Once the flight has been started, the weather information should be updated frequently.
2. File an appropriate flight plan with FAA.
3. If not instrument rated, avoid "VFR On Top" and "Special VFR." Being caught above an overcast when an emergency descent is required (or at destination) is a hazardous position for the VFR pilot. Also, accepting a clearance out of certain airport control zones with no minimum ceiling and 1-mile visibility as permitted with "Special VFR" is an invitation to disaster for a VFR pilot. The weather and/or the terrain within the control zone and beyond may be totally unsuitable for visual flight.
4. Avoid flight through or near thunderstorms. Recent research has proven beyond any doubt that all thunderstorms are potentially dangerous and should be given a wide berth.
5. Avoid flight thorough areas of known or forecast severe weather because tornadoes, squall lines, hail, and severe or extreme turbulence may be encountered. Severe or extreme clear air turbulence may be encountered frequently at low and intermediate levels up to 20 miles ahead of squall lines. The "roll cloud" ahead of a squall line is a visible sign of violent turbulence, but the absence of a roll cloud should not be interpreted as denoting the lack of turbulence.

6. Avoid flight through areas of known or forecast icing conditions unless the aircraft is well equipped with deicing/anti-icing devices. Ice accumulation through areas of freezing precipitation and wet snow can be rapid and heavy. In addition to airframe icing, carburetor icing can occur when visible moisture is present and when moisture is not visible under the right atmospheric conditions (i.e., low temperature, high humidity).

7. Avoid flight at low altitudes over mountainous terrain, particularly near the lee slopes. If the wind velocity near the level of the ridge is in excess of 40 knots and approximately perpendicular to the ridge, mountain wave conditions are likely over and near the lee slopes. If the wind velocity at the level of the ridge exceeds 50 knots, a strong mountain wave is probable with strong up and down drafts and severe or extreme turbulence. The worst turbulence will be encountered in and below the rotor zone which is usually 8 to 10 miles downward from the ridge. This zone is characterized by the presence of "roll clouds" if sufficient moisture is present. Altocumulus standing lenticular clouds are also visible signs that a mountain wave exists, but their presence is likewise dependent upon moisture. The mountain wave downdrafts may exceed the climb capability of the aircraft.

8. Avoid areas of low ceilings and restricted visibility unless the pilot is instrument proficient and utilizing an instrument equipped aircraft, then proceed with caution and have planned alternates.

9. Use caution when landing on runways that are covered by water or slush which cause hydroplaning (aquaplaning), a phenomenon that renders braking and steering ineffective because of the lack of sufficient surface friction. Snow- and ice-covered runways are also hazardous.

10. Use caution when taking off or landing during gusty wind conditions.

11. Avoid taking off or landing too close behind large aircraft. "Wake turbulence" caused by these aircraft can be hazardous.

12. When the flight has been completed, visit or call the National Weather Service airport station or FAA Flight Service Station and discuss the weather encountered en route.

Using the Aeronautical Chart

Draw the course to be flown on the sectional chart or charts. The course line should begin at the center of the airport of departure and end at the center of the destination airport. If the route is

Study available information about each airport at which a landing is intended. This should include a study of the NOTAMS and the Airport/Facility Directory. This includes location, elevation, runway and lighting facilities, available services, availability of UNICOM, types of fuel available (use to decide on refueling stops), FSS located on the airport, control tower and ground control frequencies, traffic information, remarks and other pertinent information. The NOTAMS, issued every 14 days, should be checked for additional information on hazardous conditions or changes that have been made since issuance of the Airport/Facility Directory.

The Sectional Chart Bulletin subsection should be checked for major changes that have occurred since the last publication date of each.sectional chart being used. Remember, the chart may be up to 6 months old. The published date of the chart appears at the top of the legend side of the chart.

The Airport/Facility Directory will generally have the latest information pertaining to such matters and should be used in preference to the information on the back of the chart, if there are differences.

Aircraft Flight Manual or Pilot's Operating Handbook Check these publications to determine the proper loading of the airplane (weight and balance data). The weight of the usable fuel and drainable oil aboard must be known. Also, the weight of the passengers, the weight of all baggage to be carried, and the empty weight of the airplane to be sure that the total weight does not exceed the maximum allowable. The distribution of the load must be known to tell if the resulting center of gravity is within limits. Be sure to use the latest weight and balance information in the FAA-approved Airplane Flight Manual or other permanent aircraft records, as appropriate, to obtain empty weight and empty weight center of gravity information.

Determine the takeoff and landing distances from the appropriate charts, based on the calculated load, elevation of the airport, and temperature; then compare these distances with the amount of runway available. Remember, the heavier the load and the higher the elevation, temperature, or humidity, the longer the takeoff roll and landing roll and the lower the rate of climb.

Check the fuel consumption charts to determine the rate of fuel consumption at the estimated flight altitude and power settings. Calculate the rate of fuel consumption, then compare it with the estimated time for the flight so that refueling points along the route can be included in the plan.

Using the Plotter, Computer, or Electronic Calculator, etc. When drawing a course line on the aeronautical chart, use a protractor (or plotter) to determine the true course. Then determine the magnetic variation from the

direct, the course line will consist of a single straight line. If the route is not direct, it will consist of two or more straight line segments—for example, a direct VOR station, which is off the direct route but which will make navigating easier, may be chosen.

Appropriate checkpoints should be selected along the route in some way. These should be easy-to-locate points such as large towns, large lakes and rivers, or combinations of recognizable points such as towns with an airport, towns with a network of highways and railroads entering and departing, etc. Normally choose only towns indicated by splashes of yellow on the chart. Do not choose towns represented by a small circle—these may turn out to be only a half-dozen houses. (In isolated areas, however, towns represented by a small circle can be prominent checkpoints.)

The areas on either side of the planned route should be checked for alert, warning, restricted, prohibited and military operations areas, or Air Defense Identification Zones (ADIZ). Each area will have its restrictions printed on the chart either within the area or somewhere near the border, depending on its size.

Study the terrain along the route. This is necessary for several reasons. It should be checked to determine the highest and lowest elevations to be encountered so that an appropriate altitude which will conform to FAA regulations can be selected. If the flight is to be flown at an altitude more than 3,000 feet above the terrain, conformance to the cruising altitude appropriate to the direction of flight is required. Check the route for particularly rugged terrain so it can be avoided. Areas where a takeoff or landing will be made should be carefully checked for tall obstructions. Television transmitting towers may extend to altitudes over 1,500 ft. above the surrounding terrain. It is essential that pilots be aware of their presence and location.

Make a list of the navigation aids that will be used along the route and the frequency on which they can be received. Indicate the aids that have voice facilities so that stations having weather broadcasts will be known.

It is important that the chart legend be used to determine the meaning of chart symbols or colors.

Use of the Airport/Facility Directory Make a list of the Flight Service Stations along the intended route and the frequencies which can be used for transmitting and receiving (in addition to the navigation aid frequencies selected from the chart). Check the correctness of navigation aid frequencies selected from the aeronautical chart. This can be done by checking the Sectional Chart Bulletin, NOTAMS, and the appropriate navigational aid information in the Airport/Facility Directory.

mid-isogonic line, apply it to the measured true course and obtain the magnetic course. When flying at an altitude of more than 3,000 ft. above the surface, the magnetic course must be known to decide whether to fly at an even-thousand-plus-five-hundred-foot level or an odd-thousand-plus-five-hundred-foot level. Then measure the length of the course line, using care to assure that the appropriate scale is used.

If after a thorough weather check it is decided that the flight can be made safely, the winds-aloft forecast should be obtained and altitude selected. The altitude chosen must conform to FAA regulations, and also have the most favorable winds possible. Of course, it may be desirable to sacrifice favorable winds at times in order to fly at an altitude where there is no turbulence. After determining the altitude and the forecast winds at that altitude, use this information, the estimated true airspeed, and measured true course, to compute the true heading and groundspeed. From the computed true heading, determine the compass heading by applying variation (already obtained from the mid-isogonic line on the chart) and deviation (obtained from the compass correction card). From the computed groundspeed and measured course distance, determine the total flight time. Then use the computed total time and estimated fuel consumption rate to determine the amount of fuel that will be consumed during the flight.

After making the necessary computations, a flight plan can be filed.

VFR Flight Plan An examination of en route accidents shows a striking relationship between the number of accidents by aircraft not on flight plans and those on flight plans. Filing a flight plan is not required by FAA regulations; however, it is good operating practice, since the information contained in the flight plan will be used in search and rescue operations in event of emergency.

Though flight plans can be filed in the air by radio, it is usually best to file a flight plan with the nearest FSS in person or by phone just before departing. After takeoff, contact the FSS by radio and give them the takeoff time so the flight plan can be activated. To avoid congestion of already busy communication channels, use radio for filing flight plans *only* when it is impossible to file any other way.

When a VFR flight plan is filed, it will be held by the FSS until 1 hour after the proposed departure time and then canceled unless:

(1) The actual departure time is received; or
(2) A revised proposed departure time is received; or
(3) At the time of filing, the FSS is informed that the proposed departure

time will be met, but actual time cannot be given because of inadequate communication.

The FSS specialist who accepts the flight plan will not inform the pilot of this procedure, however.

Remember, there is every advantage in filing a flight plan; but *do not forget to close the flight plan upon arrival.* Do this by telephone with the nearest FSS, if possible, to avoid radio congestion. If there is no FSS near the point of landing, the flight plan may be closed by radio with the nearest FSS upon arrival at the destination.

Figures 7-34 and 7-35 show the flight plan form a pilot files with the Flight Service Station. When filing a flight plan by telephone or radio, give the information in the order of the numbered spaces. This enables the FSS specialist to copy the information more efficiently. Most of the spaces are either self-explanatory or nonapplicable to the VFR flight plan (such as item 13). However, some spaces may need explanation.

Item 4 asks for the estimated true airspeed in knots. If the pilot is able to convert the airspeed from miles per hour to knots, there is no problem. If not, then report the airspeed in miles per hour.

Item 6 asks for the proposed departure time in Greenwich Mean Time

---

DEPARTMENT OF TRANSPORTATION—FEDERAL AVIATION ADMINISTRATION

FLIGHT PLAN

Form Approved
OMB No. 04-R0072

1. TYPE	2. AIRCRAFT IDENTIFICATION	3. AIRCRAFT TYPE/SPECIAL EQUIPMENT	4. TRUE AIRSPEED	5. DEPARTURE POINT	6. DEPARTURE TIME		7. CRUISING ALTITUDE
VFR ☒ IFR ☐ DVFR ☐	N12KA	B-315	140 KTS	ABI MUN	PROPOSED (Z) 1830Z	ACTUAL (Z)	6,500

8. ROUTE OF FLIGHT
ABI V66 MAF V94 SFL V16 ELP

9. DESTINATION (Name of airport and city)	10. EST. TIME ENROUTE		11. REMARKS
TUT'L EL PASO, TEX	HOURS 2	MINUTES 49	

12. FUEL ON BOARD		13. ALTERNATE AIRPORT(S)	14. PILOT'S NAME, ADDRESS & TELEPHONE NUMBER & AIRCRAFT HOME BASE	15. NUMBER ABOARD
HOURS 3	MINUTES 30	NR	K. ANDER — SENESCENT AIRPORT N. DAK.	3

14. COLOR OF AIRCRAFT CREAM + BLACK ON PURPLE	CLOSE VFR FLIGHT PLAN WITH ELP ____ FSS ON ARRIVAL

FAA Form 7233-1 (5-72)

☆ U.S. GOVERNMENT PRINTING OFFICE 1976 · 671-921/561/7

Figure 7-34. *Flight plan form.*

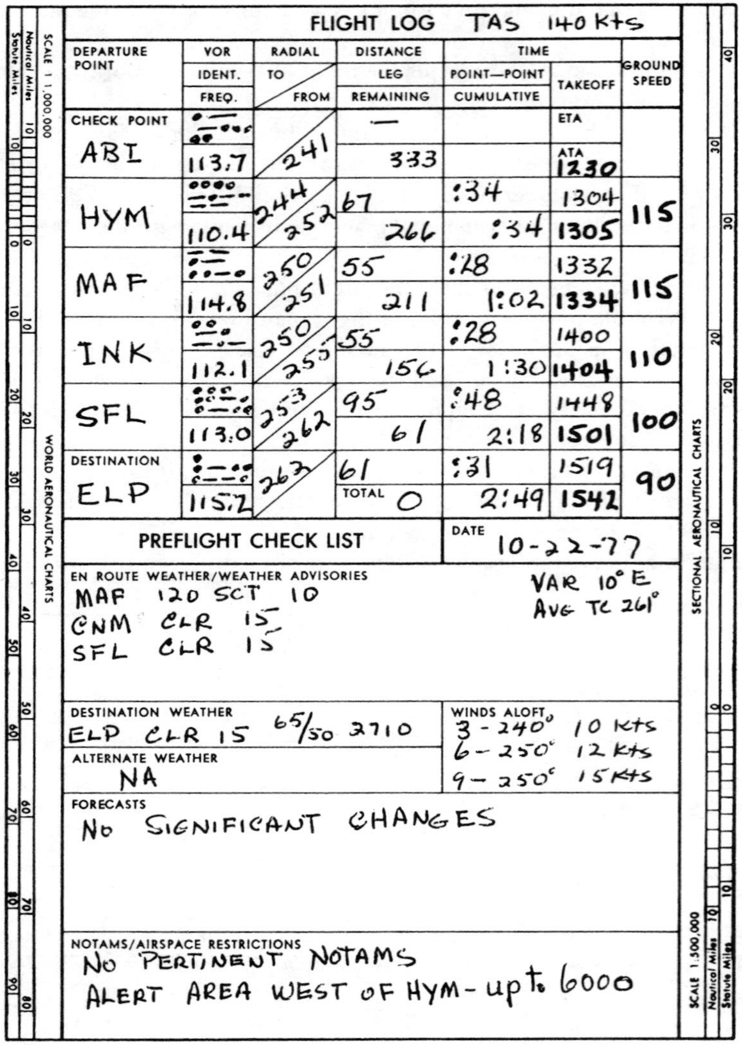

FLIGHT LOG TAS 140 Kts

DEPARTURE POINT / CHECK POINT	VOR FREQ.	RADIAL TO	RADIAL FROM	DISTANCE LEG	DISTANCE REMAINING	TIME POINT—POINT	TIME CUMULATIVE	TAKEOFF ETA	TAKEOFF ATA	GROUND SPEED
ABI	113.7		241	—	333				1230	
HYM	110.4	244	252	67	266	:34	:34	1304	1305	115
MAF	114.8	250	251	55	211	:28	1:02	1332	1334	115
INK	112.1	250	255	55	156	:28	1:30	1400	1404	110
SFL	113.0	253	262	95	61	:48	2:18	1448	1501	100
DESTINATION ELP	115.2	263		61	TOTAL 0	:31	2:49	1519	1542	90

PREFLIGHT CHECK LIST — DATE 10-22-77

EN ROUTE WEATHER/WEATHER ADVISORIES
MAF 120 SCT 10
CNM CLR 15
SFL CLR 15
VAR 10° E
AVG TC 261°

DESTINATION WEATHER
ELP CLR 15 65/50 2710

ALTERNATE WEATHER
NA

WINDS ALOFT
3 - 240° 10 Kts
6 - 250° 12 Kts
9 - 250° 15 Kts

FORECASTS
No SIGNIFICANT CHANGES

NOTAMS/AIRSPACE RESTRICTIONS
No PERTINENT NOTAMS
ALERT AREA WEST OF HYM - up to 6000

NOTE: ETAS ARE BASED ON AN ESTIMATED GROUNDSPEED OF 118 KNOTS. ATAS ARE BASED ON ACTUAL GROUNDSPEED DETERMINED DURING FLIGHT.

Figure 7–35. *Reverse side of flight plan form.*

(indicated by the "Z"). If the pilot is unable to convert local standard time to Greenwich Time, give the time as local standard and the FSS will convert it to Greenwich. To convert local standard time to Greenwich Mean Time, add 5 hours to Eastern Standard Time (EST); add 6 hours to Central Standard Time CST); add 7 hours to Mountain Standard Time (MST); and add 8 hours to Pacific Standard Time (PST). To convert local daylight time to Greenwich Mean Time, add 4 hours to Eastern Daylight Time (EDT); add 5 hours to Central Daylight Time (CDT); add 6 hours to Mountain Daylight Time (MDT); and add 7 hours to Pacific Daylight Time (PDT).

Item 7 asks for the cruising altitude. Normally "VFR" can be entered in this block, since the pilot will choose a cruising altitude to conform to FAA regulations (on IFR flights, air traffic control designates the cruising altitude).

Item 8 asks for the route of flight. If the flight is to be direct, enter the word "direct"; if not, enter the actual route to be followed such as via certain towns or navigation aids.

Item 12 asks for the fuel on board in hours and minutes. This is determined by dividing the total usable fuel aboard in gallons by the estimated rate of fuel consumption in gallons.

Figure 7–35 shows the reverse side of the flight plan. This is used as a checklist for—and a place to enter—the information pertinent to the flight. It also contains a measuring scale for both Sectional Aeronautical Charts and World Aeronautical Charts.

Even if a flight plan is not filed, regular position reports should be made to Flight Service Stations to receive altimeter setting, SIGMETS, and advisories to small aircraft. This will also enable search and rescue action to be focused in the proper area in case of an emergency. Remember, the Flight Service Stations are anxious to help in every way possible. It is only sensible to take advantage of their services.

CHAPTER VIII—FLIGHT INFORMATION PUBLICATIONS

The purpose of this chapter is to provide a means that can be used to become familiar with the contents of the Airman's Information Manual, Airport/Facility Directory, Notices to Airmen, and Graphic Notices and Supplemental Data.

The information relating primarily to the operation of aircraft in VFR conditions has been extracted from the Airman's Information Manual and placed in this chapter. A variety of subject areas are covered, the knowledge of which will be very helpful to the pilot. Also, information relating to the format, legend, and contents of the Airport/Facility Directory, Notices to Airmen and Graphic Notices and Supplemental Data has been placed in this chapter. This information will be useful in becoming familiar with these publications.

Although the information in this chapter was current at the time this handbook was developed changes may occur, because of operational necessity, which could cause obsolescence of some material. If there is reason to believe that changes have taken place, reference should be made to current publications for the updated information.

The Airman's Information Manual is designed to provide airmen with basic flight information and ATC procedures for use in the National Airspace System (NAS) of the U.S. The information contained parallels the U.S. Aeronautical Information Publication (AIP) distributed internationally.

This manual contains the basic fundamentals required in order to fly in the U.S. NAS. It also contains items of interest to pilots concerning health and medical facts, factors affecting flight safety, a pilot/controller glossary of terms used in the Air Traffic Control System, and information on safety, accident and hazard reporting.

This manual is complimented by other operational publications which are available upon separate subscription. These publications are:

Airport/Facility Directory, Alaska Supplement, Pacific Supplement—These publications contain information on airports, communications, navigational aids, instrument landing systems, VOR receiver check points, preferred routes, FSS/Weather Service telephone numbers, Air Route Traffic Control Center (ARTCC) frequencies, part-time control zones, and various other pertinent, special notices essential to air navigation. The publications are available upon subscription from the National Ocean Survey (NOS), Distribution Division (C–44), Riverdale, Maryland 20840.

Notices to Airmen (Class-II)—A publication containing current Notices to Airmen (NOTAMs) which are considered essential to the safety of flight as well as supplemental data affecting the other operational publications listed here. It also includes current FDC NOTAMs, which are regulatory in nature, issued to establish restrictions to flight or amend charts or published Instrument Approach Procedures. This publication is issued every 14 days and is available through subscription from the Superintendent of Documents.

Graphic Notices and Supplemental Data—a publication containing a tabulation of Parachute Jump Areas; Special Notice Area Graphics; Terminal Area Graphics; Terminal Radar Service Area (TRSA) Graphics; Olive Branch Routes; and other data, as required, not subject to frequent change. This publication is issued quarterly and is available through subscription from the Superintendent of Documents.

FLIGHT INFORMATION PUBLICATION POLICY

The following is, in essence, the statement issued by the FAA Administrator and published in the December 10, 1964, issue of the Federal Register, concerning the FAA policy as pertaining to the type of information that will be published as NOTAMs and in the Airman's Information Manual.

It is a pilot's inherent responsibility that he be alert at all times for and in anticipation of all circumstances, situations and conditions which affect the safe operation of his aircraft. For example, a pilot should expect to find air traffic at any time or place. At or near both civil and military airports and in the vicinity of known training areas, a pilot should expect concentrated air traffic although he should realize concentrations of air traffic are not limited to these places.

It is the general practice of the agency to advertise by NOTAM or other flight information publications such information it may deem appropriate; information which the agency may from time to time make available to pilots is solely for the purpose of assisting them in executing their regulatory responsibilities. Such information serves the aviation community as a whole and not pilots individually.

The fact that the agency under one particular situation or another may or may not furnish information does not serve as a precedent of the agency's responsibility to the aviation community; neither does it give assurance that other information of the same or similar nature will be advertised nor does it guarantee that any and all information known to the agency will be advertised.

Consistent with the foregoing, it shall be the policy of the Federal Aviation Administration to furnish information only when, in the opinion of the agency, a unique situation should be advertised and not to furnish routine information such as concentrations of air traffic, either civil or military. The Airman's Information Manual will not contain informative items concerning everyday circumstances that pilots should, either by good practice or regulation, expect to encounter or avoid.

AERONAUTICAL INFORMATION AND THE NATIONAL AIRSPACE SYSTEM

1. Aeronautical information concerning the National Airspace System is disseminated by three methods. The primary method is aeronautical charts. The second method is the Airman's Information Manual (AIM), and the third is the National Notice to Airmen System. These three systems have been designed to supplement and complement each other. The basic difference between these three systems is the frequency of issuance. To the maximum extent possible, aeronautical charts reflect the most current information available at time of printing. The AIM contains static procedural data and data changes known sufficiently in advance to permit publication.

2. Information of a time-critical nature that is required for flight planning and not known sufficiently in advance to permit publication on a chart or in the AIM receives immediate handling through the National Notice to Airmen System.

3. Information distributed by the Notice to Airmen System is categorized into two types—NOTAM (D) and NOTAM (L). It is the intent, insofar as possible, to limit to dissemination by NOTAM (D) that time-critical information which would affect a pilot's decision to make a flight; for example, an airport closed, terminal radar out of service, en route navigational aids out of service, etc. Dissemination of information in this category will include that pertaining to all navigational facilities and all IFR airports with approved instrument approach procedures and for those VFR airports which are designated as the destination point on a daily average of two or more general avia-tion VFR flight plans. All such airports are annotated in Part 2 and Part 3 of this manual by the section symbol "§".

4. Information which is primarily of an advisory or "nice-to-know" nature, plus data on airports not included above, that can be given to the pilot upon request on an "as-needed" basis before departure, while en route, or prior to landing, is classed as a NOTAM (L) and given local distribution via appropriate voice communications, local teletypewriter or telautograph circuits, telephone, etc. Examples of this type are: Men and equipment crossing a runway, taxiway closed, etc.

5. Pilots planning a flight should contact the nearest FAA Flight Service Station to obtain current flight information.

AIR NAVIGATION RADIO AIDS

GENERAL

Various types of air navigation aids are in use today, each serving a special purpose in our system of air navigation.

These aids have varied owners and operators namely: the Federal Aviation Administration, the military services, private organizations; and individual states and foreign governments.

The Federal Aviation Administration has the statutory authority to establish, operate, and maintain air navigation facilities and to prescribe standards for the operation of any of these aids which are used by both civil and military aircraft for instrument flight in federally controlled airspace. These aids are tabulated in the Airport/Facility Directory by State.

A brief description of these aids follows. Also, a composite table of normal usable altitudes and distances appears in Class of VOR/-VORTAC/TACAN.

NONDIRECTIONAL RADIO BEACON (NDB)

1. A low or medium-frequency radio beacon transmits nondirectional signals whereby the pilot of an aircraft equipped with a loop antenna can determine his bearing and "home" on the station. These facilities normally operate in the frequency band of 200 to 415 kHz and transmit a continuous carrier with 1020 Hz modulation keyed to provide identification except during voice transmission.

2. When a radio beacon is used in conjunction with the Instrument Landing System markers, it is called a Compass Locator.

3. All radio beacons except the compass locators transmit a continuous three-letter identification in code except during voice transmissions. Compass locators transmit a continuous two-letter identification in code. The first and second letters of the three-letter location identifier are assigned to the front course outer marker compass locator (LOM), and the second and third letters are assigned to the front course middle marker compass locator (LMM).

Example:

ATLANTA, ATL, LOM-AT, LMM-TL.

4. Voice transmissions are made on radio beacons unless the letter "W" (without voice) is included in the class designator (HW).

5. Radio beacons are subject to disturbances that result in ADF needle deviations, signal fades and interference from distant station during night operations. Pilots are cautioned to be on the alert for these vagaries.

VHF OMNIDIRECTIONAL RANGE (VOR)

1. VORs operate within the 108.0–117.95 MHz frequency band and have a power output necessary to provide coverage within their assigned operational service volume. The equipment is VHF, thus, it is subject to line-of-sight restriction, and its range varies proportionally to the altitude of the receiving equipment. There is some "spill over," however, and reception at an altitude of 1000 feet is about 40 to 45 miles. This distance increases with altitude.

2. Most VORs are equipped for voice transmission on the VOR frequency.

3. The effectiveness of the VOR depends upon proper use and adjustment of both ground and airborne equipment.

a. **Accuracy:** The accuracy of course alignment of the VOR is excellent, being generally plus or minus 1°.

b. **Roughness:** On some VORs, minor course roughness may be observed, evidence by course needle or brief flag alarm activity (some receivers are more subject to these irregularities than others). At a few stations, usually in mountainous terrain, the pilot may occasionally observe a brief course needle oscillation, similar to the indication of "approaching station." Pilots flying over unfamiliar routes are cautioned to be on the alert for these vagaries, and in particular, to use the "to-from" indicator to determine positive station passage.

(1) Certain propeller RPM settings can cause the VOR Course Deviation Indicator to fluctuate as much as ±6°. Slight changes to the RPM setting will normally smooth out this roughness. Helicopter rotor speeds may also cause VOR course disturbances. Pilots are urged to check for this propeller modulation phenomenon prior to reporting a VOR station or aircraft equipment for unsatisfactory operation.

4. The only positive method of identifying a VOR is by its Morse Code identification or by the recorded automatic voice identification which is always indicated by use of the word "VOR" following the range's name. Reliance on determining the identification of an omnirange should never be placed on listening to voice transmissions by the Flight Service Station (FSS) (or approach control facility) involved. Many FSS remotely operate several omniranges which have different names from each other and in some cases none have the name of the "parent" FSS. (During periods of maintenance the coded identification is removed. See MAINTENANCE OF FAA NAVAIDS.)

5. Voice identification has been added to numerous VHF omniranges. The transmission consists of a voice announcement, "AIRVILLE VOR," alternating with the usual Morse Code identification.

VOR RECEIVER CHECK

1. Periodic VOR receiver calibration is most important. If a receiver's Automatic Gain Control or modulation circuit deteriorates, it is possible for it to display acceptable accuracy and sensitivity close in to the VOR or VOT and display out-of-tolerance readings when located at greater distances where weaker signal areas exist. The likelihood of this deterioration varies between receivers, and is generally considered a function of time. The best assurance of having an accurate receiver is periodic calibration. Yearly intervals are recommended at which time an authorized repair facility should recalibrate the receiver to the manufacturer's specifications.

2. Part 91.25 of the Federal Aviation Regulations provides for certain VOR equipment accuracy checks prior to flight under instrument flight rules. To comply with this requirement and to ensure satisfactory operation of the airborne system, the FAA has provided pilots with the following means of checking VOR receiver accuracy: (1) FAA VOR test facility (VOT) or a radiated test signal from an appropriately rated radio repair station, (2) certified airborne check points, and (3) certified check points on the airport surface.

a. The FAA VOR test facility (VOT) transmits a test signal for VOR receivers which provides users of VOR a convenient and accurate means to

the antenna) is installed in the aircraft, the person checking the equipment may check one system against the other. He shall turn both systems to the same VOR ground facility and note the indicated bearing to that station. The maximum permissible variations between the two indicated bearings is 4°.

TACTICAL AIR NAVIGATION (TACAN)

1. For reasons peculiar to military or naval operations (unusual siting conditions, the pitching and rolling of a naval vessel, etc.) the civil VOR-DME system of air navigation was considered unsuitable for military air naval use. A new navigational system, Tactical Air Navigation (TACAN), was therefore developed by the military and naval forces to more readily lend itself to military and naval requirements. As a result, the FAA has been in the process of integrating TACAN facilities with the civil VOR-DME program. Although the theoretical, or technical principles of operation of TACAN equipment are quite different from those of VOR-DME facilities, the end result, as far as the navigating pilot is concerned, is the same. These integrated facilities are called VORTACs.

2. TACAN ground equipment consists of either a fixed or mobile transmitting unit. The airborne unit in conjunction with the ground unit reduces the transmitted signal to a visual presentation of both azimuth and distance information. TACAN is a pulse system and operates in the UHF band of frequencies. Its use requires TACAN airborne equipment and does not operate through conventional VOR equipment.

VHF OMNIDIRECTIONAL RANGE/TACTICAL AIR NAVIGATION (VORTAC)

1. VORTAC is a facility consisting of two components, VOR and TACAN, which provides three individual services: VOR azimuth, TACAN azimuth and TACAN distance (DME) at one site. Although consisting of more than one component, incorporating more than one operating frequency, and using more than one antenna system, a VORTAC is considered to be a unified navigational aid. Both components of a VORTAC are envisioned as operating simultaneously and providing the three services at all times.

2. Transmitted signals of VOR and TACAN are each identified by three-letter code transmission and are interlocked so that pilots using VOR azimuth with TACAN distance can be assured that both signals being received are definitely from the same ground station. The frequency channels of the VOR

determine the operational status of their receivers. The facility is designed to provide a means of checking the accuracy of a VOR receiver while the aircraft is on the ground. The radiated test signal is used by tuning the receiver to the published frequency of the test facility. With the Course Deviation Indicator (CDI) centered the omnibearing selector should read 0° with the to-from indication being "from" or the omnibearing selector should read 180° with the to-from indication reading "to." Should the VOR receiver operate an RMI (Radio Magnetic Indicator), it will indicate 180° on any OBS setting when using the VOT. Two means of identification are used with the VOR radiated test signal. In some cases a continuous series of dots is used while in others a continuous 1020 Hertz tone will identify the test signal. Information concerning an individual test signal can be obtained from the local Flight Service Station.

b. A radiated VOR test signal from an appropriately rated radio repair station serves the same purpose as an FAA VOR signal and the check is made in much the same manner with the following differences: (1) the frequency normally approved by the FCC is 108.0 MHz; (2) the repair stations are not permitted to radiate the VOR test signal continuously, consequently the owner/operator must make arrangements with the repair station to have the test signal transmitted. This service is not provided by all radio repair stations, the aircraft owner/operator must determine which repair station in his local area does provide this service. A representative of the repair station must make an entry into the aircraft logbook or other permanent record certifying to the radial accuracy which was transmitted and the date of transmission. The owner/operator or representative of the repair station may accomplish the necessary checks in the aircraft and make a logbook entry stating the results of such checks. It will be necessary to verify with the appropriate repair station the test signal radial being transmitted and whether you should get a "to," or "from" indication.

c. Airborne and ground check points consist of certified radials that should be received at specific points on the airport surface, or over specific landmarks while airborne in the immediate vicinity of the airport.

d. Should an error in excess of ±4° be indicated through use of a ground check, or ±6° using the airborne check, IFR flight shall not be attempted without first correcting the source of the error. CAUTION: no correction other than the "correction card" figures supplied by the manufacturer should be applied in making these VOR receiver checks.

e. The list of airborne check points, ground check points, and VOTs is published in the Airport/Facility Directory.

f. If dual system VOR (units independent of each other except for

DISTANCE MEASURING EQUIPMENT (DME)

1. In the operation of DME, paired pulses at a specific spacing are sent out from the aircraft (this is the interrogation) and are received at the ground station. The ground station (transponder) then transmits paired pulses back to the aircraft at the same pulse spacing but on a different frequency. The time required for the round trip of this signal exchange is measured in the airborne DME unit and is translated into distance (Nautical Miles) from the aircraft to the ground station.

2. Operating on the line-of-sight principle, DME furnishes distance information with a very high degree of accuracy. Reliable signals may be received at distances up to 199 NM at line-of-sight altitude with an accuracy of better than ½ mile or 3% of the distance, whichever is greater. Distance information received from DME equipment is SLANT RANGE distance and not actual horizontal distance.

3. DME operates on frequencies in the UHF spectrum between 962 MHz and 1213 MHz. Aircraft equipped with TACAN equipment will receive distance information from a VORTAC automatically, while aircraft equipped with VOR must have a separate DME airborne unit.

4. VOR/DME, VORTAC, ILS/DME, and LOC/DME navigation facilities established by the FAA provide course and distance information from co-located components under a frequency pairing plan. Aircraft receiving equipment which provides for automatic DME selection assures reception of azimuth and distance information from a common source whenever designated, VOR/DME, VORTAC, ILS/DME, and LOC/DME are selected.

5. Due to the limited number of available frequencies, assignment of paired frequencies has been required for certain military noncolocated VOR and TACAN facilities which serve the same area but which may be separated by distances up to a few miles. The military is presently undergoing a program to colocate VOR and TACAN facilities or to assign nonpaired frequencies to those facilities that cannot be colocated.

6. VOR/DME, VORTAC, ILS/DME, and LOC/DME facilities are identified by synchronized identifications which are transmitted on a time share basis. The VOR or localizer portion of the facility is identified by a coded tone modulated at 1020 Hz or by a combination of code and voice. The TACAN or DME is identified by a coded tone modulated at 1350 Hz. The DME or

TACAN coded identification is transmitted one time for each three or four times that the VOR or localizer coded identification is transmitted. When either the VOR or the DME is inoperative, it is important to recognize which identifier is retained for the operative facility. A single coded identification with a repetition interval of approximately 30 seconds indicates that the DME is operative.

7. Aircraft receiving equipment which provides for automatic DME selection assures reception of azimuth and distance information from a common source whenever designated VOR/DME, VORTAC and ILS/DME navigation facilities are selected. Pilots are cautioned to disregard any distance displays from automatically selected DME equipment whenever VOR or ILS facilities, which do not have the DME feature installed, are being used for position determination.

CLASS OF NAVAIDS

VOR, VORTAC, and TACAN aids are classed according to their operational use. There are three classes.

T (Terminal)
L (Low altitude)
H (High altitude)

The normal service range for the T, L, and H class aids is included in the following table. Certain operational requirements make it necessary to use some of these aids at greater service ranges than are listed in the table. Extended range is made possible through flight inspection determinations. Some aids also have lesser service range due to location, terrain, frequency protection, etc.

VOR/VORTAC/TACAN NAVAIDS
Normal Usable Altitudes and Radius Distances

Class	Altitudes	Distance (miles)
T	12,000' and below	25
L	Below 18,000'	40
H	Below 18,000'	40
H	Within the conterminous 48 states only, between 14,500' and 17,999'	100
H	18,000' – FL 450	130
H	Above FL 450	100

NON-DIRECTIONAL RADIO BEACON (NDB)

Usable Radius Distances for all Altitudes

Class	Power (watts)	Distance (miles)
Compass Locator	Under 25	15
MH	Under 50	25
H	50 – 1999	*50
HH	2000 or more	75

*Service range of individual facilities may be less than 50 miles.

MAINTENANCE OF FAA NAVAIDS

1. During periods of routine or emergency maintenance, the coded identification (or code and voice, where applicable) will be removed from certain FAA navaids; namely, ILS localizers, VHF ranges, NDB's, compass locators and 75 MHz marker beacons. The removal of identification serves as warning to pilots that the facility has been officially taken over by "Maintenance" for tune-up or repair and may be unreliable even though on the air intermittently or constantly.

NAVAIDS WITH VOICE

1. Voice equipped en route radio navigational aids are under the operational control of an FAA Flight Service Station (FSS), or an approach control facility. Most are remotely operated.

2. Unless otherwise noted on the chart, all radio navigation aids operate continuously except during interruptions for voice transmissions on the same frequencies where simultaneous transmission is not available, and during shutdowns for maintenance purposes. Hours of operation of those facilities not operating continuously are annotated on the charts.

VHF/UHF DIRECTION FINDER

1. The VHF/UHF Direction Finder (VHF/UHF/DF) is one of the Common System equipments that helps the pilot without his being aware of its operation. The VHF/UHF/DF is a ground-based radio receiver used by the operator of the ground station where it is located.

2. The equipment consists of a directional antenna system, a VHF and a UHF radio receiver. At a radar-equipped tower or center, the cathode-ray tube indications may be superimposed on the radarscope.

3. The VHF/UHF/DF display indicates the magnetic direction of the aircraft from the station each time the aircraft transmits. Where DF equipment is tied into radar, a strobe of light is flashed from the center of the radarscope in the direction of the transmitting aircraft.

4. DF equipment is of particular value in locating lost aircraft and in helping to identify aircraft on radar.

RADAR

1. Capabilities.

a. Radar is a method whereby radio waves are transmitted into the air and are then received when they have been reflected by an object in the path of the beam. *Range* is determined by measuring the time it takes (at the speed of light) for the radio wave to go out to the object and then return to the receiving antenna. The *direction* of a detected object from a radar site is determined by the position of the rotating antenna when the reflected portion of the radio wave is received.

b. More reliable maintenance and improved equipment have reduced radar system failures to a negligible factor. Most facilities actually have some components duplicated—one operating and another which immediately takes over when a malfunction occurs to the primary component.

2. Limitations.

It is very important for the aviation community to recognize the fact that there are limitations to radar service and that ATC controllers may not always be able to issue traffic advisories concerning aircraft which are not under ATC control and cannot be seen on radar.

(1) The characteristics of radio waves are such that they normally travel in a continuous straight line unless they are:

(a) "Bent" by abnormal atmospheric phenomena such as temperature inversions;

(b) Reflected or attenuated by dense objects such as heavy clouds, precipitation, ground obstacles, mountains, etc.; or

(c) Screened by high terrain features.

(2) The bending of radar pulses, often called anomalous propagation or ducting, may cause many extraneous blips to appear on the radar

operator's display if the beam has not been bent toward the ground or may decrease the detection range if the wave is bent upward. It is difficult to solve the effects of anomalous propagation, but using beacon radar and electronically eliminating stationary and slow moving targets by a method called moving target indicator (MTI) usually negate the problem.

(3) Radar energy that strikes dense objects will be reflected and displayed on the operator's scope thereby blocking out aircraft at the same range and greatly weakening or completely eliminating the display of targets at a greater range. Again, radar beacon and MTI are very effectively used to combat ground clutter and weather phenomena, and a method of circularly polarizing the radar beam will eliminate some weather returns. A negative characteristic of MTI is that an aircraft flying a speed that coincides with the canceling signal of the MTI (tangential or "blind" speed) may not be displayed to the radar controller.

(4) Relatively low altitude aircraft will not be seen if they are screened by mountains or are below the radar beam due to earth curvature. The only solution to screening is the installation of strategically placed multiple radars which has been done in some areas.

(5) There are several other factors which affect radar control. The amount of reflective surface of an aircraft will determine the size of the radar return. Therefore, a small light airplane or a sleek jet fighter will be more difficult to see on radar than a large commercial jet or military bomber. Here again, the use of radar beacon is invaluable if the aircraft is equipped with an airborne transponder. All ARTCC radars in the conterminous U.S. and many airport surveillance radars have the capability to interrogate Mode C and display altitude information to the controller from appropriately equipped aircraft. However, there are a number of airport surveillance radars that are still two dimensional (range and azimuth) only and altitude information must be obtained from the plot.

(6) The controllers' ability to advise a pilot flying on instruments or in visual conditions of his proximity to another aircraft will be limited if the unknown aircraft is not observed on radar, if no flight plan information is available, or if the volume of traffic and workload prevent his issuing traffic information. First priority is given to establishing vertical, lateral, or longitudinal separation between aircraft flying IFR under the control of ATC.

3. FAA radar units operate continuously at the locations shown in the Airport/Facility Directory, and their services are available to all pilots, both civil and military. Contact the associated FAA control tower or ARTCC on any frequency guarded for initial instructions, or in an emergency, any FAA facility for information on the nearest radar service.

AIR TRAFFIC CONTROL RADAR BEACON SYSTEM (ATCRBS)

1. The Air Traffic Control Radar Beacon System (ATCRBS), sometimes referred to as secondary surveillance radar, consists of three main components:

a. Interrogator. Primary radar relies on a signal being transmitted from the radar antenna site and for this signal to be reflected or "bounced back" from an object (such as an aircraft). This reflected signal is then displayed as a "target" on the controller's radarscope. In the ATCRBS, the *Interrogator*, a ground based radar beacon transmitter-receiver, scans in synchronism with the primary radar and transmits discrete radio signals which repetitiously requests all transponders, on the mode being used, to reply. The replies received are then mixed with the primary returns and both are displayed on the same radarscope.

b. Transponder. This airborne radar beacon transmitter-receiver automatically receives the signals from the interrogator and selectively replies with a specific pulse group (code) only to those interrogations being received on the mode to which it is set. These replies are independent of, and much stronger than a primary radar return.

c. Radarscope. The radarscope used by the controller displays returns from both the primary radar system and the ATCRBS. These returns, called targets, are what the controller refers to in the control and separation of traffic.

2. The job of identifying and maintaining identification of primary radar targets is a long and tedious task for the controller. Some of the advantages of ATCRBS over primary radar are:

a. Reinforcement of radar targets.

b. Rapid target identification.

c. Unique display of selected codes.

3. A part of the ATCRBS ground equipment is the decoder. This equipment enables the controller to assign discrete transponder codes to each aircraft under his control. Normally only one code will be assigned for the entire flight. Assignments are made by the ARTCC computer on the basis of the National Beacon Code Allocation Plan. The equipment is also designed to receive Mode C altitude information from the aircraft.

4. It should be emphasized that aircraft transponders greatly improve the effectiveness of radar systems.

AIRPORT, AIR NAVIGATION LIGHTING AND MARKING AIDS

AERONAUTICAL (LIGHT) BEACONS

1. An aeronautical light beacon is a visual NAVAID displaying flashes of white and/or colored light to indicate the location of an airport, a heliport, a landmark, a certain point of a Federal airway in mountainous terrain, or a hazard. The light used may be a rotating beacon or one or more flashing lights. The flashing lights may be supplemented by steady burning lights of lesser intensity.

2. The color or color combination displayed by a particular beacon and/or its auxiliary lights tell whether the beacon is indicating a landing place, landmark, point of the Federal airways, or hazard. Coded flashes of the auxiliary lights, if employed, further identify the beacon site.

ROTATING BEACON

1. The rotating beacon has a vertical light distribution such as to make it most effective at angles of one to three degrees above the horizontal from its site; however, it can be seen well above and below this peak spread. Rotation is in clockwise direction when viewed from above. It is always rotated at a constant speed which produces the visual effect of flashes at regular intervals. Flashes may be one or two colors alternately. The total number of flashes are:

12 to 30 per minute for beacons marking airports, landmarks, and points on Federal airways.

30 to 60 per minute for beacons marking heliports.

12 to 60 per minute for hazard beacons.

COLOR

2. The colors and color combinations of rotating beacons and auxiliary lights are basically:

White and Green......	Lighted land airport.
*Green alone.........	Lighted land airport.
White and Yellow.....	Lighted water airport.
*Yellow alone........	Lighted water airport.
White and red........	Landmark or navigational point.
White alone..........	Unlighted land airport (rare installation).
Red alone............	Hazard.
Green, Yellow, and White.............	Lighted heliport.
White...............	Hazard.

*Green alone or yellow alone is used only in connection with a not far distant white-and-green or white-and-yellow beacon display, respectively.

3. Military airport beacons flash alternately white and green, but are differentiated from civil beacons by dual-peaked (two quick) white flashes between the green flashes.

4. Pilots should not rely solely on the operation of the rotating beacon to indicate weather conditions, IFR versus VFR. In control zones, operation of the rotating beacon during the hours of daylight may indicate that the ground visibility is less than 3 miles and/or the ceiling is less than 1,000 feet. ATC clearance in accordance with FAR Part 91 would be required for landing, takeoff and flight in the traffic pattern. At locations with control towers and if controls are provided, ATC personnel turn the beacon on. However, at many airports throughout the country, the rotating beacon is turned on by a photoelectric cell or time clocks and ATC personnel have no control as to when it shall be turned on. Also, there is no regulatory requirement for daylight operation and pilots are reminded that it remains their responsibility for complying with proper pre-flight planning in accordance with FAR Part 91.5.

AUXILIARY LIGHTS

1. The auxiliary lights are of two general kinds: code beacons and course lights. The code beacon, which can be seen from all directions, is used to identify airports and landmarks and to mark hazards. The number of code beacon flashes are:

a. Green coded flashes not exceeding 40 flashes or character elements per minute, or constant flashes 12 to 15 per minute, for identifying land airports.

b. Yellow coded flashes not exceeding 40 flashes or character elements per minute, or constant flashes 12 to 15 per minute, for identifying water airports.

c. Red flashes, constant rate, 12 to 40 flashes per minute, for marking hazards.

2. The course light, which can be seen clearly from only one direction, is used only with rotating beacons of the Federal Airway System; two course lights, back to back, direct coded flashing beams of light in either direction along the course of airway.

OBSTRUCTIONS

1. Obstructions are marked/lighted to warn airmen of its presence during daytime and nighttime conditions. They may be marked/lighted in any of the following combinations:

a. Aviation Red Obstruction Lights. Flashing aviation red beacons and steady burning aviation red lights during nighttime operation. Aviation orange and white paint is used for daytime marking.

b. High Intensity White Obstruction Lights. Flashing high intensity white lights during daytime with reduced intensity for twilight and nighttime operation. When this type system is used, the marking of structures with red obstruction lights and aviation orange and white paint may be omitted.

c. Dual Lighting. A combination of flashing aviation red beacons and steady burning aviation red lights for nighttime operation and flashing high intensity white lights for daytime operation. Aviation orange and white paint may be omitted.

2. High intensity flashing white lights are being used to identify some supporting structures of overhead transmission lines located across rivers, chasms, gorges, etc. These lights flash in a middle, top, lower light sequence at approximately 60 flashes per minute. The top light is normally installed near the top of the supporting structure, while the lower light indicates the approximate lower portion of the wire span. The lights are beamed towards the companion structure and identify the area of the wire span.

3. High intensity flashing white lights are also employed to identify tall structures, such as chimneys and towers, as obstructions to air navigation. The lights provide a 360 degree coverage about the structure at 40 flashes per minute and consist of from one to seven levels of lights depending upon the height of the structure. Where more than one level is used the vertical banks flash simultaneously.

AIRWAY BEACONS

Airway beacons are remnants of the "lighted" airways which antedated the present electronically equipped Federal Airways System. Only a few of these beacons exist today to mark airway segments in remote mountain areas. Flashes in Morse Code identify the beacon site.

CONTROL OF LIGHTING SYSTEMS

1. Operation of approach light systems and runway lighting is controlled by the control tower. At some locations the FSS may control the lights where there is no control tower in operation.

2. Pilots may request that lights be turned on or off. Runway edge lights, in-pavement lights and approach lights also have intensity controls which may be varied to meet the pilot's request. Sequenced flashing lights may be turned on and off. Some sequenced flashing systems also have intensity control.

3. The Medium Intensity Approach Lighting System with Runway Alignment Indicators (MALSR) has been installed at many airports. The control of MALSR is now being transferred from the runway light circuits to Air Traffic Control towers and/or radio control from approaching aircraft. As soon as the transfer becomes effective, operational procedures for control from aircraft will be published. In the interim, with few exceptions, MALSR will operate only when runway edge lights are turned on and its intensity will vary directly as the runway edge light intensity is varied.

4. Omnidirectional flashing light lead-in approach and runway end identifier light systems are now being installed at several airports. This system consists of five flashing lights located on the extended runway centerline and two located on either side of the runway threshold. The lights flash toward the threshold in sequence.

PILOT CONTROL OF AIRPORT LIGHTING

1. The Federal Aviation Administration is installing controls on selected airport lights to provide pilots with the ability to control lights by keying the microphone. These controls will be available at all times at selected locations that do not have a tower or flight service station. Airports served by part-time towers or stations will have the control system activated when the tower or station is not operating. Control of the lights will be possible when aircraft are within 15 miles of the airport. Only one lighting system per runway may be operated by a pilot control system. Where a single runway is served by both approach lights and runway edge lights, priority for pilot control will be given to the approach light system. If no approach lights are installed, priority will be given to runway edge lights over other lighting systems such as REIL and VASI.

2. FAA approved control systems provide for the installation of three types of radio controls. These types are: a three step system that provides low, medium, or high intensity; a two step system that provides medium or high intensity; and a control to turn on a light system without regard to intensity. Each activation or change of intensity will start a timer to maintain the selected light intensity step of 15 minutes (which should be adequate time to complete an approach, landing, and necessary taxiing). A new 15 minute period may be obtained by repeating the (desired) microphone keying to the appropriate step.

3. The two step control must be activated to the highest intensity setting (key mike 5 times) before low intensity may be selected.

4. The three step control may be activated to provide either low, medium or high intensity initially.

5. Suggested useage would be to always activate the control by keying the mike 5 times to ensure the lights are activated. All controls, regardless of the system can be activated by keying the mike 5 times. Adjustment can then be made to high or low intensity as appropriate or desired at a later time. Each microphone keying resets the timer to maintain the lights for an additional 15 minutes from each keying.

LEGEND FOR LIGHT CONTROLS

Radio Control System	Key Mike	Intensity
3-step light system	7 times in 5 seconds	High
	5 times in 5 seconds	Medium
	3 times in 5 seconds	Low
*2-step light system	5 times in 5 seconds	High
	3 times in 5 seconds	Medium
ACTIVATE (Rwy lights, REIL, or VASI)	5 times in 5 seconds	Lights on

*Must be activated to High intensity before MEDIUM may be selected.

6. Where the airport is not served by an instrument approach procedure, it may have either the standard FAA approved control system or an independent type system of different specification installed by the airport sponsor. The Airport/Facility Directory contains descriptions of pilot controlled lighting systems for each airport having these systems and explains the type of lights, method of control, and operating frequency in clear text.

7. Where the airport is served by one or more instrument approach procedures, the instrument approach chart will include sufficient data to identify the control device, light system, and the control frequency(s). For example: a three step control for a MALSR installed on runway 25 that is controlled on frequency 122.8 MHz would be noted on the instrument approach chart as follows: 3 step MALSR Rwy 25—122.8. To turn on Runway 25 edge lights using 122.3 or 281.6 with a control that does not have the intensity steps, the instrument approach chart would be annotated: Activate Runway 25 MIRL 122.3/281.6.

VISUAL APPROACH SLOPE INDICATOR (VASI)

1. The VASI is a system of lights so arranged to provide visual descent guidance information during the approach to a runway. These lights are visible from 3–5 miles during the day and up to 20 miles or more at night. The visual glide path of the VASI provides safe obstruction clearance within ±10 degrees of the extended runway centerline and to 4 nautical miles from the runway threshold. Descent, using the VASI, should not be initiated until the aircraft is visually aligned with the runway. Lateral course guidance is provided by the runway or runway lights.

2. VASI installations may consist of either 2, 4, 6, 12, or 16-light units arranged in bars referred to as near, middle, and far bars. Most VASI installations consist of two bars, near and far, and may consist of 2, 4, or 12-light units. Some airports have VASIs consisting of three bars, near, middle, and far, which provide an additional visual glide path for use by high cockpit aircraft. This installation may consist of either 6- or 16-light units. VASI installations consisting of 2, 4, or 6-light units are located on one side of the runway, usually the left. Where the installation consists of 12- or 16-light units, the light units are located on both sides of the runway.

3. Two bar VASI installations provide one visual glide path which is normally set at 3 degrees. Three bar VASI installations provide two visual glide paths. The lower glide path is provided by the near and middle bars and is normally set at 3 degrees while the upper glide path, provided by the middle and far bars, is normally ¼ degree higher. This higher glide path is intended for use only by high cockpit aircraft to provide a sufficient threshold crossing height. Although normal glide path angles are three degrees, angles at some locations may be as high as 4.5 degrees to give proper obstacle clearance. Pilots of high performance aircraft are cautioned that use of VASI angles in excess of 3.5 degrees may cause an increase in runway length required for landing and rollout.

4. The following information is provided for pilots as yet unfamiliar with the principles and operation of this system and pilot technique required. The basic principle of the VASI is that of color differentiation between red and white. Each light unit projects a beam of light having a white segment in the upper part of the beam and red segment in the lower part of the beam. The light units are arranged so that the pilot using the VASIs during an approach will see the combination of lights in Figure 8–1.

2-BAR VASI

Light Bar		Color
(a) Below glide path	Far	Red
	Near	Red
(b) On glide path	Far	Red
	Near	White
(c) Above glide path	Far	White
	Near	White

3-BAR VASI

Light Bar		Color
(a) Below both glide paths	Far	Red
	Middle	Red
	Near	Red
(b) On lower glide path	Far	Red
	Middle	Red
	Near	White
(c) On upper glide path	Far	Red
	Middle	White
	Near	White
(d) Above both glide paths	Far	White
	Middle	White
	Near	White

Figure 8–1. *VASI.*

5. When on the proper glide path of a 2-bar VASI, the pilot will see the near bar as white and the far bar as red. From a position below the glide path, the pilot will see both bars as red. In moving up to the glide path, the pilot will see the color of the near bar change from red to white. From a position above the glide slope the pilot will see both bars as white. In moving down to

the glide path, the pilot will see the color of the far bar change from white to pink to red. When the pilot is below the glide path the red bars tend to merge into one distinct red signal and a safe obstruction clearance may not exist under this condition.

6. When using a 3–bar VASI it is not necessary to use all three bars. The near and middle bars constitute a two bar VASI for using the lower glide path. Also, the middle and far bars constitute a 2-bar VASI for using the upper glide path. A simple rule of thumb when using a two-bar VASI is:

All Red		Too Low
All White		Too High
Red & White		On Glide Path

7. In haze or dust conditions or when the approach is made into the sun, the white lights may appear yellowish. This is also true at night when the VASI is operated at a low intensity. Certain atmospheric debris may give the white lights an orange or brownish tint; however, the red lights are not affected and the principle of color differentiation is still applicable.

TRI-COLOR VISUAL APPROACH SLOPE INDICATOR

Tri-color Visual Approach Indicators have been installed at general aviation and air carrier airports. The Tri-color Approach Slope Indicator normally consists of a single light unit, projecting a three-color visual approach path into the final approach area of the runway upon which the system is installed. In all of these systems, a below glide path indication is red, the above glide path indication is amber and the on path indication is green.

Presently installed Tri-color Visual Approach Slope Indicators are low candlepower projector-type systems. Research tests indicate that these systems generally have a daytime useful range of approximately ½ to 1 mile. Nighttime useful range, depending upon visibility conditions, varies from 1 to 5 miles. Projector-type Visual Approach Slope Indicators may be initially difficult to locate in flight due to their small light source. Once the light source is acquired, however, it will provide accurate vertical guidance to the runway. Pilots should be aware that this yellow-green-red configuration produces a yellow-green transition light beam between the yellow and green primary light segments and an anomalous yellow transition light beam between the green and red primary light segments. This anomalous yellow signal could cause confusion with the primary yellow too-high signal.

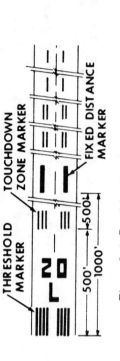

Figure 8-4. *Precision instrument runway.*

d. Threshold—A line perpendicular to the runway centerline designating the beginning of that portion of a runway usable for landing.

e. Displaced Threshold—A threshold that is not at the beginning of the full strength runway pavement.

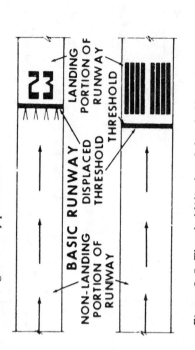

Figure 8-5. *Threshold/displaced threshold markings.*

f. Closed or Overrun/Stopway Areas—Any surface or area which appears usable but which, due to the nature of its structure, is unusable.

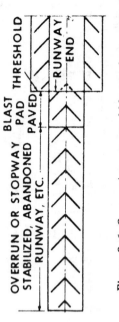

Figure 8-6. *Overrun/stopway and blast pad area.*

MARKINGS

1. In the interest of safety, regularity, or efficiency of aircraft operations, the FAA has recommended for the guidance of the public the following airport marking. (Runway numbers and letters are determined from the approach direction. The number is the whole number nearest one-tenth the magnetic azimuth of the centerline of the runway, measured clockwise from the magnetic north.) The letter or letters differentiate between parallel runways:

For two parallel runways "L" "R"
For three parallel runways "L" "C" "R"

a. Basic Runway Marking—markings used for operations under Visual Flight Rules: centerline marking and runway direction numbers.

Figure 8-2. *Basic runway.*

b. Nonprecision Instrument Runway Marking—markings on runways served by a nonvisual navigation aid and intended for landings under instrument weather conditions: basic runway markings plus threshold marking.

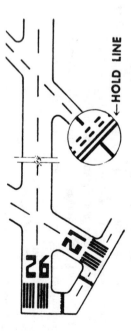

Figure 8-3. *Nonprecision instrument runway.*

c. Precision Instrument Runway Marking—markings on runway served by non-visual precision approach aids and on runways having special operational requirements, non-precision instrument runway marking, touchdown zone marking, fixed distance marking, plus side stripe.

Figure 8-7. *Closed runway or taxiway.*

g. Fixed Distance Marker—To provide a fixed distance marker for landing of turbojet aircraft on other than a precision instrument runway. This marking is similar to the fixed distance marking on a precision instrument runway and located 1,000 feet from the threshold.

h. STOL (Short Take Off and Landing) Runway—In addition to the normal runway number marking, the letters STOL are painted on the approach end of the runway and a touchdown aim point is shown.

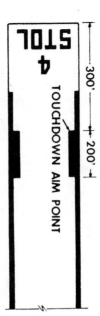

Figure 8-8. *STOL runway.*

i. Taxiway Marking—The taxiway centerline is marked with a continuous yellow line. The taxiway edge may be marked with two continuous yellow lines six inches apart. Taxiway HOLDING LINES consist of two continuous and two dashed lines, spaced six inches between lines, perpendicular to the centerline. When instructed by ATC "HOLD SHORT OF (runway, ILS critical area, etc.)" the pilot should stop so no part of the aircraft extends beyond the holding line. When approaching the holding line from the side with the continuous lines, a pilot should not cross the holding line without ATC clearance at a controlled airport or without making sure of adequate separation from other aircraft at uncontrolled airports. An aircraft exiting the runway is not clear until all parts of the aircraft have crossed the holding line.

AIRSPACE

Airspace users' operations and needs are varied. Because of the nature of some operations, restrictions must be placed upon others for safety reasons. The complexity or density of aircraft movements in other airspace areas may result in additional aircraft and pilot requirements for operation within such airspace. It is of the utmost importance that pilots be familiar with the operational requirements for the various airspace segments.

UNCONTROLLED AIRSPACE

GENERAL

Uncontrolled airspace is that portion of the airspace that has not been designated as continental control area, control area, control zone, terminal control area, or transition area and within which ATC has neither the authority nor the responsibility for exercising control over air traffic.

VFR REQUIREMENTS

Rules governing VFR flight have been adopted to assist the pilot in meeting his responsibility to see and avoid other aircraft. Minimum weather conditions and distance from clouds required for VFR flight are contained in these rules. (FAR 91.105) See Figures 8-9 and 8-10.

ALTITUDE	UNCONTROLLED AIRSPACE		CONTROLLED AIRSPACE	
	Flight Visibility	Distance From Clouds	**Flight Visibility	**Distance From Clouds
1200' or less above the surface, regardless of MSL Altitude	*1 statute mile	Clear of clouds	3 statute miles	500' below 1000' above 2000' horizontal
More than 1200' above the surface, but less than 10,000' MSL	1 statute mile	500' below 1000' above 2000' horizontal	3 statute miles	500' below 1000' above 2000' horizontal
More than 1200' above the surface and at or above 10,000' MSL	5 statute miles	1000' below 1000' above 1 statute mile horizontal	5 statute miles	1000' below 1000' above 1 statute mile horizontal

* Helicopters may operate with less than 1 mile visibility, outside controlled airspace at 1200 feet or less above the surface, provided they are operated at a speed that allows the pilot adequate opportunity to see any air traffic or obstructions in time to avoid collisions.

** In addition, when operating within a control zone beneath a ceiling, the ceiling must not be less than 1000'. If the pilot intends to land or takeoff or enter a traffic pattern within a control zone, the ground visibility must be at least 3 miles at that airport. If ground visibility is not reported at the airport, 3 miles flight visibility is required. (FAR 91.105)

Figure 8-9. *Minimum visibility and distance from clouds—VFR.*

CONTROLLED AND UNCONTROLLED AIRSPACE VFR ALTITUDES AND FLIGHT LEVELS			
If your magnetic course (ground track) is	More than 3000' above the surface but below 18,000' MSL fly	Above 18,000' MSL to FL 290 (except within Positive Control Area, FAR 71.193) fly	Above FL 290 (except within Positive Control Area, FAR 71.193) fly 4000' intervals
0° to 179°	Odd thousands, MSL, plus 500' (3500, 5500, 7500, etc)	Odd Flight Levels plus 500' (FL 195, 215, 235, etc)	Beginning at FL 300 (FL 300, 340, 380, etc)
180° to 359°	Even thousands, MSL, plus 500' (4500, 6500, 8500, etc)	Even Flight Levels plus 500' (FL 185, FL 205, 225, etc)	Beginning at FL 320 (FL 320, 360, 400, etc)

Figure 8-10. *Altitudes and flight levels.*

CONTROLLED AIRSPACE

GENERAL

1. Safety, users' needs, and volume of flight operations are some of the factors considered in the designation of controlled airspace. When so designated, the airspace is supported by ground/air communications, navigation aids, and air traffic services.

2. Controlled airspace consists of those areas designated as Continental Control Area, Control Area, Control Zones, Terminal Control Areas and Transition Areas, within which some or all aircraft may be subject to Air Traffic Control.

CONTINENTAL CONTROL AREA

The continental control area consists of the airspace of the 48 contiguous States, the District of Columbia and Alaska, excluding the Alaska peninsula west of Longitude 160°00'00"W, at and above 14,500 feet MSL, but does not include:

1. The airspace less than 1,500 feet above the surface of the earth; or
2. Prohibited and restricted areas, other than the restricted areas listed in FAR Part 71 Subpart D.

CONTROL AREAS

Control areas consist of the airspace designated as Colored Federal airways, VOR Federal airways, Additional Control Areas, and Control Area Extensions, but do not include the Continental Control Area. Unless otherwise designated, control areas also include the airspace between a segment of a main VOR airway and its associated alternate segments. The vertical extent of the various categories of airspace contained in control area is defined in FAR Part 71.

POSITIVE CONTROL AREA

Positive control area is airspace so designated in Part 71.193 of the Federal Aviation Regulations. This area includes specified airspace within the conterminous United States from 18,000 feet to and including FL600, excluding Santa Barbara Island, Farallon Island, and that portion south of latitude 25°04'N. In Alaska, it includes the airspace over the State of Alaska from 18,000 feet to and including FL600, but not including the airspace less than 1,500 feet above the surface of the earth and the Alaskan Peninsula west of longitude 160°00'W. Rules for operating in positive control area are found in FARs 91.97 and 91.24.

TRANSITION AREAS

1. Controlled airspace extending upward from 700 feet or more above the surface when designated in conjunction with an airport for which an instrument approach procedure has been prescribed; or from 1,200 feet or more above the surface when designated in conjunction with airway route structures or segments. Unless specified otherwise, transition areas terminate at the base of overlying controlled airspace.

2. Transition areas are designated to contain IFR operations in controlled airspace during portions of the terminal operation and while transitioning between the terminal and en route environment.

CONTROL ZONES

1. Controlled airspace which extends upward from the surface and terminates at the base of the continental control area. Control zones that do not underlie the continental control area have no upper limit. A control zone may include one or more airports and is normally a circular area within a radius of 5 statute miles and any extensions necessary to include instrument departure and arrival paths.

2. Control zones are depicted on charts (for example—on the sectional charts the zone is outlined by a broken blue line) and if a control zone is effec-

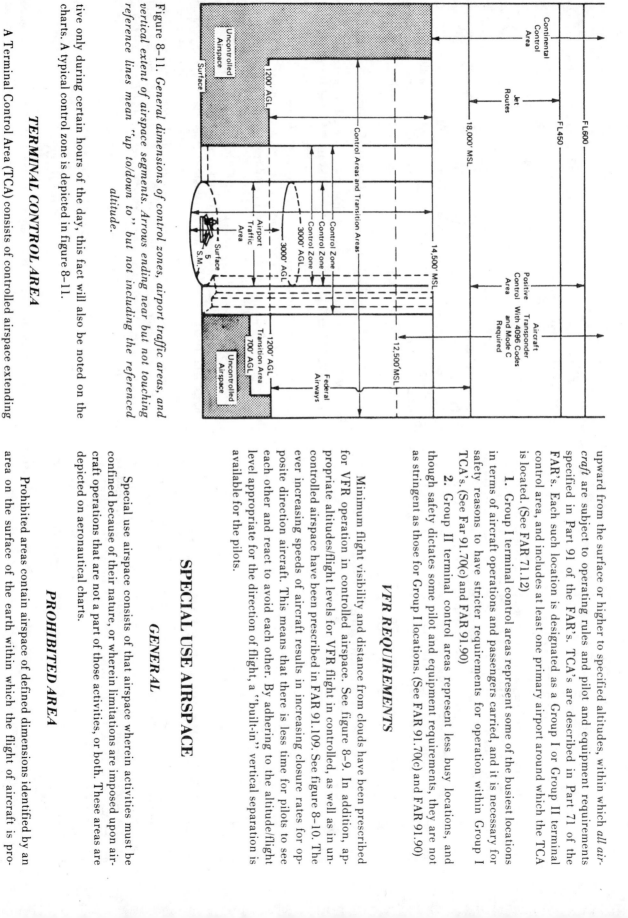

Figure 8–11. *General dimensions of control zones, airport traffic areas, and vertical extent of airspace segments. Arrows ending near but not touching reference lines mean "up to/down to" but not including the referenced altitude.*

TERMINAL CONTROL AREA

A Terminal Control Area (TCA) consists of controlled airspace extending upward from the surface or higher to specified altitudes, within which *all aircraft* are subject to operating rules and pilot and equipment requirements specified in Part 91 of the FAR's. Each such location is designated as a Group I or Group II terminal control area, and includes at least one primary airport around which the TCA is located. (See FAR 71.12)

1. Group I terminal control areas represent some of the busiest locations in terms of aircraft operations and passengers carried, and it is necessary for safety reasons to have stricter requirements for operation within Group I TCA's. (See Far 91.70(c) and FAR 91.90)

2. Group II terminal control areas represent less busy locations, and though safety dictates some pilot and equipment requirements, they are not as stringent as those for Group I locations. (See FAR 91.70(c) and FAR 91.90)

VFR REQUIREMENTS

Minimum flight visibility and distance from clouds have been prescribed for VFR operation in controlled airspace. See figure 8–9. In addition, appropriate altitudes/flight levels for VFR flight in controlled, as well as in uncontrolled airspace have been prescribed in FAR 91.109. See figure 8–10. The ever increasing speeds of aircraft results in increasing closure rates for opposite direction aircraft. This means that there is less time for pilots to see each other and react to avoid each other. By adhering to the altitude/flight level appropriate for the direction of flight, a "built-in" vertical separation is available for the pilots.

SPECIAL USE AIRSPACE

GENERAL

Special use airspace consists of that airspace wherein activities must be confined because of their nature, or wherein limitations are imposed upon aircraft operations that are not a part of those activities, or both. These areas are depicted on aeronautical charts.

PROHIBITED AREA

Prohibited areas contain airspace of defined dimensions identified by an area on the surface of the earth within which the flight of aircraft is pro-

tive only during certain hours of the day, this fact will also be noted on the charts. A typical control zone is depicted in figure 8–11.

211

hibited. Such areas are established for security or other reasons associated with the national welfare. These areas are published in the Federal Register and depicted on aeronautical charts.

RESTRICTED AREA

Restricted areas contain airspace identified by an area on the surface of the earth within which the flight of aircraft, while not wholly prohibited, is subject to restrictions. Activities within these areas must be confined because of their nature or limitations imposed upon aircraft operations that are not a part of those activities or both. Restricted areas denote the existence of unusual often invisible, hazards to aircraft such as artillery firing, aerial gunnery, or guided missiles. Penetration of restricted areas without authorization from the using or controlling agency may be extremely hazardous to the aircraft and its occupants. Restricted areas are published in the Federal Register and constitute Part 73 of the Federal Aviation Regulations.

WARNING AREA

Warning areas are airspace which may contain hazards to non-participating aircraft in international airspace. Warning areas are established beyond the 3-mile limit. Though the activities conducted within warning areas may be as hazardous as those in Restricted areas, Warning areas cannot be legally designated because they are over international waters. Penetration of Warning areas during periods of activity may be hazardous to the aircraft and its occupants. Official descriptions of Warning areas may be obtained on request to the FAA, Washington, D.C. 20591.

MILITARY OPERATIONS AREAS (MOA)

Military Operations Areas consist of airspace of defined vertical and lateral limits established for the purpose of separating certain military training activities from IFR traffic. Whenever an MOA is being used, non-participating IFR traffic may be cleared through an MOA if IFR separation can be provided by ATC. Otherwise, ATC will reroute or restrict non-participating IFR traffic.

Some training activities may necessitate acrobatic maneuvers, and the USAF is exempted from the regulation prohibiting acrobatic flight on airways within MOAs.

Pilots operating under VFR should exercise extreme caution while flying within an MOA when military activity is being conducted. Information regard-

ing activity in MOA's may be obtained from any FSS within 200 miles of the area.

These areas will be depicted on Sectional, VFR Terminal, and Low Altitude En Route Charts.

ALERT AREA

Alert areas are depicted on aeronautical charts to inform non-participating pilots of areas that may contain a high volume of pilot training or an unusual type of aerial activity. Pilots should be particularly alert when flying in these areas. All activity within an Alert Area shall be conducted in accordance with Federal Aviation Regulations, without waiver, and pilots of participating aircraft as well as pilots transiting the area shall be equally responsible for collision avoidance. Information concerning these areas may be obtained upon request to the FAA, Washington, D.C. 20591.

VFR LOW ALTITUDE TRAINING ROUTES

VFR Low Altitude Training Routes are developed in accordance with mutually acceptable procedures established by the FAA and the military services for use by the military services in conducting VFR low altitude navigation and tactical training as well as flight testing. Flights are conducted on established routes when the forecasts and weather conditions are equal to or better than a ceiling of 3000 feet and a visibility of five miles. The routes are flown in one direction only by aircraft whose speeds exceed 250 Kts. Flights will be conducted at or below 1500 feet AGL. Unless current information indicates otherwise, it should be assumed that routes are active on a continuous basis. The latest information may be obtained from the FAA FSS located nearest the route of flight. The flights are conducted in a see-and-avoid environment and pose no restrictions to other aircraft. However, because of the high speeds under which the routes are flown, *the most effective means of avoiding potential conflict is to fly above the 1500 feet AGL limit established for the routes.*

OTHER AIRSPACE AREAS

AIRPORT TRAFFIC AREAS

1. Unless otherwise specifically designated (FAR Part 93), that airspace within a horizontal radius of five statute miles from the geographical center of

212

any airport at which a control tower is operating, extending from the surface up to, but not including, an altitude of 3,000 feet above the elevation of the airport.

2. The rules prescribed for airport traffic areas are established in FAR 91.70, 91.85 and 91.87. They require, in effect, that unless a pilot is landing or taking off from an airport within the airport traffic area, he must avoid the area unless otherwise authorized by ATC. If operating to, from or on the airport served by the control tower, he must also establish and maintain radio communications with the tower. Maximum indicated airspeeds are prescribed. Airport traffic areas are indicated on sectional charts by the blue airport symbol, but the actual boundary is not depicted. See figure 8-11.

AIRPORT ADVISORY AREA

1. The area within five statute miles of an airport where a control tower is not operating but where a Flight Service Station is located. At such locations, the FSS provides advisory service to arriving and departing aircraft.

2. It is not mandatory that pilots participate in the airport advisory service program, but it is strongly recommended that they do.

TEMPORARY FLIGHT RESTRICTIONS

1. Temporary flight restrictions may be put into effect in the vicinity of any incident or event which by its nature may generate such a high degree of public interest that the likelihood of a hazardous congestion of air traffic exists. FAR 91.91, as amended 1 March, 1971, prohibits the operation of nonessential aircraft in airspace that has been designated in a NOTAM as an area within which temporary flight restrictions apply. The revised rule will continue to be implemented in the case of disasters of substantial magnitude. It will also be implemented as necessary in the case of demonstrations, riots, pageants, and other civil disturbances, as well as major sporting events, parades, and similar functions which are likely to attract large crowds and encouraging viewing from the air.

2. NOTAM's implementing temporary flight restrictions will contain a description of the area in which the restrictions apply. Normally the area will include the airspace below 2,000 feet above the surface within 5 miles of the site of the incident. However, the exact dimensions will be included in the NOTAM.

3. Pilots are not to operate aircraft within such an area described in the NOTAM unless they are one of the following: (1) That aircraft is participating in disaster relief activities and is being operated under the direction of the agency responsible for relief activities; (2) They are operating to or from an airport within the area and such operation will not hamper or endanger relief activities; (3) Their operation is authorized under an IFR ATC clearance; (4) Flight around the area is impracticable because of weather or other considerations and advance notice is given to the Air Traffic facility specified in the NOTAM, and enroute flight through the area will not hamper or endanger relief activities; or (5) They are carrying accredited news representatives or persons on official business concerning the incident, and the flight is conducted in accordance with FAR 91.79 and a flight plan is filed with the Air Traffic facility specified in the NOTAM.

AIR TRAFFIC CONTROL

SERVICES AVAILABLE TO PILOTS

Centers are established primarily to provide air traffic service to aircraft operating on IFR flight plans within controlled airspace, and principally during the enroute phase of flight.

CONTROL TOWERS

Towers have been established to provide for a safe, orderly and expeditious flow of traffic on and in the vicinity of an airport. When the responsibility has been so delegated, towers also provide for the separation of IFR aircraft in the terminal areas (Approach Control).

FLIGHT SERVICE STATIONS

Flight Service Stations are the Air Traffic Service facilities within the National Airspace System which have the prime responsibility for preflight pilot briefing, en route communications with VFR flights, assisting lost VFR aircraft, originating NOTAMS, broadcasting aviation weather information, accepting and closing flight plans, monitoring radio NAVAIDS, participating with search and rescue units in locating missing VFR aircraft, and operating the national weather teletypewriter systems. In addition, at selected locations, FSSs take weather observations, issue airport advisories, administer airman written examinations, and advise Customs and Immigration of transborder flight.

PILOT VISITS

Pilots are encouraged to visit air traffic facilities—Towers, Centers and Flight Service Stations and participate in "Operation Raincheck". Operation Raincheck is conducted at these facilities and is designed to familiarize pilots with the ATC system, its functions, responsibilities and benefits. On rare occasions, facilities may not be able to approve a visit because of workload or other reasons. It is therefore requested that pilots contact the facility prior to the visit—give the number of persons in the group, the time and date of the proposed visit, the primary interest of the group. With this information available, the facility can prepare an itinerary and have someone available to guide the group through the facility.

VFR ADVISORY SERVICE

1. VFR advisory service is provided by numerous nonradar *Approach Control* facilities to those pilots intending to land at an airport served by an approach control tower. This service includes: wind, runway, traffic and NOTAM information, unless this information is contained in the ATIS broadcast and the pilot indicates he has received the ATIS information.

2. Such information will be furnished upon initial contact with concerned approach control facility. The pilot will be requested to change to the *tower* frequency at a predetermined time or point, to receive further landing information.

3. Where available, use of this procedure will not hinder the operation of VFR flights by requiring excessive spacing between aircraft or devious routing. Radio contact points will be based on time or distance rather than on landmarks.

4. Compliance with this procedure is not mandatory but pilot participation is encouraged.

AIRPORT ADVISORY PRACTICES AT NONTOWER AIRPORTS

There is no substitute for alertness while in the vicinity of an airport. An airport may have a flight service station, UNICOM operator, or no facility at all. Pilots should predetermine what, if any, service is available at a particular airport. Combining an aural/visual alertness and complying with the following

recommended practices will enhance safety of flight into and out of uncontrolled airports.

1. Recommended Traffic Advisory Practices—As standard operating practice all inbound traffic should continuously monitor the appropriate field facility frequency from 15 miles to landing. Departure aircraft should monitor the appropriate frequency either prior to or when ready to taxi.

a. Inbound Aircraft

Airport	Frequency	Broadcast Position Altitude, Intentions	Broadcast Position
Part-Time Tower (when closed)	Tower Local Control	5 Miles	Downwind, Base, Final
Part-Time Tower (closed) but Full-Time FSS	Tower Local Control	*15 Miles	*5 Miles
Part-Time FSS	123.6	5 Miles	Downwind, Base, Final
Full-Time or Part-Time FSS (closed)	123.6	*15 Miles	*5 Miles
UNICOM (open)	122.8	*	
UNICOM (if unable establish contact)	122.8	5 Miles	Downwind, Base, Final
No Facility on Airport	122.9	5 Miles	Downwind, Base, Final

b. Outbound Aircraft

Airport	Frequency	Broadcast Position And Intentions
Part-Time Tower (closed)	Tower Local Control	When ready to taxi; and before taking runway for takeoff
Part-Time Tower (closed) but Full-Time FSS	Tower Local Control	*When ready to taxi; and before taking runway for takeoff
Part-Time FSS (closed)	123.6	When ready to taxi; and before taking runway for takeoff

Continued

Airport	Frequency	Broadcast Position And Intentions
Full-Time or Part-Time FSS (open)	123.6	*When ready to taxi; and before taking runway for takeoff
UNICOM	122.8	*When ready to taxi; and before taking runway for takeoff
UNICOM (if unable establish contact)	122.8	When ready to taxi; and before taking runway for takeoff
No facility on airport	122.9	When ready to taxi; and before taking runway for takeoff

*Contact appropriate facility first; e.g., "Zanesville Radio, this is Cessna 12345, Over," before announcing arrival/departure intentions. Except for scheduled air carriers and other civil operators having authorized company call signs, departure aircraft should state the aircraft type, identification number, type of flight planned, i.e., VFR or IFR, and the planned destination.

NOTE.—FSS at part-time non-FAA tower locations do not have tower local control frequency. Use 123.6.

c. *Information Furnished by FSS or UNICOM*

(1) FSSs provide airport advisory service at airports where there is no control tower or when the tower is not in operation (part-time tower location). Advisories provide: wind direction and velocity, favored or designated runway, altimeter setting, known traffic (caution: all aircraft in the airport vicinity may not be communicating with the FSS), notices to airmen, airport traffic patterns, and instrument approach procedures. These elements are varied so as to best serve the current traffic situation. Some airport managers have specified that under certain wind or other conditions, designated runways are to be used. Pilots using other than the favored or designated runways should advise the FSS immediately.

NOTE.—*Airport Advisory Service is offered to enhance safety; CONTROL IS NOT EXERCISED.*

2. **Recommended Phraseologies**

   a. Departures

Example:

Aircraft: JOHNSON RADIO, COMANCHE SIX ONE THREE EIGHT, ON TERMINAL BUILDING RAMP, READY TO TAXI, VFR TO DULUTH, OVER

FSS: COMANCHE SIX ONE THREE EIGHT, JOHNSON RADIO, ROGER, WIND THREE TWO ZERO DEGREES AT TWO FIVE, FAVORING RUNWAY THREE ONE, ALTIMETER THREE ZERO ZERO ONE, CESSNA ONE-SEVENTY ON DOWNWIND LEG MAKING TOUCH AND GO LANDINGS ON RUNWAY THREE ONE

NOTE.—*The takeoff time should be reported to the FSS as soon as practicable. If the aircraft has limited equipment and it is necessary to use the navigational feature of the radio aid immediately after takeoff, advise the FSS of this before changing frequency; In such cases, advisories will be transmitted over 123.6 or the tower local control frequency, as appropriate, and the aid frequency.*

   b. Arrivals

Example:

Aircraft: JOHNSON RADIO, TRIPACER ONE SIX EIGHT NINER, OVER KEY WEST, TWO THOUSAND, LANDING GRAND FORKS, OVER

FSS: TRIPACER ONE SIX EIGHT NINER, JOHNSON RADIO, OVER KEY WEST AT TWO THOUSAND, WIND ONE FIVE ZERO DEGREES AT FOUR, DESIGNATED RUNWAY FIVE, ALTIMETER THREE ZERO ZERO ONE, DC-3 TAKING OFF RUNWAY FIVE, BONANZA ON DOWNWIND LEG RUNWAY FIVE MAKING TOUCH AND GO LANDINGS, COMANCHE DEPARTED RUNWAY ONE SEVEN AT ONE SIX PROCEEDING EASTBOUND, OVER

NOTE.—*Pilots should guard 123.6 or the tower local control frequency, as appropriate, until clear of the runway after landing and report leaving the runway to the FSS.*

   c. Transmissions (blind broadcasts) When Not Communicating With an FSS or UNICOM Operator

Example:

   (1) **Inbound**

THIS IS APACHE TWO TWO FIVE ZULU, FIVE MILES EAST OF STRAWN AIRPORT, TWO THOUSAND DESCENDING TO ENTER DOWNWIND FOR RUNWAY ONE SEVEN STRAWN

   (2) **Outbound**

Example:

THIS IS QUEENAIRE SEVEN ONE FIVE FIVE BRAVO, TAXIING ONTO RUNWAY TWO SIX AT STRAWN AIRPORT FOR TAKEOFF

AUTOMATIC TERMINAL INFORMATION SERVICE (ATIS)

Automatic Terminal Information Service (ATIS) is the continuous broadcast of recorded noncontrol information in selected high activity terminal areas. Its purpose is to improve controller effectiveness and to relieve frequency congestion by automating the repetitive transmission of essential but routine information.

Information to include the time of the latest weather sequence, ceiling, visibility (sky conditions/ceilings below 5,000 feet and visibility less than 5 miles will be broadcast; if conditions are at or better than 5,000 and 5, sky condition/ceiling and visibility may be omitted) obstructions to visibility, temperature, wind direction (magnetic) and velocity, altimeter, other pertinent remarks, instrument approach and runways in use is continuously broadcast on the voice feature of a TVOR/VOR/VORTAC located on or near the airport, or in a discrete UHF/VHF frequency. Where VFR arrival aircraft are expected to make initial contact with approach control, this fact and the appropriate frequencies may be broadcast on ATIS. Pilots of aircraft arriving or departing the terminal area can receive the continuous ATIS broadcasts at times when cockpit duties are least pressing and listen to as many repeats as desired. ATIS broadcasts shall be updated upon the receipt of any official weather, regardless of content change and reported values. A new recording will also be made when there is a change in other pertinent data such as runway change, instrument approach in use, etc.

Sample Broadcast:

DULLES INTERNATIONAL INFORMATION SIERRA. 1300 GREENWICH WEATHER. MEASURED CEILING THREE THOUSAND OVERCAST. VISIBILITY THREE, SMOKE. TEMPERATURE SIX EIGHT. WIND THREE FIVE ZERO AT EIGHT. ALTIMETER TWO NINER NINER TWO. ILS RUNWAY ONE RIGHT APPROACH IN USE. LANDING RUNWAY ONE RIGHT AND LEFT. DEPARTURE RUNWAY THREE ZERO. ARMEL VORTAC OUT OF SERVICE. ADVISE YOU HAVE SIERRA.

1. Pilots should listen to ATIS broadcasts whenever ATIS is in operation.

2. Pilots should notify controllers that they have received the ATIS broadcast by repeating the alphabetical code word appended to the broadcast.

EXAMPLE: "INFORMATION SIERRA RECEIVED."

AERONAUTICAL ADVISORY STATIONS (UNICOM)

The frequency available is indicated.

Frequency allocation for UNICOM use has been revised as reflected in the following table. This change will be accomplished over approximately a two year period as license are granted or renewed. All frequency changes will take place by February 1, 1979.

122.8, 123.0, 122.75 MHz for Landing Areas (except heliports) without an ATC Tower or FSS.

122.95 MHz for Landing Areas (except Heliports) with an ATC Tower or FSS.

122.75, 122.750 MHz for Landing Areas not open to the public.
123.05, 123.075 MHz for Heliports with or without ATC Tower or FSS.
122.85 MHz Multicom.

Its use is limited to the necessities of safe and expeditious operation of private aircraft pertaining to runway and wind conditions, types of fuel available, weather, and dispatching. Secondarily, communications may be transmitted concerning ground transportation, food and lodging during transit.

THIS SERVICE SHALL NOT BE USED FOR AIR TRAFFIC CONTROL PURPOSES, except for the verbatim relay of ATC information limited to the following:

a. Revision of proposed departure time.
b. Takeoff, arrival, or flight plan cancellation time.
c. ATC clearances *provided* arrangements are made between the ATC facility and UNICOM licensee to handle such messages.

AERONAUTICAL MULTICOM SERVICE 122.9MHz

1. A mobile service used to provide communications essential to the conduct of activities being performed by or directed from private aircraft.

Example:

Ground/air communications pertaining to agriculture, ranching, conservation activities, forest fire fighting, aerial advertising and parachute jumping. THIS SERVICE SHALL NOT BE USED FOR AIR TRAFFIC CONTROL PURPOSES, except for the verbatim relay of ATC information limited to the following:

a. Revision of proposed departure time.
b. Takeoff, arrival, or flight plan cancellation time.
c. ATC clearances *provided* arrangements are made between the ATC facility and UNICOM licensee to handle such messages.

3. When the pilot acknowledges that he has received the ATIS broadcast, controllers may omit those items contained on the broadcast if they are current. Rapidly changing conditions will be issued by Air Traffic Control and the ATIS will contain words as follows:

"LATEST CEILING/VISIBILITY/ALTIMETER/WIND/(OTHER CONDITIONS) WILL BE ISSUED BY APPROACH CONTROL/TOWER."

The absence of a sky condition/ceiling and/or visibility on ATIS indicates a sky condition/ceiling of 5,000 feet or above and visibility of 5 miles or more. A remark may be made on the broadcast, "The weather is better than 5,000 and 5," or the existing weather may be broadcast.

4. Controllers will issue pertinent information to pilots who do not acknowledge receipt of a broadcast or who acknowledge receipt of a broadcast which is not current.

5. To serve frequency-limited aircraft, Flight Service Stations (FSS) are equipped to transmit on the omnirange frequency at most en route VORs used as ATIS voice outlets. Such communication interrupts the ATIS broadcast. Pilots of aircraft equipped to receive on other FSS frequencies are encouraged to do so in order that these override transmissions may be kept to an absolute minimum.

Pilots are urged to cooperate in the ATIS program as it relieves frequency congestion on approach control, ground control, and local control frequencies.

6. Some pilots use the phrase "Have Numbers" in communications with the control tower. Use of this phrase means that the pilot has received wind and runway information ONLY and the tower does not have to repeat this information. It does not indicate receipt of the ATIS broadcast and should never be used for this purpose.

Airport/Facility Directory indicates airports for which ATIS is provided.

RADAR TRAFFIC INFORMATION SERVICE

1. A service provided by radar air traffic control facilities. Pilots receiving this service are advised of any radar target observed on the radar display which may be in such proximity to the position of their aircraft or its intended route of flight that it warrants their attention. This service is not intended to relieve the pilot of his responsibility for continual vigilance to see and avoid other aircraft.

a. Purpose of the Service—The issuance of traffic information as observed on a radar display is based on the principle of assisting and advising a pilot that a particular radar target's position and track indicates it may intersect or pass in such proximity to his intended flight path that it warrants his attention. This is to alert the pilot to the traffic so that he can be on the lookout for it and thereby be in a better position to take appropriate action should the need arise.

Pilots are reminded that the surveillance radar used by ATC does not provide altitude information unless the aircraft is equipped with Mode C and the Radar Facility is capable of displaying altitude information.

b. Provision of the Service—Many factors, such as limitations of the radar, volume of traffic, controller workload and communications frequency congestion, could prevent the controller from providing this service. The controller possesses complete discretion for determining whether he is able to provide or continue to provide this service in a specific case. His reason against providing or continuing to provide the service in a particular case is not subject to question nor need it be communicated to the pilot. In other words, the provision of this service is entirely dependent upon whether the controller believes he is in a position to provide it. Traffic information is routinely provided to all aircraft operating on IFR Flight Plans except when the pilot advises he does not desire the service, or the pilot is operating within positive controlled airspace. Traffic information may be provided to flights not operating on IFR Flight Plans when requested by pilots of such flight.

When receiving VFR radar advisory service, pilots should monitor the assigned frequency at all times. This it to preclude controllers' concern for radio failure or emergency assistance to aircraft under his jurisdiction. VFR radar advisory service does not include vectors away from conflicting traffic unless requested by the pilot. When advisory service is no longer desired, advise the controller before changing frequencies then change your transponder code to 1200 if applicable. Except in programs where radar service is automatically terminated, the controller will advise the aircraft when radar is terminated.

NOTE.—Participation by VFR pilots in formal programs implemented at certain terminal locations constitutes pilot request. This also applies to participating pilots at those locations where arriving VFR flights are encouraged to make their first contact with the tower on the approach control frequency.

c. Issuance of Traffic Information—Traffic information will include the following concerning a target which may constitute traffic for an aircraft that is:

217

(1) Radar identified:

   (a) Azimuth from the aircraft in terms of the twelve hour clock;

   (b) Distance from the aircraft in nautical miles;

   (c) Direction in which the target is proceeding; and

   (d) Type of aircraft and altitude if known.

Example:

Traffic 10 o'clock, 3 miles, west-bound (type aircraft and altitude, if known, of the observed traffic). The pilot may, upon receipt of traffic information, request a vector (heading) to avoid such traffic. The vector will be provided to the extent possible as determined by the controller provided the aircraft to be vectored is within the airspace under the jurisdiction of the controller.

(2) Not radar identified:

   (a) Distance and direction with respect to a fix;

   (b) Direction in which the target is proceeding; and

   (c) Type of aircraft and altitude if known.

Example:

Traffic 8 miles south of the airport northeast-bound, (type aircraft and altitude if known).

(d) The examples depicted in figures 8–12 and 8–13 point out the possible error in the position of this traffic when it is necessary for a pilot to apply drift correction to maintain this track. This error could also occur in the event a change in course is made at the time radar traffic information is issued.

Figure 8–12. *Radar traffic information.*

In figure 8–12 traffic information would be issued to the pilot of aircraft "A" as 12 o'clock. The actual position of the traffic as seen by the pilot of aircraft "A" would be one o'clock. Traffic information issued to aircraft "B" would also be given as 12 o'clock, but in this case, the pilot of "B" would see his traffic at 11 o'clock.

Figure 8–13. *Radar traffic information.*

In figure 8–13 traffic information would be issued to the pilot of aircraft "C" as two o'clock. The actual position of the traffic as seen by the pilot of aircraft "C" would be at three o'clock. Traffic information issued to aircraft "D" would be at an 11 o'clock position. Since it is not necessary for the pilot of aircraft "D" to apply wind correction (crab) to make good his track, the actual position of the traffic issued would be correct. Since the radar controller can only observe aircraft track (course) on his radar display, he must issue traffic advisories accordingly, and pilots should give due consideration to this fact when looking for reported traffic.

SAFETY ADVISORY

A safety advisory will be issued to pilots of aircraft being controlled by ATC if the controller is aware the aircraft is at an altitude which, in the controller's judgment, places the aircraft in unsafe proximity to terrain/obstructions. The primary method of detecting unsafe proximity is through Mode C.

1. Terrain/Obstruction Advisory.

a. The controller will immediately issue an advisory to the pilot of an aircraft under his control if he is aware the aircraft is at an altitude which, in the controller's judgment, places the aircraft in unsafe proximity to terrain/obstructions or other aircraft.

Example:

LOW ALTITUDE ALERT, CHECK YOUR ALTITUDE IMMEDIATELY.

b. Some automated terminal facilities (ARTS III) are equipped with a computer function which, if operating, generates an alert to the controller when a tracked Mode C equipped aircraft under his control is below or is predicted, by the computer, to go below a predetermined minimum safe altitude. This function is called Minimum Safe Altitude Warning (MSAW). It is designed solely to aid the controller in detecting tracked aircraft with an operating Mode C transponder which may be in unsafe proximity to terrain/obstructions. The automated facility (ARTS III) which is equipped with this function will, when it is operating, process:

(1) All IFR aircraft, with an operating transponder and altitude encoder, which are being tracked.

(2) All VFR aircraft, with an operating transponder and altitude encoder, which are being tracked and have requested MSAW.

NOTE:—Pilots operating on VFR flight plans may request MSAW processing by the ARTS III computer if their aircraft has an operating transponder with altitude encoding (Mode C).

Example:

APACHE THREE THREE PAPA REQUEST MSAW.

2. Aircraft Conflict Advisory.

The controller will immediately issue an advisory to the pilot of an aircraft under his control if he is aware of an aircraft that is not under his control at an altitude which, in the controller's judgment, places both aircraft in unsafe proximity to each other. With the alert the controller will offer the pilot an alternate course(s) of action when feasible. Any alternate course(s) of action the controller may recommend to the pilot will be predicated only on other traffic under his control.

Example:

AMERICAN THREE, TRAFFIC ALERT, ADVISE YOU TURN RIGHT/LEFT HEADING (DEGREES) AND/OR CLIMB/DESCEND TO (ALTITUDE) IMMEDIATELY.

The provision of this service is contingent upon the capability of the controller to have an awareness of situation(s) involving unsafe proximity to terrain, obstructions and uncontrolled aircraft. The issuance of a safety advisory cannot be mandated, but it can be expected on a reasonable, though intermittent basis. Once the advisory is issued, it is solely the pilot's prerogative to determine what course of action, if any, he will take. This procedure is intended for use in time critical situations where aircraft safety is in question. Non-critical situations should be handled via the normal traffic advisory procedures.

3. In many cases, the controller will be unable to determine if flight into instrument conditions will result from his instructions. To avoid possible hazards resulting from being vectored into IFR conditions, pilots should keep the controller advised of the weather conditions in which he is operating and along the course ahead.

4. Radar navigation assistance (vectors) may be initiated by the controller when one of the following conditions exist:

a. The controller suggests the vector and the pilot concurs.

b. A special program has been established and vectoring service has been advertised.

c. In the controller's judgment the vector is necessary for air safety.

5. Radar navigation assistance (vectors) and other radar derived information may be provided in response to pilot requests. Many factors, such as limitations of radar, volume of traffic, communications frequency, congestion, and controller workload could prevent the controller from providing it. The controller has complete discretion for determining if he is able to provide the service in a particular case. His decision not to provide the service in a particular case is not subject to question.

TERMINAL RADAR PROGRAMS FOR VFR AIRCRAFT

1. Stage I Service (Radar Advisory Service for VFR Aircraft)

a. In addition to the use of radar for the control of IFR aircraft, Stage I facilities provide traffic information and limited vectoring to VFR aircraft on a workload permitting basis.

b. Vectoring service may be provided when requested by the pilot or with pilot concurrence when suggested by ATC.

c. Pilots of arriving aircraft should contact approach control on the publicized frequency, give their position, altitude, radar beacon code (if transponder equipped), destination, and request traffic information.

d. Approach control will issue wind and runway, except when the pilot states "HAVE NUMBERS" or this information is contained in the ATIS broadcast and the pilot indicates he has received the ATIS information. Traffic information is provided on a workload permitting basis. Approach control will specify the time or place at which the pilot is to contact the tower on local control frequency for further landing information. Upon being told to contact the tower, radar service is automatically terminated.

2. Stage II Service (Radar Advisory and Sequencing for VFR Aircraft).

a. This service has been implemented at certain terminal locations. The purpose of the service is to adjust the flow of arriving VFR and IFR aircraft into the traffic pattern in a safe and orderly manner and to provide radar traffic information to departing VFR aircraft. Pilot participation is urged but it is not mandatory.

b. Pilots of arriving VFR aircraft should initiate radio contact with approach control when approximately 25 miles from the airport at which Stage II services are being provided. On initial contact by VFR aircraft, approach control will assume that Stage II service is requested. Approach control will provide the pilot with wind and runway (except when the pilot states "Have Numbers" or that he has received the ATIS information), routings, etc., as necessary for proper sequencing with other participating VFR and IFR traffic en route to the airport. Traffic information will be provided on a workload permitting basis. If an arriving aircraft does not want the service, the pilot should state NEGATIVE STAGE II, or make a similar comment, on initial contact with approach control.

c. After radar contact is established, the pilot may navigate on his own into the traffic pattern or, depending on traffic conditions, he may be directed to fly specific headings to position the flight behind a preceding aircraft in the approach sequence. When a flight is positioned behind the preceding aircraft and the pilot reports having that aircraft in sight, he will be directed to follow it. If other "non-participating" or "local" aircraft are in the traffic pattern, the tower will issue a landing sequence.

d. Standard radar separation will be provided between IFR aircraft until such time as the aircraft is sequenced and the pilot sees the traffic he is to follow. Standard radar separation between VFR or between VFR and IFR aircraft will not be provided.

e. Pilots of departing VFR aircraft are encouraged to request radar traffic information by notifying ground control on initial contact with their request and proposed direction of flight.

Example:

"XRAY GROUND CONTROL, N18 AT HANGAR 6, READY TO TAXI, VFR SOUTHBOUND, HAVE INFORMATION BRAVO AND REQUEST RADAR TRAFFIC INFORMATION."

Following takeoff, the tower will advise when to contact departure control.

f. Pilots of aircraft transiting the area and in radar contact/communication with approach control will receive traffic information on a controller workload permitting basis. Pilots of such aircraft should give their position, altitude, radar beacon code (if transponder equipped), destination, and/or route of flight.

3. Stage III Service (Radar Sequencing and Separation Service for VFR Aircraft).

a. This service has been implemented at certain terminal locations. The purpose of this service is to provide separation between all participating VFR aircraft and all IFR aircraft operating within the airspace defined as the Terminal Radar Service Area (TRSA). Pilot participation is urged but it is not mandatory.

b. If any aircraft does not want the service, the pilot should state NEGATIVE STAGE III, or make a similar comment, on initial contact with approach control or ground control, as appropriate.

c. TRSA charts and a further description of the Services Provided, Flight Procedures, and ATC Procedures are contained in certain flight publications.

d. The TRSA is contained in a radar environment and control is predicated thereon; however, this does not preclude application of nonradar separation when required or deemed appropriate. The type of separation used will depend on prevailing conditions (traffic volume, type of aircraft, ceiling and visibility, etc.).

e. Visual separation is used when prevailing conditions permit and it will be applied as follows:

(1) When a VFR flight is positioned behind the preceding aircraft and the pilot reports having that aircraft in sight, he will be directed to follow it.

(2) When IFR flights are being sequenced with other traffic and the pilot reports the aircraft he is to follow in sight, the pilot may be directed to follow it and will be cleared for a "visual approach."

(3) If other "non-participating" or "local" aircraft are in the traffic pattern, the tower will issue a landing sequence.

(4) Departing VFR aircraft may be asked if they can visually follow a preceding departure out of the TRSA. If the pilot concurs, he will be directed to follow it until leaving the TRSA.

f. Until visual separation is obtained, standard vertical or radar separation will be provided.

(1) 1,000 feet vertical separation may be used between IFR aircraft.

(2) 500 feet vertical separation may be used between VFR aircraft, or between a VFR and an IFR aircraft.

(3) Radar separation varies depending on size of aircraft and aircraft distance from the radar antenna. The minimum separation used will be 1½ miles for most VFR aircraft under 12,500 pounds GWT. If being separated from larger aircraft, the minimum is increased appropriately.

g. Pilots operating VFR in a TRSA—

(1) Must maintain an altitude when assigned by ATC.

(2) When not assigned an altitude should coordinate with ATC prior to any altitude change.

h. Within the TRSA, traffic information on observed but unidentified targets will, to the extent possible, be provided all IFR and participating VFR aircraft. At the request of the pilot, he will be vectored to avoid the observed traffic, insofar as possible, provided the aircraft to be vectored is within the airspace under the jurisdiction of the controller.

i. Departing aircraft should inform ATC of their intended destination and/or route of flight and proposed cruising altitude.

4. PILOTS RESPONSIBILITY: THESE PROGRAMS ARE NOT TO BE INTERPRETED AS RELIEVING PILOTS OF THEIR RESPONSIBILITIES TO SEE AND AVOID OTHER TRAFFIC OPERATING IN BASIC VFR WEATHER CONDITIONS, TO MAINTAIN APPROPRIATE TERRAIN AND OBSTRUCTION CLEARANCE, OR TO REMAIN IN WEATHER CONDITIONS EQUAL TO OR BETTER THAN THE MINIMA REQUIRED BY FAR 91.105. WHENEVER COMPLIANCE WITH AN ASSIGNED ROUTE, HEADING AND/OR ALTITUDE IS LIKELY TO COMPROMISE SAID PILOT RESPONSIBILITY RESPECTING TERRAIN AND OBSTRUCTION CLEARANCE AND WEATHER MINIMA, APPROACH CONTROL SHOULD BE SO ADVISED AND A REVISED CLEARANCE OR INSTRUCTION OBTAINED.

TERMINAL CONTROL AREA OPERATION

1. Operating Rules and Equipment Requirements. Regardless of weather conditions, ATC authorization is required prior to operating within a TCA. Pilots should not request such authorization unless the requirements of FAR 91.24 and 91.90 are met. Included among these requirements are:

Group I TCAs

a. A two-way radio capable of communicating with ATC on appropriate frequencies.

b. A VOR or TACAN receiver, except for helicopters.

c. A 4096 code transponder with Mode C automatic altitude reporting equipment, except for helicopters operating at or below 1,000 feet AGL under a Letter of Agreement. (ATC may authorize a deviation from the altitude reporting equipment requirement immediately; however, request for

a deviation from the 4096 transponder equipment requirement must be submitted to the controlling ATC facility at least 4 hours before the proposed operation.)

d. A private pilot certificate or better in order to land or takeoff from an airport within the TCA.

e. Unless otherwise authorized by ATC, each person operating a large turbine engine powered airplane to or from a primary airport shall operate at or above the designated floors while within the lateral limits of the terminal control area.

f. No person may operate an aircraft in the airspace underlying a terminal control area, at an indicated airspeed of more than 200 knots (230 m.p.h.)(FAR 91.70).

Group II TCAs

a. A two-way radio capable of communicating with ATC on appropriate frequencies.

b. A VOR or TACAN receiver, except for helicopters.

c. A 4096 code transponder, except for helicopters operating at or below 1,000 feet under a letter of agreement, or for IFR flights operating to or from an airport outside of but in close proximity to the TCA when the commonly used transition, approach, or departure procedures to such airport require flight within the TCA. (ATC may authorize deviations from the transponder requirements. Requests for deviation should be submitted to the controlling ATC facility at least four hours before the proposed operation.)

d. Unless otherwise authorized by ATC, large turbine-powered aircraft must operate at or above the floor of the TCA while operating to or from the primary airport.

e. No person may operate an aircraft in the airspace underlying a Terminal Control Area, at an indicated airspeed of more than 200 knots (230 m.p.h.)(FAR 91.70).

2. Flight Procedures.

a. *IFR Flights.* Aircraft operating within the TCA shall be operated in accordance with current IFR procedures. A clearance for a visual approach is not authorization for an aircraft to operate below the designated floors of the TCA.

b. *VFR Flights.*

(1) Arriving VFR flights should contact ATC on the appropriate frequency and in relation to geographical fixes shown on local

charts. Although a pilot may be operating beneath the floor of the TCA on initial contact, communications with ATC should be established in relation to the points indicated for spacing and sequencing purposes.

(2) Departing VFR aircraft should advise the ground controller of the intended altitude and route of flight to depart the TCA.

(3) Aircraft not landing/departing the primary airport may obtain ATC clearance to transit the TCA when traffic conditions permit and provided the requirements of FAR 91.90 are met. Such VFR transiting aircraft are encouraged, to the extent possible, to transit through VFR corridors or above or below the TCA.

3. ATC Clearances and Separations. While operating within a TCA, pilots are provided the service and separation as in the Stage III, Terminal Radar Programs For VFR Aircraft in this chapter. In the event of a radar outage, separation and sequencing of VFR aircraft will be suspended as this service is dependent on radar. The pilot will be advised that the service is not available and issued wind, runway information and the time or place to contact the tower. Traffic information will be provided on a workload permitting basis.

4. Assignment of radar headings and/or altitudes are based on the provision that a pilot operating in accordance with visual flight rules is expected to advise ATC if compliance with an assigned route, radar heading or altitude will cause the pilot to violate such rules.

RADAR SERVICE FOR VFR AIRCRAFT IN DIFFICULTY

1. Radar equipped FAA Air Traffic Control facilities provide radar assistance and navigation service (vectors) to VFR aircraft in difficulty provided the aircraft can communicate with the facility, are within radar coverage, and can be radar identified. Pilots should clearly understand that authorization to proceed in accordance with such radar navigational assistance does not constitute authorization for the pilot to violate Federal Aviation Regulations. In effect, assistance provided is on the basis that navigational guidance information issued is advisory in nature and the job of flying the aircraft safely, remains with the pilot.

2. Experience has shown that many pilots who are not qualified for instrument flight cannot maintain control of their aircraft when clouds or other reduced visibility conditions are encountered. In many cases, the controller

will be unable to determine if flight into instrument conditions will result from his instructions. To avoid possible hazards resulting from being vectored into IFR conditions, a pilot in difficulty should keep the controller advised of the weather conditions in which he is operating and along the course ahead, and should observe the following:

a. If an alternative course of action is available which will permit flight in VFR weather conditions, noninstrument rated pilots should choose the alternative rather than requesting a vector or approach into IFR weather conditions; or,

b. If no alternative course of action is available, the noninstrument rated pilot should so advise the controller and 'declare an emergency.'

c. If the pilot is instrument rated and the aircraft is instrument equipped, the pilot should so indicate by filing an IFR flight plan. Assistance will be provided on the basis that the flight can operate safely in IFR weather conditions.

3. Some 'DO's' and 'DONT's:'

a. DO let ATC know of your difficulty immediately. DON'T wait until the situation becomes an emergency.

b. DO give as much information as possible on initial contact with ATC-nature of difficulty, position (in relation to a navaid if possible), altitude, radar beacon code (if transponder equipped), weather conditions, if instrument rated or not, destination, service requested.

c. DON'T change radio frequency without informing the controller.

d. DO adhere to ATC instructions or information, or if not possible, DO advise ATC immediately that you cannot comply.

TRANSPONDER OPERATION

1. General

a. Air Traffic Control Radar Beacon System (ATC-RBS) is similar to and compatible with military coded radar beacon equipment. Civil Mode A is identical to military Mode 3.

b. Civil and military transponders should be adjusted to the "on" or normal operating position as late as practicable prior to takeoff and to "off" or "standby" as soon as practicable after completing landing roll unless the change to "standby" has been accomplished previously at the request of ATC.

c. If entering a U.S. domestic control area from outside the U.S., the pilot should advise on first radio contact with a U.S. radar air traffic control facility that such equipment is available by adding "transponder" to the air-craft identification.

d. It should be noted by all users of the ATC Transponders that the coverage they can expect is limited to "line of sight." Low altitude or aircraft antenna shielding by the aircraft itself may result in reduced range. Range can be improved by climbing to a higher altitude. It may be possible to minimize antenna shielding by locating the antenna where dead spots are only noticed during abnormal flight attitudes.

e. For ATC to utilize one or a combination of the 4096 discrete codes FOUR DIGIT CODE DESIGNATION will be used, e.g., code 2100 will be expressed as TWO ONE ZERO ZERO.

f. Pilots should be particularly sure to abide by the provisions of sub-paragraph **b** above. Additionally, due to the operational characteristics of the rapidly expanding automated air traffic control system, THE LAST TWO DIGITS OF THE SELECTED TRANSPONDER CODE SHOULD ALWAYS READ '00' UNLESS SPECIFICALLY REQUESTED BY ATC TO BE OTHERWISE.

g. Some transponders are equipped with a Mode C automatic altitude reporting capability. This system converts aircraft altitude in 100 feet in-crements, to coded digital information which is transmitted together with MODE C framing pulses to the interrogating radar facility. The manner in which transponder panels are designed differs, therefore, a pilot should be thoroughly familiar with the operation of his transponder so that ATC may realize its full capabilities.

h. Adjust transponder to reply on the Mode A/3 code specified by ATC and, if equipped, to reply on Mode C with altitude reporting *capability ac-tivated* unless deactivation is directed by ATC or unless the installed aircraft equipment has not been tested and calibrated as required by FAR 91.36. If deactivation is required and your transponder is so designed, turn off the altitude reporting switch and continue to transmit MODE C framing pulses. If this capability does not exist, turn off MODE C.

i. Pilots of aircraft with operating Mode C altitude reporting transponders should report exact altitude/flight level to the nearest hundred foot increment when establishing initial contact with an air traffic control facility. Exact altitude/flight level reports on initial contact provide air traffic control with information that is required prior to using Mode C altitude in-formation for separation purposes. This will significantly reduce altitude verification requests.

j. The transponder shall be operated only as specified by ATC. Activate the "IDENT" feature only upon request of the ATC controller.

k. Under no circumstances should a pilot of a civil aircraft operate the transponder on Code 0000. This code is reserved for military interceptor operations.

l. Military pilots operating VFR or IFR within restricted/warning areas should adjust their transponders to code 4000 unless another code has been assigned by ATC.

m. When making routine code changes, pilots should avoid inadvertent selection of codes 7500, 7600 or 7700 thereby causing momentary false alarms at automated ground facilities. For example when switching from code 2700 to code 7200, switch first to 2200 then 7200, NOT to 7700 and then 7200. This procedure applies to nondiscrete code 7500 and all discrete codes in the 7600 and 7700 series (i.e., 7600-7677, 7700-7777) which will trigger special in-dicators in automated facilities. Only nondiscrete code 7500 will be decoded as the hijack code. An aircraft's transponder code (when available) is utilized to enhance the tracking capabilities of the ATC facility, therefore, pilots should not turn the transponder to standby when making routine code changes.

n. Specific details concerning requirements, exceptions and ATC authorized deviations for transponder and Mode C operation above 12,500' and below 18,000' MSL are found in FAR 91.24. In general, the FAR requires aircraft to be equipped with Mode A/3 (4096 codes) and Mode C altitude re-porting capability when operating in controlled airspace of the 48 contiguous States and the District of Columbia above 12,500' MSL, excluding airspace at and below 2,500' AGL. Pilots should insure that their aircraft transponder is operating on an appropriate or ATC assigned VFR/IFR code and Mode C when operating in such airspace. If in doubt about the operational status of either feature of your transponder while airborne, contact the nearest ATC facility or Flight Service Station and they will advise you what facility you should contact for determining the status of your equipment. In-flight re-quests for "immediate" deviation may be approved by controllers only when the flight will continue IFR or when weather conditions prevent VFR descent and continued VFR flight in airspace not affected by the FAR. All other re-quests should be made by contacting the nearest Flight Ser-vice/Air Traffic facility in person or by telephone. The nearest ARTC Center will normally be the controlling agency and is responsible for coordinating re-quests involving deviations in other ARTCC areas.

NOTE: *Positive Control Area (PCA) and Terminal Control Area (TCA) devia-tion requests are handled as they have been in the past.*

o. Pilots should be aware that proper application of these procedures will provide both VFR and IFR aircraft with a higher degree of safety in the environment where high-speed closure rates are possible. Transponders substantially increase the capability of radar to see an aircraft and the Mode C feature enables the controller to quickly determine where potential traffic conflicts may exist. Even VFR pilots who are not in contact with ATC will be afforded greater protection from IFR aircraft and VFR aircraft which are receiving traffic advisories. Nevertheless, pilots should never relax their visual scanning vigilance from other aircraft.

2. Instrument Flight Rules (IFR) Flight Plan

a. If the pilot cancels his IFR flight plan prior to reaching the terminal area of destination, the transponder should be adjusted according to the instructions below for VFR flight.

b. The transponder shall be operated only as specified by ATC. Activate the "IDENT" feature only upon request of the ATC controller.

3. Visual Flight Rules (VFR)

a. Unless otherwise instructed by an Air Traffic Control Facility adjust Transponder to reply on Mode 3/A code 1200 regardless of altitude.

b. Adjust transponder to reply on Mode C, with altitude reporting *capability activated* if the aircraft is so equipped, unless deactivation is directed by ATC or unless the installed equipment has not been tested and calibrated as required by FAR 91.36. If deactivation is required and your transponder is so designed, turn off the altitude reporting switch and continue to transmit MODE C framing pulses. If this capability does not exist, turn off MODE C.

4. Emergency Operation

a. When an emergency occurs, the pilot of an aircraft equipped with a coded radar beacon transponder, who desires to alert a ground radar facility to an emergency condition and who cannot establish communications without delay with an air traffic control facility may adjust the transponder to reply on Mode A/3, Code 7700.

b. Pilots should understand that they may not be within a radar coverage area and that, even if they are certain radar facilities are not yet equipped to automatically recognize Code 7700 as an emergency signal. Therefore, they should establish radio communications with an air traffic control facility as soon as possible.

5. Radio Failure

a. Should the pilot of an aircraft equipped with a coded radar beacon transponder experience a loss of two-way radio capability the pilot should:

(1) Adjust his transponder to reply on Mode A/3, code 7700 for a period of 1 minute,

(2) then change to code 7600 and remain on 7600 for a period of 15 minutes or the remainder of the flight, whichever occurs first.

(3) repeat steps 1 and 2, as practicable.

b. Pilots should understand that they may not be in an area of radar coverage. Also many radar facilities are not presently equipped to automatically display code 7600 and will interrogate 7600 only when the aircraft is under direct radar control at the time of radio failure. However, replying on code 7700 first increases the probability of early detection of a radio failure condition.

6. Radar Beacon Phraseology

Air traffic controllers, both civil and military, will use the following phraseology when referring to operation of the Air Traffic Control Radar Beacon System (ATCRBS) Mark XIFF (SIF). Instructions by air traffic control refer only to Mode A/3 or Mode C operations and do not affect the operation of the transponder on other Modes.

SQUAWK (number)—Operate radar beacon transponder on designated code in Mode A/3.

IDENT—Engage the "IDENT" feature (military I/P) of the transponder.

SQUAWK (number) AND IDENT—Operate transponder on specified code in Mode A/3 and engage the "IDENT" (military I/P) feature.

SQUAWK STANDBY—Switch transponder to standby position.

SQUAWK LOW/NORMAL—Operate transponder on low or normal sensitivity as specified. Transponder is operated in "NORMAL" position unless ATC specified "LOW" ("ON" is used instead of "NORMAL" as a master control label on some types of transponders).

SQUAWK ALTITUDE—Activate MODE C with automatic altitude reporting.

STOP ALTITUDE SQUAWK—Turn off altitude reporting switch and continue transmitting MODE C framing pulses. If your equipment does not have this capability, turn off MODE C.

STOP SQUAWK (mode in use)—Switch off specified mode. (Use for military aircraft when the controller is unaware if a military service requires the aircraft to continue operating on another MODE.)

STOP SQUAWK—Switch off transponder.

SQUAWK MAYDAY—Operate transponder in the emergency position. (Mode A Code 7700 for civil transponder. Mode 3 Code 7700 and emergency feature for military transponder.)

SQUAWK VFR—Operate transponder on code 1200 regardless of altitude.

RADIO COMMUNICATIONS

PHRASEOLOGY AND TECHNIQUES

Radio communications are a critical link in the ATC system. The link can be a strong bond between pilot and controller—or it can be broken with surprising speed and disastrous results. Discussion herein provides basic procedures for new pilots and also highlights safe operating concepts for all pilots.

CLARITY AND BREVITY

The single, most important thought in pilot-controller communications is understanding. Brevity is important, and contacts should be kept as brief as possible, but the controller must know what you want to do before he can properly carry out his control duties. And you, the pilot, must know exactly what he wants you to do. Since concise phraseology may not always be adequate, use whatever words are necessary to get your message across.

PROCEDURAL WORDS AND PHRASES

All pilots will find the Pilot/Controller Glossary very helpful in learning what certain words or phrases mean. Good phraseology enhances safety and is the mark of a professional pilot. Jargon, chatter and "CB" slang have no place in ATC communications. The Pilot/Controller Glossary is the same glossary used in the ATC controller's handbook. We recommend that it be studied and reviewed from time to time to sharpen your communication skills.

RADIO TECHNIQUE

1. *Listen* before you transmit. Many times you can get the information you want through ATIS or by monitoring the frequency. Except for a few situations where some frequency overlap occurs, if you hear someone else talk-

ing, the keying of your transmitter will be futile and you will probably jam their receivers causing them to repeat their call. If you have just changed frequencies, pause for your receiver to tune, listen and make sure the frequency is clear.

2. *Think before you key* your transmitter. Know what you want to say and if it is lengthy, e.g., a flight plan or IFR position report, jot it down. (But do not lock your head in the cockpit.)

3. The microphone should be very close to your lips and after pressing the mike button, a slight pause may be necessary to be sure the first word is transmitted. Speak in a normal conversational tone.

4. When you release the button, wait a few seconds before calling again. The controller or FSS specialist may be jotting down your number, looking for your flight plan, transmitting on a different frequency, or selecting his transmitter to your frequency.

5. Be alert to the sounds *or lack of sounds* in your receiver. Check your volume, recheck your frequency and *make sure that your microphone is not stuck* in the transmit position. Frequency blockage can, and has, occurred for extended periods of time due to unintentional transmitter operation. This type of interference is commonly referred to as a "stuck mike," and controllers may refer to it in this manner when attempting to assign an alternate frequency.

6. Be sure that you are within the performance range of your radio equipment and the ground station equipment. Remote radio sites do not always transmit and receive on all of a facilities available frequencies, particularly with regard to VOR sites where you can hear but not reach a ground station's receiver. Remember that higher altitude increases the range of VHF "line of sight" communications.

CONTACT PROCEDURES

1. *Initial Contact.* The term "initial contact" or "initial callup" means the first radio call you make to a given facility, or the first call to a different controller/FSS specialist within a facility. *Use the following format:* (a) name of facility being called, (b) your *full* aircraft identification as filed in the flight plan or as discussed under Aircraft Call Signs below, (c) type of message to follow or your request if it is short, and (d) the word "Over."

Examples:

"NEW YORK RADIO, MOONEY THREE ONE ONE ECHO, OVER." "COLUMBIA GROUND CONTROL, CESSNA THREE ONE SIX ZERO FOXTROT, IFR MEMPHIS, OVER."

If radio reception is reasonably assured, inclusion of your request, your position or altitude, the phrase "Have numbers" or "Information Charlie received" (for ATIS) in the initial contact helps decrease radio frequency congestion. Use discretion and do not overload the controller with information he does not need. When you do not get a response from the ground station, recheck your radios or use another transmitter and keep the next contact short.

2. *Initial contact when your transmitting and receiving frequencies are different.* If you are attempting to establish contact with a ground station and you are receiving on a different frequency than that transmitted, indicate the VOR name or the frequency on which you expect a reply. Most FSSs and control facilities can transmit on several VOR stations in the area. Use the appropriate FSS call sign as indicated on charts.

Example:

New York FSS transmits on the Kennedy, Deer Park and Riverhead VORTACs. If you are in the Riverhead area, your callup should be "NEW YORK RADIO, CESSNA THREE ONE SIX ZERO FOXTROT, RECEIVING RIVERHEAD VOR, OVER."

If the chart indicates FSS frequencies above the VORTAC or in FSS communications boxes, transmit or receive on those frequencies nearest your location.

When unable to establish contact and you wish to call *any* ground station, use the phrase "ANY RADIO (TOWER) (STATION), GIVE THE CESSNA THREE ONE SIX ZERO FOXTROT A CALL ON (FREQUENCY) OR (VOR)." If an emergency exists or you need assistance, so state.

3. *Subsequent Contacts and Responses to Callup from a Ground Facility.* Use the same format as used for initial contact except you should state your message or request with the callup in one transmission. The ground station name and the word "Over" may be omitted if the message requires an obvious reply and there is no possibility for misunderstandings. *You should acknowledge all callups or clearances* unless the controller or FSS specialist advises otherwise. There are some occasions when the controller must issue time-critical instructions to other aircraft and he may be in a position to observe your response, either visually or on radar. If the situation demands your response, take appropriate action or immediately advise the facility of any problem. Acknowledgement is made with one of the words "Wilco, Roger, Affirmative, Negative," or other appropriate remarks; e.g., "Piper Two one Four Lima, Roger." If you have been receiving services, e.g., VFR traffic ad-

visories and you are leaving the area or changing frequencies, advise the ATC facility and terminate contact.

4. *Acknowledgement of Frequency Changes.* When advised by ATC to change frequencies, acknowledge the instruction. If you select the new frequency without an acknowledgement, the controller's workload is increased because he has no way of knowing whether you received the instruction or lost your radios.

5. *Precautions in the Use of Call Signs.* Improper use of call signs can result in pilots executing a clearance intended for another aircraft. Call signs *should never be abbreviated on an initial contact or at any time when other aircraft call signs have similar numbers/sounds or identical letters/numbers,* (e.g., Cessna 6132F, Cessna 1622F, Baron 123F, Cherokee 7732F, etc.). As an example, assume that a controller issues an approach clearance to an aircraft at the bottom of a holding stack and an aircraft with a similar call sign (at the top of the stack) acknowledges the clearance with the last two or three numbers of his call sign. If the aircraft at the bottom of the stack did not hear the clearance and intervene, flight safety would be affected, and there would be no reason for either the controller or pilot to suspect that anything is wrong. This kind of "human factors" error can strike swiftly and is extremely difficult to rectify. *Pilot's therefore, must be certain that aircraft identification is complete and clearly identified before taking action on an ATC clearance.* FAA personel will not abbreviate call signs of air carrier or other civil aircraft having authorized call signs. *FAA* may initiate abbreviated call signs of other aircraft by using the *prefix and the last three digits/letters of* the aircraft identification after communications are established. Controllers, when aware of similar/identical call signs, will take action to minimize errors by emphasizing certain numbers/letters, by repeating the entire call sign, repeating the prefix, or by asking pilots to use a different call sign temporarily. Pilots should use the phrase "Verify clearance for (your complete call sign)" if doubt exists concerning proper identity.

AIRCRAFT CALL SIGNS

1. Civil aircraft pilots should state the aircraft type, model or manufacturer's name, or if none of these are known, the digits/letters of the registration number. Generally, the prefix "N" is dropped and the aircraft manufacturer's name or model is stated instead.

Examples:

"BONANZA SIX FIVE FIVE GOLF," "DOUGLAS ONE ONE ZERO," "BREEZY SIX ONE THREE ROMEO EXPERIMENTAL." (Omit "Experimental" after initial contact).

2. Air Taxi or other commercial operators *not* having FAA authorized call signs should prefix their normal identification with the phonetic word "Tango." For example, Tango Aztec Two Four Six Four Alpha.

3. Air carriers and commuter air carriers having FAA authorized call signs should identify themselves by stating the complete call sign, using group form for the numbers.

Examples:

UNITED TWENTY-FIVE, MIDWEST COMMUTER SEVEN ELEVEN.

4. Military aircraft use a variety of systems including serial numbers, word call signs and combinations of letters/numbers. Examples include Army Copter 48931, Air Force 61782, MAC 31792, Pat 157, Air Evac 17652, Navy Golf Alfa Kilo 21, Marine 4 Charlie 36, etc.

5. Civilian airborne ambulance flights (aircraft carrying ambulatory or litter patients, organ donors, or organs for transplant) will be expedited and necessary notification will be made when the pilot requests. When filing flight plans for such flights, add the word "Lifeguard" in the remarks section. In radio communications use the call sign "Lifeguard" followed by the type, digits, and letters of the registration number. Pilots should use discretion in the use of this term. It should be used for those missions of an urgent nature.

Example:

LIFEGUARD CESSNA TWO SIX FOUR SIX.

GROUND STATION CALL SIGNS

Examples are self-explanatory:

Airport Unicom	"Shannon Unicom"
FAA Flight Service Station	"Shannon Radio"
FAA Flight Service Station (En Route Flight Advisory Service (Weather)	"Seattle Flight Watch"
Airport Traffic Control Tower	"Augusta Tower"
Clearance Delivery Position (IFR)	"Dallas Clearance Delivery"
Ground Control Position in Tower	"Miami Ground"
Radar or Nonradar Approach Control Position	"Oklahoma City Approach"
Radar Departure Control Position	"St. Louis Departure"
FAA Air Route Traffic Control Center	"Washington Center"

TIME

1. FAA uses Greenwich Mean Time (GMT) (or "Z") for all operations.

To Convert From:	To Greenwich Mean Time:
Eastern Standard Time	Add 5 hours*
Central Standard Time	Add 6 hours*
Mountain Standard Time	Add 7 hours*
Pacific Standard Time	Add 8 hours*

* For Daylight Time subtract 1 hour.

2. The 24-hour clock system is used in radiotelephone transmissions. The hour is indicated by the first two figures and the minutes by the last two figures.

Examples:

0000 . . . . ZERO ZERO ZERO ZERO
0920 . . . . ZERO NINER TWO ZERO

3. Time may be stated in minutes only (two figures) in radio telephone communications when no misunderstanding is likely to occur.

4. Current time in use at a station is stated in the nearest quarter minute in order that pilots may use this information for time checks. Fractions of a quarter minute less than eight seconds are stated as the preceding quarter minute; fractions of a quarter minute of eight seconds or more are stated as the succeeding quarter minute.

Examples:

Time	
0929:05	TIME, ZERO NINER TWO NINER
0929:10	TIME, ZERO NINER TWO NINER AND ONE-QUARTER

FIGURES

1. Figures indicating hundreds and thousands in round numbers, as for ceiling heights, and upper wind levels up to 9900 shall be spoken in accordance with the following examples:

500 . . . . FIVE HUNDRED
4500 . . . . FOUR THOUSAND FIVE HUNDRED

2. Numbers above 9900 shall be spoken by separating the digits preceding the word "thousand." Examples:

10000 . . . . ONE ZERO THOUSAND
13500 . . . . ONE THREE THOUSAND FIVE HUNDRED

SPEEDS

The separate digits of the speed followed by the word 'knots'. The controller may omit the word "knots" when using speed adjustment procedures, "Reduce/Increase Speed To One Five Zero."

Examples:
250 . . . . TWO FIVE ZERO KNOTS
185 . . . . ONE EIGHT FIVE KNOTS
95 . . . . NINER FIVE KNOTS

PHONETIC ALPHABET

1. The International Civil Aviation Organization (ICAO) phonetic alphabet is used by FAA personnel when communications conditions are such that the information cannot be readily received without their use. Air traffic control facilities may also request pilots to use phonetic letter equivalents when aircraft with similar sounding identifications are receiving communications on the same frequency.

2. Pilots should use the phonetic alphabet when identifying their aircraft during initial contact with air traffic control facilities. Additionally use the phonetic equivalents for single letters and to spell out groups of letters or difficult words during adverse communications conditions. (See Figure 8-14.)

COMMUNICATIONS WITH TOWER WHEN AIRCRAFT TRANSMITTER/RECEIVER OR BOTH ARE INOPERATIVE
(See FAR 91.87 and 91.77)

1. Arrival

a. Receiver inoperative—If you have reason to believe your receiver is inoperative, remain outside or above the airport traffic area until the direction and flow of traffic has been determined, then advise the tower of your type aircraft, position, altitude, intention to land and request that you be controlled with light signals. (The color, type, and meanings of light signals are published in FAR 91.77 and the AIM.) When you are approximately 3 to 5 miles from the airport, advise the tower of your position and join the airport traffic pattern. From this point on, watch the tower for light signals. Thereafter, if a complete pattern is made, transmit your position when downwind and/or turning base leg.

3. Transmit airway or jet route numbers as follows:

Examples:
V12 . . . . VICTOR TWELVE
J533 . . . . J FIVE THIRTY THREE
4. All other numbers shall be transmitted by pronouncing each digit.

Example:
10 . . . . ONE ZERO
5. When a radio frequency contains a decimal point, the decimal point is spoken as "POINT."

Example:
122.1 . . . . ONE TWO TWO POINT ONE
(ICAO Procedures require the decimal point be spoken as "DECIMAL" and FAA will honor such usage by military aircraft and all other aircraft required to use ICAO Procedures.)

ALTITUDES AND FLIGHT LEVELS

1. Up to but not including 18,000' MSL—by stating the separate digits of the thousands, plus the hundreds, if appropriate.

Examples:
12,000 . . . . ONE TWO THOUSAND
12,500 . . . . ONE TWO THOUSAND FIVE HUNDRED
2. At and above 18,000' MSL (FL 180) by stating the words "flight level" followed by the separate digits of the flight level.

Example:
190 . . . . FLIGHT LEVEL ONE NINER ZERO

DIRECTIONS

The three digits of the magnetic course, bearing, heading or wind direction. All of the above should always be magnetic. The word "true" must be added when it applies.

Examples:
(magnetic course) 005 . . . . ZERO ZERO FIVE
(true course) 050 . . . . ZERO FIVE ZERO TRUE
(magnetic bearing) 360 . . . . THREE SIX ZERO
(magnetic heading) 100 . . . . ONE ZERO ZERO
(wind direction) 220 . . . . TWO TWO ZERO

Letter	Word	Pronunciation
A	Alfa	(AL-FAH)
B	Bravo	(BRAH-VOH)
C	Charlie	(CHAR-LEE) (or SHAR LEE)
D	Delta	(DELL-TAH)
E	Echo	(ECK-OH)
F	Foxtrot	(FOKS-TROT)
G	Golf	(GOLF)
H	Hotel	(HOH-TEL)
I	India	(IN-DEE-AH)
J	Juliett	(JEW-LEE-ETT)
K	Kilo	(KEY-LOH)
L	Lima	(LEE-MAH)
M	Mike	(MIKE)
N	November	(NO-VEM-BER)
O	Oscar	(OSS-CAH)
P	Papa	(PAH-PAH)
Q	Quebec	(KEH-BECK)
R	Romeo	(ROW-ME-OH)
S	Sierra	(SEE-AIR-RAH)
T	Tango	(TANG-GO)
U	Uniform	(YOU-NEE-FORM) (or OO-NEE-FORM)
V	Victor	(VIK-TAH)
W	Whiskey	(WISS-KEY)
X	Xray	(ECKS-RAY)
Y	Yankee	(YANG-KEY)
Z	Zulu	(ZOO-LOO)
1	One	(WUN)
2	Two	(TOO)
3	Three	(TREE)
4	Four	(FOW-ER)
5	Five	(FIFE)
6	Six	(SIX)
7	Seven	(SEV-EN)
8	Eight	(AIT)
9	Nine	(NIN-ER)
0	Zero	(ZEE-RO)

Figure 8-14. *Phonetic alphabet and Morse code.*

b. Transmitter inoperative—Remain outside or above the airport traffic area until the direction and flow of traffic has been determined, then join the airport traffic pattern. Monitor the primary local control frequency as depicted on Sectional Charts for landing or traffic information, and look for a light signal which may be addressed to your aircraft. During hours of daylight, acknowledge tower transmissions or light signals by rocking your wings. At night, acknowledge by blinking the landing or navigation lights.

c. Transmitter and receiver inoperative—Remain outside or above the airport traffic area until the direction and flow of traffic has been determined, then join the airport traffic pattern and maintain visual contact with the tower to receive light signals. Acknowledge light signals in accordance with 1.b. above.

2. Departures.

If you experience radio failure prior to leaving the parking area, make every effort to have the equipment repaired. If you are unable to have the malfunction repaired, call the tower by telephone and request authorization to depart without two-way radio communications. If tower authorization is granted, you will be given departure information and requested to monitor the tower frequency or watch for light signals, as appropriate. During daylight hours, acknowledge tower transmissions or light signals by moving the ailerons or rudder. At night, acknowledge by blinking the landing or navigation lights. If radio malfunction occurs after departing the parking area, watch the tower for light signals or monitor tower frequency.

COMMUNICATIONS GUARD ON VFR FLIGHTS

On VFR flights, guard the voice channel of VORs for broadcasts and calls from FAA Flight Service Stations. (FSS). Where the VOR voice channel is being utilized for ATIS broadcasts, pilots of VFR flights are urged to guard the voice channel of an adjacent VOR.

When in contact with a control facility, notify the controller if you plan to leave the frequency. This could save the controller time by not trying to call you on that frequency.

AIRPORT OPERATIONS

Increased traffic congestion, aircraft in climb and descent attitudes, and pilots preoccupation with cockpit duties are some factors that increase the hazardous accident potential near the airport. The situation is further compounded when the weather is marginal—that is, just meeting VFR requirements. Pilots must be particularly alert when operating in the vicinity of an airport. This section defines some rules, practices and procedures that pilots should be familiar with, and adhere to, for safe airport operations.

TOWER-CONTROLLED AIRPORTS

1. When operating to an airport where traffic control is being exercised by a control tower, pilots are required to maintain two-way radio contact with the tower while operating within the airport traffic area unless the tower authorizes otherwise. Initial call-up should be made about 15 miles from the airport.

2. When necessary, the tower controller will issue clearances or other information for aircraft to generally follow the desired flight path (traffic patterns) when flying in the airport traffic area/control zone, and the proper taxi routes when operating on the ground. If not otherwise authorized or directed by the tower, pilots approaching to land in an airplane must circle the airport to the left, and pilots approaching to land in a helicopter must avoid the flow of fixed wing traffic. However, an appropriate clearance must be received from the tower before landing. (See Figure 8–15.)

NOTE.—This diagram is intended only to illustrate terminology used in identifying various components of a traffic pattern. It should not be used as a reference or guide on how to enter a traffic pattern.

3. The following terminology for the various components of a traffic pattern has been adopted as standard for use by control towers and pilots:

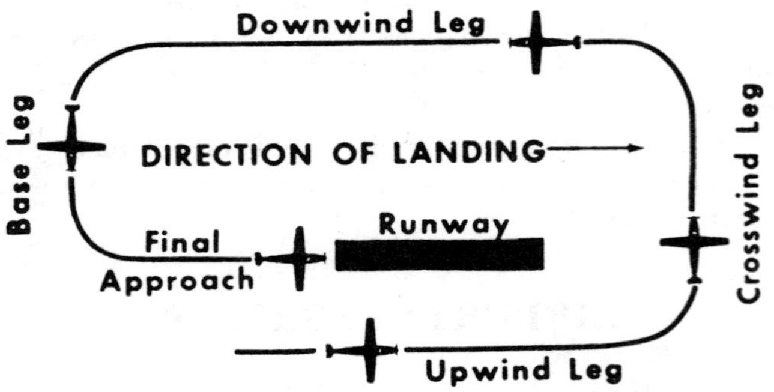

Upwind leg—A flight path parallel to the landing runway in the direction of landing.

Crosswind leg—A flight path at right angles to the landing runway off its takeoff end.

Downwind Leg—A flight path parallel to the landing runway in the direction opposite to landing.

Base leg—A flight path at right angles to the landing runway off its approach end and extending from the downwind leg to the intersection of the extended runway center line.

Final approach—A flight path in the direction of landing along the extended runway center line from the base leg to the runway.

4. The tower controller will consider that pilots of turbinepowered aircraft are ready for takeoff when they reach the runway/warm-up block unless they advise otherwise.

NONTOWER AIRPORTS

1. Preparatory to landing at an airport without an operating control tower, *but at which either an FSS or a UNICOM is located,* pilots should contact the FSS or UNICOM for traffic advisories, wind, runway in use, and traffic flow information. CAUTION—ALL AIRCRAFT MAY NOT BE COMMUNICATING WITH THE FSS OR UNICOM. THEY CAN ONLY ISSUE TRAFFIC ADVISORIES ON THOSE THEY ARE AWARE OF.

2. At those airports *not having a tower, FSS or UNICOM,* visual indicators, if installed, provide the following information:

a. The segmented circle system is designed to provide traffic pattern information at airports without operating control towers. The system consists of the following components:

The segmented circle—Located in a position affording maximum visibility to pilots in the air and on the ground and providing a centralized location for other elements of the system.

The wind direction indicator—A wind cone installed at the center of the circle and used to indicate wind direction and velocity. The large end of the wind cone points into the wind.

The landing direction indicator—A tetrahedron or a tee installed when conditions at the airport warrant its use, located at the center of the circle, and used to indicate the direction in which landings and takeoffs should be made. The large end (cross bar) of a tee is in the direction of landing. The small end of a tetrahedron points in the direction of landing.

Figure 8–15. *Traffic pattern.*

Landing strip indicators—Installed in pairs as shown in the segmented circle diagram and used to show the alignment of landing strips.

Traffic pattern indicators—Arranged in pairs in conjunction with landing strip indicators and used to indicate the direction of turns when there is a variation from the normal left traffic pattern. (If there is no segmented circle installed at the airport, traffic pattern indicators may be installed on or near the end of the runway.)

b. Where installed, a flashing amber light near the center of the segmented circle (or on top of the control tower or adjoining building) indicates that a right traffic pattern is in effect at the time.

c. Preparatory to landing at an airport without a control tower, or when the control tower is not in operation, the pilot should concern himself with the indicator for the approach end of the runway to be used. When approaching for landing all turns must be made to the left unless a light signal or traffic pattern indicator indicates that turns should be made to the right. If the pilot will mentally enlarge the indicator for the runway to be used, the base and final approach legs of the traffic pattern to be flown immediately become apparent. Similar treatment of the indicator at the departure end of the runway will clearly indicate the direction of turn after takeoff.

d. When two or more aircraft are approaching an airport for the purpose of landing, the aircraft at the lower altitude has the right of way, but it shall not take advantage of this rule to cut in front of another which is on final approach to land, or to overtake that aircraft. (Ref: FAR 91.67(f))

TRAFFIC PATTERNS

At most airports and military air bases, traffic pattern altitudes for propeller driven aircraft generally extend from 600 feet to as high as 1,500 feet above the ground. Also traffic pattern altitudes for military turbojet aircraft sometimes extend up to 2,500 feet above the ground. Therefore, pilots of en route aircraft should be constantly on the alert for other aircraft in traffic patterns and avoid these areas whenever possible.

Traffic pattern altitudes should be maintained unless otherwise required by the applicable distance from cloud criteria (FAR 91.105) (See Figure 8-16.)

UNEXPECTED MANEUVERS IN THE AIRPORT TRAFFIC PATTERN

There have been several incidents in the vicinity of controlled airports that were caused primarily by aircraft executing unexpected maneuvers. Air-

port traffic control service is based upon observed or known traffic and airport conditions. Controllers establish the sequence of arriving and departing aircraft by requiring them to adjust flight as necessary to achieve proper spacing. These adjustments can only be based on observed traffic, accurate pilot reports, and anticipated aircraft maneuvers. Pilots are expected to cooperate so as to preclude disruption of traffic flow or creation of conflicting patterns.

The pilot in command of an aircraft is directly responsible for and is the final authority as to the operation of that aircraft. On occasion it may be necessary for a pilot to maneuver his aircraft to maintain spacing with the traffic he has been sequenced to follow. The controller can anticipate minor maneuvering such as shallow "S" turns. The controller cannot however, anticipate a major maneuver such as a 360 degree turn. If a pilot makes a 360 degree turn after he has obtained a landing sequence the result is usually a gap in the landing interval and more importantly it causes a chain reaction which may result in a conflict with following traffic and interruption of the sequence established by the tower or approach controller. Should a pilot decide he needs to make maneuvering turns to maintain spacing behind a preceding aircraft, he should always advise the controller if at all possible. Except when requested by the controller or in emergency situations, a 360 degree turn should never be executed in the traffic pattern or when receiving radar service without first advising the controller.

USE OF RUNWAYS

Runways are numbered to correspond to their magnetic bearing. Runway 27, for example, has a bearing of 270 degrees. Wind direction issued by the tower is also magnetic.

1. At airports where an informal or formal runway use program is not established, ATC clearances may specify: (1) the runway most nearly aligned with the wind when it is five knots or more, (2) the "calm wind" runway when wind is less than five knots, or (3) another runway if operationally advantageous. It is not necessary for a controller to specifically inquire if the pilot will use a specific runway or to offer him a choice of runways. If a pilot prefers to use a different runway than that specified or the one most nearly aligned with the wind, he is expected to inform ATC accordingly.

2. At airports where an informal runway use program is established for airplanes over 12,500 pounds and all turbojet airplanes, ATC will specify runways on the basis of noise sensitivity starting with those that provide the greatest noise abatement benefits and subsequently in order of decreasing

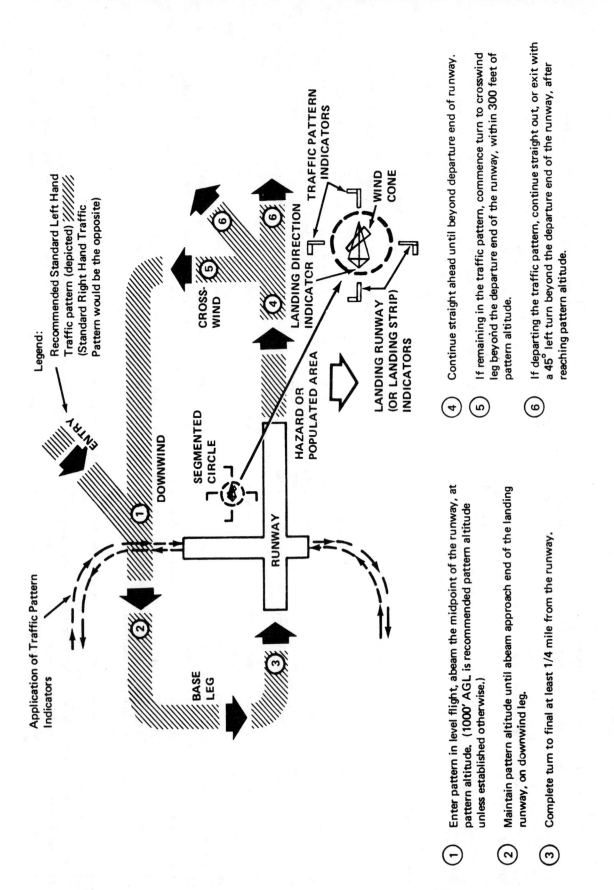

Application of Traffic Pattern Indicators

Legend:

Recommended Standard Left Hand Traffic pattern (depicted) ⫽⫽⫽
(Standard Right Hand Traffic Pattern would be the opposite)

ENTRY

DOWNWIND

SEGMENTED CIRCLE

BASE LEG

RUNWAY

CROSS-WIND

LANDING DIRECTION INDICATOR

HAZARD OR POPULATED AREA

TRAFFIC PATTERN INDICATORS

WIND CONE

LANDING RUNWAY (OR LANDING STRIP) INDICATORS

① Enter pattern in level flight, abeam the midpoint of the runway, at pattern altitude. (1000' AGL is recommended pattern altitude unless established otherwise.)

② Maintain pattern altitude until abeam approach end of the landing runway, on downwind leg.

③ Complete turn to final at least 1/4 mile from the runway.

④ Continue straight ahead until beyond departure end of runway.

⑤ If remaining in the traffic pattern, commence turn to crosswind leg beyond the departure end of the runway, within 300 feet of pattern altitude.

⑥ If departing the traffic pattern, continue straight out, or exit with a 45° left turn beyond the departure end of the runway, after reaching pattern altitude.

preference, those that provide lesser advantages from a noise benefits standpoint. Both wind direction and velocity will be given. The wind will not be described as "calm" unless the velocity is less than 3 knots. If the pilot of any aircraft subject to the informal runway use program prefers to use a different runway than that specified, he is expected to inform ATC accordingly. Such requests from pilots of large or turbojet airplanes will be honored and the pilot will be informed if the requested runway is "noise sensitive."

3. At airports where a formal runway use program is established for airplanes over 12,500 pounds and all turbojet airplanes, ATC will assign noise abatement runways, when acceptable to the pilot, if: (1) runways are clear and dry; i.e., there is no ice, slush, etc., (2) wind velocity does not exceed 15 knots, and (3) any crosswind does not exceed 80 degrees from either side of the centerline of the runway in the direction of use. The pilot of an aircraft subject to the formal runway use program will be informed that the runway specified is the noise abatement runway only when he requests the use of another runway which is more noise sensitive.

4. When it is determined that turboprop airplanes of less than 12,500 pounds create a noise problem such airplanes will be subject to the informal or formal runway use program established for that airport.

5. If a pilot prefers to use a different runway than that specified, he is expected to advise ATC accordingly. When use of a different runway is requested, pilot cooperation is solicited to preclude disruption of the traffic flow or creation of conflicting patterns.

INTERSECTION TAKEOFFS

1. In order to enhance airport capacities, reduce taxing distances, minimize departure delays, and provide for more efficient movement of air traffic, controllers may initiate intersection takeoffs as well as approve them when the pilot requests. If for ANY reason a pilot prefers to use a different intersection or the full length of the runway or desires to obtain the distance between the intersection and the runway end, HE IS EXPECTED TO INFORM ATC ACCORDINGLY.

2. Controllers are required to separate small propeller driven aircraft (less than 12,500 pounds) taking off from an intersection on the same runway (same or opposite direction takeoff) following a Category III aircraft (12,500 or more) by ensuring that at least a 3-minute interval exists between the time that the preceeding Category III aircraft has taken off and the succeeding air-

craft begins takeoff roll. To inform the pilot of the required 3-minute wait, the controller will state "Hold for Wake Turbulence." If after considering wake turbulence hazards the pilot feels that a lesser time interval is appropriate, he may request a waiver to the 3-minute interval. Pilots must initiate such a request by stating "REQUEST WAIVER TO 3 MINUTE INTERVAL," or by making a similar statement. Controllers may then issue a takeoff clearance if other traffic permits, since the pilot has accepted responsibility for his own wake turbulence separation. A pilot may not waive the 3 minute interval when departing behind a heavy jet.

SIMULTANEOUS LANDINGS ON INTERSECTING RUNWAYS

1. Despite the many new and lengthened runways which have been added to the Nation's airports in recent years, limited runway availability remains a major contributing factor to airport congestion. Many high-density airports have gained operational experience with intersecting runways which clearly indicates that simultaneous landings are safe and feasible. Tower controllers may authorize simultaneous landings on intersecting runways when the following conditions are met:

a. The controller has received no reports that braking action is less than good.

b. Operations are conducted in VFR conditions unless visual separation is applied.

c. Instructions are issued to restrict one aircraft from entering the intersecting runway being used by another aircraft.

d. Traffic information is issued to and an acknowledgement received from one or both pilots as appropriate to the situation.

e. The measured distance from runway threshold to intersection is issued if the pilot requests it.

f. The conditions specified in c., d., and e. are met at or before issuance of the landing clearance.

g. The distance from landing threshold to the intersection is adequate for the category of aircraft being held short. Controllers are provided a general table of aircraft category/minimum runway length requirements as a general guide. IT IS INCUMBENT ON THE PILOT TO DETERMINE HIS ABILITY TO HOLD SHORT OF AN INTERSECTION AFTER LANDING, WHEN SO INSTRUCTED.

3. THE SAFETY AND OPERATION OF AN AIRCRAFT REMAIN

Color and Type of Signal	On the Ground	In Flight
STEADY GREEN	Cleared for take-off	Cleared to land
FLASHING GREEN	Cleared to taxi	Return for landing (to be followed by steady green at proper time)
STEADY RED	Stop	Give way to other aircraft and continue circling
FLASHING RED	Taxi clear of landing area (runway) in use	Airport unsafe—do not land
FLASHING WHITE	Return to starting point on airport	
ALTERNATING RED & GREEN	General Warning Signal—Exercise Extreme Caution	

Figure 8-17. *Light signals.*

THE RESPONSIBILITY OF THE PILOT. IF FOR ANY REASON (e.g. DIFFICULTY IN DISCERNING LOCATION OF AN INTERSECTION AT NIGHT, INABILITY TO HOLD SHORT OF AN INTERSECTION, WIND FACTORS, ETC.) A PILOT ELECTS TO USE THE FULL LENGTH OF THE RUNWAY, A DIFFERENT RUNWAY OR DESIRES TO OBTAIN THE DISTANCE FROM THE LANDING THRESHOLD TO THE INTERSECTION, HE IS EXPECTED TO PROMPTLY INFORM ATC ACCORDINGLY.

LIGHT SIGNALS

1. The following procedures are used by airport traffic control towers in the control of aircraft not equipped with radio. These same procedures will be used to control aircraft equipped with radio if radio contact cannot be established. Airport traffic control personnel use a directive traffic control signal which emits an intense narrow beam of a selected color (either red, white, or green) when controlling traffic by light signals. Although the traffic signal light offers the advantage that some control may be exercised over nonradio equipped aircraft, pilots should be cognizant of the disadvantages which are:

a. The pilot may not be looking at the control tower at the time a signal is directed toward him.

b. The directions transmitted by a light signal are very limited since only approval or disapproval of a pilot's anticipated actions may be transmitted. No supplement or explanatory information may be transmitted except by the use of the "General Warning Signal" which advises the pilot to be on the alert.

2. Portable traffic control light signals.

3. Between sunset and sunrise, a pilot wishing to attract the attention of the control tower should turn on a landing light and taxi the aircraft into a position, clear of the active runway, so that light is visible to the tower. The landing light should remain on until appropriate signals are received from the tower.

4. Pilots should acknowledge light signals by moving the ailerons or rudder during the hours of daylight or by blinking the landing or navigation lights during the hours of darkness.

5. During the hours of daylight the lighting of the rotating beacon will mean that ground visibility is less than three miles and/or that the ceiling is less than 1,000 feet. The operation of the rotating beacon indicates that a clearance from air traffic control is necessary for landing, takeoff, or flight in the traffic pattern if the airport is within a control zone. (See Figure 8-17.)

COMMUNICATIONS

1. Pilots of departing aircraft should communicate with the control tower on the appropriate ground control/clearance delivery frequency prior to starting engines to receive engine start time, taxi and/or clearance information. Unless otherwise advised by the tower, remain on that frequency during taxiing and runup, then change to local control frequency when ready to request takeoff clearance. Note—Refer to AUTOMATIC TERMINAL INFORMATION SERVICE (ATIS) for continuous broadcast of terminal information.

2. Ground control frequencies are provided in the 121.6–121.9 MHz band to eliminate frequency congestion on the tower (local control) channel. These ground control frequencies, whose use is limited to communications between the tower and aircraft on the ground and between the tower and utility vehicles on the airport, provide a clear VHF channel for arriving and departing aircraft. They are used for issuance of taxi information, clearances, and other necessary contacts between the tower and aircraft or other vehicles operated on the airport. A pilot who has just landed should not change from the tower frequency to the ground control frequency until he is directed to do so by the controller. Normally, only one ground control frequency is assigned at an airport; however, at locations where the amount of traffic so warrants, a second ground control frequency and/or another frequency designated as a clearance delivery frequency, may be assigned.

3. The controller may omit the frequency or the numbers preceding the decimal point in the frequency when directing the pilot to change to a VHF ground control frequency if, in the controller's opinion, this usage will be clearly understood by the pilot; e.g. 121.7–'Contact ground' or 'Contact ground point seven'.

DEPARTURE DELAYS

Pilots should contact ground control/clearance delivery prior to starting engines as gate hold procedures will be in effect whenever departure delays exceed or are anticipated to exceed 5 minutes. The sequence for departure will be maintained in accordance with initial call up unless modified by flow control restrictions. Pilots should monitor the ground control/clearance delivery frequency for engine startup advisories or new proposed start time if the delay changes.

TAXIING

1. Approval must be obtained prior to moving an aircraft or vehicle onto the movement area during the hours an airport traffic control tower is in operation. Always state your position on the airport when calling the tower for taxi instructions. The movement area is normally described in local bulletins issued by the airport manager or control tower. These bulletins may be found in FSSs, fixed base operators offices, air carrier offices and operations offices. The control tower also issues bulletins describing areas where they cannot provide airport traffic control service due to nonvisibility or other reasons. In addition, a clearance must be obtained prior to taxiing on a runway, taking off or landing during the hours an airport traffic control tower is in operation. Authorization to taxi "to" a runway is authorization to cross runways that intersect the taxi route unless instructions to the contrary are received. Authorization to taxi "to" a runway does not constitute a clearance to taxi "on" that runway.

2. ATC clearances or instructions pertaining to taxiing are predicated on known traffic and known physical airport conditions. Therefore, it is important that pilots clearly understand the clearance or instruction. Although an ATC clearance is issued for taxiing purposes, when operating in accordance with the FARs, it is the responsibility of the pilot to avoid collision with other aircraft. Since "the pilot in command of an aircraft is directly responsible for, and is the final authority as to, the operation of that aircraft" the pilot should obtain clarification of any clearance or instruction which is not understood.

3. At those airports where the United States Government operates the control tower and ATC has authorized non-compliance with the requirement for two-way radio communications while operating within the airport traffic area, or at those airports where the United States Government does not operate the control tower and radio communications cannot be established, pilots shall obtain a clearance by visual light signal prior to taxiing on a runway and prior to take-off and landing.

4. The following phraseologies and procedures are used in radio-telephone communications with aeronautical ground stations.

a. Aircraft identification, location, type of operation planned (VFR or IFR) and the point of first intended landing.

Example:
Aircraft: "WASHINGTON GROUND BEECHCRAFT ONE THREE ONE FIVE NINER AT HANGAR EIGHT, READY TO TAXI, IFR TO CHICAGO, OVER"

Tower: "BEECHCRAFT ONE THREE ONE FIVE NINER, RUNWAY THREE SIX, WIND ZERO THREE ZERO DEGREES AT TWO FIVE, ALTIMETER THREE ZERO ZERO FOUR, HOLD SHORT OF RUNWAY THREE."

b. Air route traffic control clearances are relayed to pilots by airport traffic controllers in the following manner.

Example:
Tower: BEECHCRAFT ONE THREE ONE FIVE NINER CLEARED TO THE CHICAGO MIDWAY AIRPORT, VIA VICTOR EIGHT, MAINTAIN EIGHT THOUSAND, OVER.

Aircraft: "BEECHCRAFT ONE THREE ONE FIVE NINER CLEARED TO THE CHICAGO MIDWAY AIRPORT, VIA VICTOR EIGHT, MAINTAIN EIGHT THOUSAND, OVER."

NOTE.—Normally, an ATC IFR clearance is relayed to a pilot by the ground controller. At busy locations, however, pilots may be instructed by the ground controller to "CONTACT CLEARANCE DELIVERY" on a frequency designated for this purpose. No surveillance or control over the movement of traffic is exercised by this position of operation. See Clearance Readback in ATC Clearances/Separations.

c. Aircraft identification, location and request for taxi instructions after landing.

Example:

Aircraft: "DULLES GROUND BEECHCRAFT ONE FOUR TWO SIX ONE CLEARING RUNWAY ONE RIGHT ON TAXIWAY E3, REQUEST CLEARANCE TO PAGE."

Tower: "BEECHCRAFT ONE FOUR TWO SIX ONE, TAXI TO PAGE VIA TAXIWAYS E3, E1, AND E9."

TAXI DURING LOW VISIBILITY

1. Pilots and aircraft operators should be constantly aware that during certain low visibility conditons the movement of aircraft and vehicles on airports may not be visible to the tower controller. This may prevent visual confirmation of an aircraft's adherence to taxi instructions. Pilots should, therefore, exercise extreme vigilance and proceed cautiously under such conditions. Of vital importance is the need for pilots to notify the controller when such difficulties are anticipated and to immediately inform the controller at the first indication of becoming disoriented.

2. Pilots should proceed with extreme caution when taxiing toward the sun. When vision difficulties are encountered pilots should immediately inform the controller.

CLEARING THE RUNWAY AFTER LANDING

After landing, unless otherwise instructed by the control tower, aircraft should continue to taxi in the landing direction, proceed to the nearest turnoff and exit the runway without delay. Do not change to the ground control frequency while on an active runway or make a 180 degree turn to taxi back on an active runway without authorization from the control tower.

SPECIAL VFR CLEARANCES

Special VFR Flight Clearance Procedures (F.A.R. Part 91.107)

1. An ATC clearance must be obtained *prior* to operating within a control zone when the weather is less than that required for VFR flight. A VFR pilot may request and be given a clearance to enter, leave or operate within most control zones in special VFR conditions, traffic permitting, and providing such flight will not delay IFR operations. The visibility requirements for Special VFR fixed-wing aircraft are: 1 mile flight visibility for operations within the control zone and 1 mile ground visibility if taking off or landing. All special VFR flights must remain clear of clouds. When a control tower is located within the control zone, requests for clearances should be to the tower. If no tower is located within the control zone, a clearance may be obtained from the nearest tower, flight service station or center.

2. It is not necessary to file a complete flight plan with the request for clearance but the pilot should state his intentions in sufficient detail to permit air traffic control to fit his flight into the traffic flow. The clearance will not contain a specific altitude as the pilot must remain clear of clouds. The controller may require the pilot to fly at or below a certain altitude due to other traffic, but the altitude specified will permit flight at or above the minimum safe altitude. In addition, at radar locations, flights may be vectored if necessary for control purposes or on pilot request.

3. ATC provides separation between special VFR flights and between them and other IFR flights.

4. Special VFR operations by fixed-wing aircraft are prohibited in some control zones due to the volume of IFR traffic. A list of these control zones is contained in FAR 93.113, and also depicted on Sectional Aeronautical Charts.

5. Special VFR operations by fixed-wing aircraft are prohibited between sunset and sunrise unless the pilot is instrument rated and the aircraft is equipped for IFR flight.

6. Special VFR clearances are effective within control zones only. ATC does not provide separation after an aircraft leaves the control zone on a special VFR clearance.

ATC CLEARANCE/INSTRUCTION READBACK

Pilots of airborne aircraft should read back *those parts* of ATC clearances/instrucions containing altitude assignments or vectors, as a means of mutual verification. The readback of the "numbers" serves as a double check between pilots and controllers, and such, it is an invaluable aid in reducing the kinds of communications errors that occur when a number is either "misheard" or is incorrect.

1. Precede all readbacks/acknowledgments with the aircraft identification. This is the only way that controllers can determine that the correct aircraft received the clearance/instruction. The requirement to include aircraft identification in all readbacks/acknowledgments becomes more important as frequency congestion increases and when aircraft with similar call signs are on the same frequency.

2. Read back altitudes, altitude restrictions, and vectors in the same sequence as they are given in the clearance/instruction.

3. Altitudes contained in charted procedures such as SIDs, instrument approaches, etc., should not be read back unless they are specifically stated by the controller.

ADHERENCE TO CLEARANCES

When air traffic clearance has been obtained under either the Visual or Instrument Flight Rules, the pilot in command of the aircraft shall not deviate from the provisions thereof unless an amended clearance is obtained. The addition of a VFR or other restriction, i.e., climb/descent point or time, crossing altitude, etc., does not authorize a pilot to deviate from the route of flight or any other provision of the air traffic control clearance.

RUNWAY SEPARATION

Tower controllers establish the sequence of arriving and departing aircraft by requiring them to adjust flight or ground operation as necessary to achieve proper spacing. They may "HOLD" an aircraft short of the runway to achieve spacing between it and another arriving aircraft; the controller may instruct a pilot to "EXTEND DOWNWIND" in order to establish spacing from another arriving or departing aircraft. At times a clearance may include the word "IMMEDIATE." For example: "CLEARED FOR IMMEDIATE TAKEOFF." In such cases "IMMEDIATE" is used for purposes of *air traffic separation*. It is up to the pilot to refuse the clearance if, in his opinion, compliance would adversely affect his operation.

PREFLIGHT
GENERAL

1. Every pilot is urged to receive a preflight briefing and to file a flight plan. This briefing would consist of weather, airport, and enroute navaid information. Briefing service may be obtained from a Flight Service Station either by telephone/interphone, by radio when airborne, or by a personal visit to the Station.

2. In addition to the filing of a flight plan, if the flight will traverse or land in one or more foreign countries, it is particularly important that pilots leave a complete itinerary with someone directly concerned, keep that person advised of the flight's progress and inform him that, if serious doubt arises as to the safety of the flight, he should first contact the FSS.

3. Pilots operating under provisions of FAR Part 135, ATCO, certificate and not having an FAA assigned 3-letter designator, are urged to prefix the normal registration (N) number with the letter "T" on flight plan filing.

Example:

TN1234B.

4. Pilots are urged to use only the latest issue of aeronautical charts in planning and conducting flight operations. Aeronautical charts are revised and reissued on a periodic basis to ensure that depicted data are current and reliable. In the conterminous United States, sectional charts are updated each 6 months, IFR en route charts each 56 days, and amendments to civil IFR approach charts are accomplished on a 7-day revision cycle. Charts that have been superseded by those of a more recent date may contain obsolete or incomplete flight information.

WEATHER BRIEFING

1. Consult your local flight service station (FSS), combined station/tower (CS/T), or National Weather Service Office (NWSO) for preflight weather briefing. FSS and NWSO personnel are certificated pilot weather briefers; however, since CS/T personnel are not certificated pilot weather briefers, weather briefings they furnish are limited to factual data derived directly from weather sequence and forecast information.

2. When telephoning for information, use the following procedure:

a. Identify yourself as a pilot. (Many persons calling WB stations want information for purposes other than flying.)

b. State your intended route, destination, proposed departure time, estimated time en route and type of aircraft.

c. Advise if you intend to fly only VFR.

d. When talking to an FSS, you will be asked your aircraft identification for activity record purposes.

3. Telephone briefings by the specialist may be monitored and/or recorded by management personnel on an unscheduled basis for purposes of quality control, training or evaluation of specialist performance.

4. You are urged to use the Pilot's Preflight Check List which is on the reverse of the flight plan form. The Check List is a reminder of items you should be aware of before beginning flight. Also provided beneath the Check List is a Flight Log for your use if desired.

5. FSSs are required to advise of pertinent NOTAMs, but if they are overlooked, don't hesitate to remind the specialist that you have not received NOTAM information. Additionally, NOTAMs which are known in sufficient time for publication and are of 7 days duration or longer are normally incorporated in Notices to Airmen and carried there until cancellation time. FDC NOTAMs, which apply to instrument flight procedures, are also included in Notices to Airmen up to and including the number indicated in the FDC NOTAM legend. These NOTAMs are not provided during a briefing unless specifically requested by the pilot since the FSS specialist has no way of knowing whether the pilot has already checked NOTAMs prior to calling. Remember to *ask* for NOTAMs contained in the Notices to Airmen—they are not normally furnished during your briefing.

FLIGHT PLAN—VFR

GENERAL

a. Except for operation in or penetrating a Coastal or Domestic ADIZ or DEWIZ, a flight plan is not required for VFR flight; however, it is strongly recommended that one be filed.

b. Indicate aircraft equipment capabilities when filling VFR flight plans by appending the appropriate suffix to aircraft type in the same manner as that prescribed for IFR flight.

Under some circumstances, ATC computer tapes can be useful in constructing the radar history of a downed or crashed aircraft. In each case, knowledge of the aircraft's transponder equipment is necessary in determining whether or not such computer tapes might prove effective.

c. To obtain maximum benefits of the flight plan program, flight plans should be filed directly with the nearest flight service station. For your convenience, FSSs provide one-call (telephone/interphone) or one-stop (personal) aeronautical and meteorological briefings while accepting flight plans. Radio may be used to file if no other means are available. Also, some states operate aeronautical communications facilities which will accept and forward flight plans to the FSS for further handling.

d. Pilots are encouraged to give their departure times directly to the flight service station with which the flight plan was filed. This will ensure more efficient flight plan service and permit the FSS to advise you of significant changes in aeronautical facilities or meteorological conditions. The following procedures are in effect: when a VFR flight plan is filed, it will be held until one hour after the proposed departure time and then canceled unless:

1. The actual departure time is received.
2. A revised proposed departure time is received.
3. At a time of filing, the FSS is informed that the proposed departure time will be met, but actual time cannot be given because of inadequate communications.

e. On pilot's request, at a location having an active tower, the aircraft identification will be forwarded to the tower for reporting the actual departure time. This procedure should be avoided at busy airports.

f. Although position reports are not required for VFR flight plans, periodic reports to FAA Flight Service Stations along the route are good practice. Such contacts permit significant information to be passed to the transmitting aircraft and also serve to check the progress of the flight should it be necessary for any reason to locate the aircraft or its occupants.

Example 1:
Bonanza 314K, over Kingfisher at (time), VFR flight plan, Tulsa to Amarillo.

Example 2:
Cherokee 5123J, over Oklahoma City at (time), Shreveport to Denver, no flight plan.

g. When a "stopover" flight is anticipated to cover an extended period of time, it is recommended that a separate flight plan be filed for each "leg" when the stop is expected to be more than one hour duration.

CLOSING VFR/DVFR FLIGHT PLANS

A pilot is responsible for ensuring that his VFR or DVFR flight plan is cancelled (See FAR 91.83). You should close your flight plan with the nearest Flight Service Station, or if one is not available you may request any ATC facility to relay your cancellation to the FSS. *Control towers do not automatically close VFR or DVFR flight plans* as they may not be aware that a particular VFR aircraft is on a flight plan. If you fail to report or cancel your flight plan within ½ hour (15 minutes for jets) after your ETA, Search and Rescue procedures are started.

EMERGENCY PROCEDURES

GENERAL

1. A pilot in any emergency phase (uncertainty, alert, or distress) should do three things to obtain assistance:

a. *If equipped with a radar beacon transponder (civil) or IFF/SIF (military), and if unable to establish voice communications with an air traffic control facility, switch to Mode 3/A and Code 7700. Military transponder should also be placed in the Emergency position. If crash is imminent and equipped with a Locator Beacon, actuate the emergency signal.*

b. *Contact controlling agency and give nature of distress and pilot's intentions. If unable to contact controlling agencies, attempt to contact any agency or assigned frequency or any of the following frequencies (transmit and receive):*

Frequency	Emission	Effective Range in Nautical Miles	Guarded by
121.5 MHz	Voice	Generally limited to radio line-of-sight.	All military towers, most civil towers, VHF direction finding stns, radar facilities. Flight Service Stns. Ocean Station Vessels.
243.0 MHz	Voice	Generally limited to radio-line-of-sight.	All military towers, most civil towers, UHF direction finding stns, radar facilities. Flight Service Stns. Ocean Station Vessels.
2182 kHz	Voice	Generally less than 300 miles for average aircraft installations.	Some ships and boats, Coast Guard stations, most commercial coast stations.
500 kHz	CW	Generally less than 100 miles for average aircraft installations.	Most large ships, most Coast Guard radio stations, most commercial coast stations.
8364 kHz	CW	Up to several thousand miles, depending upon propagation conditions. Subject to "skip."	U.S.N. Direction Finding Stations, Ocean Station Vessels, most Coast Guard radio stations and some FAA International Flight Service Stations (IFSS).

Transmit as much of the following as possible:

(1) MAYDAY, MAYDAY, MAYDAY (if distress), or PAN, PAN, PAN (if uncertainty or alert). If CW transmission, use SOS (distress) or XXX (uncertainty or alert).

(2) Aircraft identification repeated three times.

(3) Type of aircraft.

(4) Position or estimated position (stating which).

(5) Heading (true or magnetic) (stating which).

(6) True airspeed or estimated true airspeed (stating which).

(7) Altitude.

(8) Fuel remaining in hours and minutes.

(9) Nature of distress.

(10) Pilot's intentions (bailout, ditch, crash landing, etc.).

(11) Assistance desired (fix, steer, bearing, escort, etc.).

(12) Two 10-second dashes with mike button (voice) or key (CW) followed by aircraft identification (once) and OVER (voice) or K (CW).

NOTE.—ARTCC emergency frequency capability normally does not extend to radar coverage limits. If the ARTCC does not respond to transmission on emergency frequency 121.5 MHz or 243.0 MHz pilots should initiate a call to the nearest Flight Service Station or Airport Traffic Control Tower.

EMERGENCY LOCATOR TRANSMITTERS

Emergency Locator Transmitters of various types which are independently powered and of incalculable value in an emergency have been developed as a means of locating downed aircraft and their occupants. These electronic, battery operated transmitters are not a fire hazard. They are designed to emit a distinctive downward swept audio tone for homing purposes on 121.5 MHz and/or 243.0 MHz, preferably on both emergency frequencies. The power source shall be capable of providing power for continuous operation for at least 48 hours or more at a very wide range of ambient temperatures and can expedite search and rescue operations as well as facilitate accident investigation.

This equipment is required for most general aviation and small private aircraft. The pilot and other occupants could survive the crash impact only to die of exposure before they are located. These transmitters are made by several manufacturers for civil aviation use.

Once the transmitter has been activated and the signal detected, it will be a simple matter for the search aircraft with homing equipment to locate the scene. Search patterns have been developed that enable search aircraft, equipped with but one receiver, to locate the transmitter site.

Some models of locator transmitter may be of value for transmitting a signal for inflight emergencies. If an emergency situation occurs in conjunction with a radio failure, the pilot could actuate his transmitter signal to supplement other actions taken to declare the emergency. DF stations hearing the signal would take bearings on the aircraft in distress and notify search and rescue authorities.

Caution should be exercised to prevent the inadvertent actuation of locator transmitters in the air or while they are being handled on the ground. Operational testing of transmitters should be carried out only in shielded areas under controlled conditions. False signals on the distress frequencies can interfere with actual distress transmissions as well as decrease the degree of urgency that should be attached to such signals.

Aircraft operational testing is authorized on 121.5 MHz as follows:

a. Tests should be no longer than three audio sweeps.

b. If the antenna is removable, a dummy load should be substituted during test procedures.

c. Tests shall be conducted *only* within the time period made up of the first five minutes after every hour. Emergency tests outside of this time have to be coordinated with the nearest FSS or Control Tower. Airborne ELT tests are not authorized.

c. *Comply with information and clearance received.* Accept the communications control offered to you by the ground radio station, silence interfering radio stations, and do not shift frequency or shift to another ground station unless absolutely necessary.

2. Pilots of IFR flights experiencing two-way radio failure are expected to adhere to the procedures prescribed under "RADIO COMMUNICATIONS FAILURE" (FAR Part 91.127).

a. The pilot should remember that he has two means of declaring an emergency:

    (1) Emergency SQUAWK (Code 7700) from transponders

    (2) Sending emergency message

b. Some ground stations have *three* electronic means of assisting:

    (1) Receipt of emergency message

    (2) DF bearings; and

    (3) Detection of transponder emergency SQUAWK (Code 7700).

c. **Pilots should remember the FOUR C's**

    (1) **Confess** your predicament to any ground radio station. Do not wait too long. Give SAR a chance!

    (2) **Communicate** with your ground link and pass as much of the distress message on first transmission as possible. We need information for best SAR action!

    (3) **Climb** if possible for better radar and DF detection. If flying at low altitude, the chance for establishing radio contact is improved by climbing.

NOTE.—Unauthorized climb or descent under IFR conditions within controlled airspace is not permitted except in emergency. Any variation in altitude will be unknown to Air Traffic Control except at radar locations having height finding capabilities. Air Traffic Control will operate on the assumption that the provisions of FAR 91.127 are being followed by the pilot.

    (4) **Comply**—especially *Comply*—with advice and instructions received, if you really want help. Assist the ground "communications control" station to control communications on the distress frequency on which you are working (as that is the distress frequency for your case). Tell interfering stations to maintain silence until you call. Cooperate!

3. For bailout, set radio for continuous emission. For ditching or crash landing, radio should, if it is considered that there is no additional risk of fire and if circumstances permit, be set for continuous transmission.

SEARCH AND RESCUE

1. General

a. Search and Rescue is a life-saving service provided through the combined efforts of the FAA, Air Force, Coast Guard, State Aeronautic Commissions or other similar state agencies who are assisted by other organizations such as the Civil Air Patrol, Sheriffs Air Patrol, State Police, etc. It provides search, survival aid, and rescue of personnel of missing or crashed aircraft.

b. Prior to departure on every flight, local or otherwise, someone at the departure point should be advised of your destination and the route of flight if other than direct. Search efforts are often wasted and rescue is often delayed because of pilots who thoughtlessly take off without telling anyone where they are going.

c. All you need to remember to obtain this valuable protection is:

(1) File a Flight Plan with an FAA Flight Service Station in person or by telephone or radio.

(2) Close your flight plan with the appropriate authority immediately upon landing.

(3) If you land at a location other than the intended destination, report the landing to the nearest FAA Flight Service Station.

(4) If you land en route and are delayed more than 30 min. (15 min. for jets), report this information to the nearest FSS.

(5) Remember that if you fail to report within one-half hour after your ETA, a search will be started to locate you.

d. If a crashed aircraft is observed:

(1) Determine if crash is marked with yellow cross; if so, crash has already been reported and identified.

(2) Determine, if possible, type and number of aircraft and whether there is evidence of survivors.

(3) Fix, as accurately as possible, exact location of crash.

(4) If circumstances permit, orbit scene to guide in other assisting units or until relieved by another aircraft.

(5) Transmit information to nearest FAA or other appropriate aircraft.

(6) Immediately after landing, make a complete report to nearest FAA, Air Force, or Coast Guard installation. Report may be made by long distance collect telephone.

e. To assist survival and rescue in the event of a crash landing the following advice is given:

(1) For flight over uninhabited land areas, it is wise to take suitable survival equipment depending on type of climate and terrain.

(2) If forced landing occurs at sea, chances for survival are governed by degree of crew proficiency in emergency procedures and by effectiveness of water survival equipment.

(3) If it becomes necessary to ditch, distressed aircraft should make every effort to ditch near a surface vessel. If time permits, the position of the nearest vessel can be obtained from a Coast Guard Rescue Coordination Center through the FAA facility.

(4) The rapidity of rescue on land or water will depend on how accurately your position may be determined. If flight plan has been followed and your position is on course, rescue will be expedited.

(5) Unless you have good reason to believe that you will not be located by search aircraft, it is better to remain near your aircraft and prepare means for signalling whenever aircraft approach your position.

f. Search and Rescue facilities include:

(1) Rescue Coordination Centers.

(2) Search and Rescue aircraft.

(3) Rescue vessels.

(4) Pararescue and ground rescue teams.

(5) Emergency radio fixing.

2. Close Your Flight Plan

The control tower does not automatically close VFR flight plans since many of the landing aircraft are not operating on flight plans. It remains the responsibility of a pilot to close his own flight plan. This will prevent a needless search.

3. National Search and Rescue Plan

Under the National Search and Rescue Plan, the U.S. Coast Guard is responsible for coordination of search and rescue for the Maritime Region, and the U.S. Air Force is responsible for coordination of search and rescue for the Inland Region. In order to carry out this responsibility the Air Force and the Coast Guard have established Rescue Coordination Centers to direct search and rescue activities within their regions. This service is available to all persons and property in distress, both civilian and military. Normally, for aircraft incidents, information will be passed to the Rescue Coordination Centers through the appropriate Air Route Traffic Control Center or Flight Service Station.

WAKE TURBULENCE

GENERAL

Every airplane generates a wake while in flight. Initially, when pilots encountered this wake in flight, the disturbance was attributed to "prop wash." It is known, however, that this disturbance is caused by a pair of counter rotating vortices trailing from the wing tips. The vortices from large aircraft pose problems to encountering aircraft. For instance, the wake of these aircraft can impose rolling moments exceeding the roll control capability of some aircraft. Further, turbulence generated within the vortices can damage aircraft components and equipment if encountered at close range. The pilot must learn to envision the location of the vortex wake generated by large aircraft and adjust his flight path accordingly.

During ground operations, jet engine blast (thrust stream turbulence) can cause damage and upsets if encountered at close range. Exhaust velocity versus distance studies at various thrust levels have shown a need for light aircraft to maintain an adequate separation during ground operations. Below are examples of the distance requirements to avoid exhaust velocities of greater than 25 mph:

25 MPH VELOCITY	B-727	DC-8	DC-10
Takeoff Thrust	550 Ft.	700 Ft.	2,100 Ft.
Breakaway Thrust	200 Ft.	400 Ft.	850 Ft.
Idle Thrust	150 Ft.	35 Ft.	350 Ft.

Engine exhaust velocities generated by large jet aircraft during initial takeoff roll and the drifting of the turbulence in relation to the crosswind component dictate the desirability of lighter aircraft awaiting takeoff to hold well back of the runway edge of taxiway hold line; also, the desirability of aligning the aircraft to face the possible jet engine blast movement. Additionally, in the course of running up engines and taxiing on the ground, pilots of large aircraft in particular should consider the effects of their jet blasts on other aircraft.

The FAA has established new standards for the location of taxiway hold lines at airports served by air carriers as follows:

Taxiway holding lines will be established at 100 feet from the edge of the runway, except at locations where "heavy jets" will be operating, the taxiway holding line markings will be established at 150 feet. (The "heavy" category can include some B-707 and DC-8 type aircraft.)

VORTEX GENERATION

Lift is generated by the creation of a pressure differential over the wing surface. The lowest pressure occurs over the upper wing surface and the highest pressure under the wing. This pressure differential triggers the roll up of the airflow aft of the wing resulting in swirling air masses trailing downstream of the wing tips. After the roll up is completed, the wake consists of two counter rotating cylindrical vortices. (See Figure 8-18.)

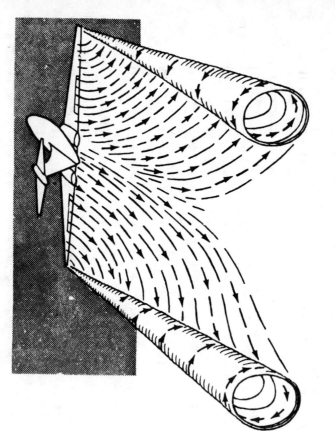

Figure 8-18. *Vortex generation.*

VORTEX STRENGTH

1. The strength of the vortex is governed by the weight, speed and shape of the wing of the generating aircraft. The vortex characteristics of any given aircraft can also be changed by extension of flaps or other wing configuring devices as well as by change in speed. However, as the basic factor is weight, the vortex strength increases proportionately. During a recent test, peak vortex tangential velocities were recorded at 224 feet per second, or about 133

knots. The greatest vortex strength occurs when the generating aircraft is HEAVY—CLEAN—SLOW.

2. Induced Roll. In rare instances a wake encounter could cause in-flight structural damage of catastrophic proportions. However, the usual hazard is associated with induced rolling moments which can exceed the rolling capability of the encountering aircraft. In flight experiments, aircraft have been intentionally flown directly up trailing vortex cores of large aircraft. It was shown that the capability of an aircraft to counteract the roll imposed by the wake vortex primarily depends on the wing span and counter-control responsiveness of the encountering aircraft.

Counter control is usually effective and induced roll minimal in cases where the wing span and ailerons of the encountering aircraft extend beyond the rotational flow field of the vortex. It is more difficult for aircraft with short wing span (relative to the generating aircraft) to counter the imposed roll induced by vortex flow. Pilots of short span aircraft, even of the high performance type, must be especially alert to vortex encounters.

The wake of large aircraft requires the respect of all pilots. (See Figure 8-19.)

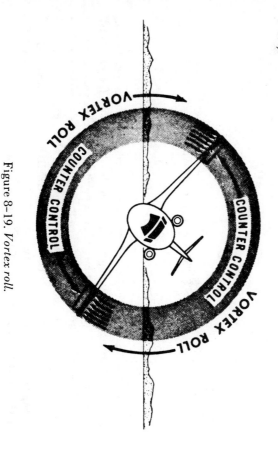

Figure 8-19. *Vortex roll.*

VORTEX BEHAVIOR

Trailing vortices have certain behavioral characteristics which can help a pilot visualize the wake location and thereby take avoidance precautions.

1. Vortices are generated from the moment aircraft leave the ground, since trailing vortices are a by-product of wing lift.

Prior to takeoff or touchdown pilots should note the rotation or touchdown point of the preceding aircraft. (See Figure 8-20.)

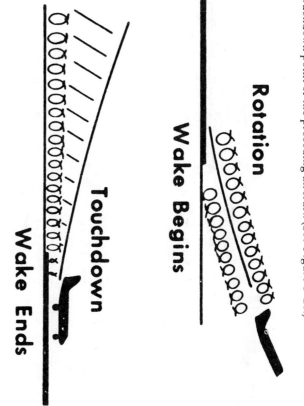

Figure 8-20. *Vortices.*

2. The vortex circulation is outward, upward and around the wing tips when viewed from either ahead or behind the aircraft. Tests with large aircraft have shown that the vortex flow field in a plane cutting thru the wake at any point downstream, covers an area about 2 wing spans in width and one wing span in depth. The vortices remain so spaced (about a wing span apart even drifting with the wind, at altitudes greater than a wing span from the ground. In view of this, if persistent vortex turbulence is encountered, a slight change of altitude and lateral position (preferably upwind) will provide a flight path clear of the turbulence.

3. Flight tests have shown that the vortices from large aircraft sink at a rate of about 400 to 500 feet per minute. They tend to level off at a distance about 900 feet below the flight path of the generating aircraft. Vortex strength diminishes with time and distance behind the generating aircraft. Atmospheric turbulence hastens breakup.

Pilots should fly at or above the large aircraft's flight path, altering course as necessary to avoid the area behind and below the generating aircraft. (See Figure 8–21.)

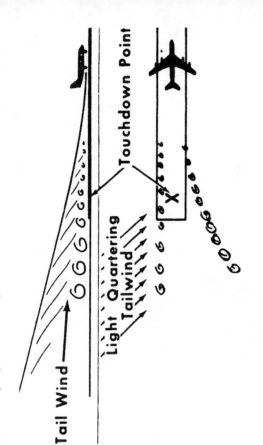

Figure 8-21. *Vortex sink.*

4. When the vortices of large aircraft sink close to the ground (within about 200 feet), they tend to move laterally over the ground at a speed of about 5 knots. (See Figure 8–22.)

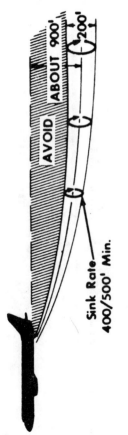

Figure 8-22. *Vortex—no wind.*

A crosswind will decrease the lateral movement of the upwind vortex and increase the movement of the downwind vortex. Thus a light wind of 3 to 7 knots could result in the upwind vortex remaining in the touchdown zone for a period of time and hasten the drift of the downwind vortex toward another runway. Similarly, a tailwind condition can move the vortices of the preceding aircraft forward into the touchdown zone.

THE LIGHT QUARTERING TAILWIND REQUIRES MAXIMUM CAUTION.

Pilots should be alert to large aircraft upwind from their approach and takeoff flight paths. (See Figure 8–23.)

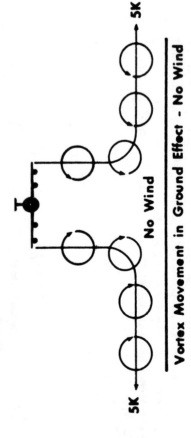

Figure 8-23. *Vortex movement in wind.*

OPERATIONS PROBLEM AREAS

1. A wake encounter is not necessarily hazardous. It can be one or more jolts with varying severity depending upon the direction of the encounter, distance from the generating aircraft, and point of vortex encounter. The probability of induced roll increases when the encountering aircraft's heading is generally aligned with the vortex trail or flight path of the generating aircraft. AVOID THE AREA BELOW AND BEHIND THE GENERATING AIRCRAFT, ESPECIALLY AT LOW ALTITUDE WHERE EVEN A MOMENTARY WAKE ENCOUNTER COULD BE HAZARDOUS.

Pilots should be particularly alert in calm wind conditions and situations where the vortices could:

a. Remain in the touchdown area.
b. Drift from aircraft operating on a nearby runway.

c. Sink into takeoff or landing path from a crossing runway.

d. Sink into the traffic patterns from other airport operations.

e. Sink into the flight path of **VFR** flight operating at the hemispheric altitude 500 feet below.

2. Pilots of all aircraft should visualize the location of the vortex trail behind large aircraft and use proper vortex avoidance procedures to achieve safe operation. It is equally important that pilots of large aircraft plan or adjust their flight paths to minimize vortex exposure to other aircraft.

VORTEX AVOIDANCE PROCEDURES

1. *GENERAL.* Under certain conditions, airport traffic controllers apply procedures for separating aircraft from heavy jet aircraft. The controllers will also provide VFR aircraft, with whom they are in communication and which in the tower's opinion may be adversely affected by wake turbulence from a large aircraft, the position, altitude and direction of flight of the large aircraft followed by the phrase "CAUTION—WAKE TURBULENCE." WHETHER OR NOT A WARNING HAS BEEN GIVEN, HOWEVER, THE PILOT IS EXPECTED TO ADJUST HIS OPERATIONS AND FLIGHT PATH AS NECESSARY TO PRECLUDE SERIOUS WAKE ENCOUNTERS.

2. The following vortex avoidance procedures are recommended for the various situations:

a. Landing behind a large aircraft—same runway: Stay at or above the large aircraft's final approach flight path—note his touchdown point—land beyond it.

b. Land'ng behind a large aircraft—when parallel runway is closer than 2,500 feet: Consider possible drift to your runway. Stay at or above the large aircraft's final approach flight path—note his touchdown point.

c. Landing behind a large aircraft—crossing runway: Cross above the large aircraft's flight path.

d. Landing behind a departing large aircraft—same runway: Note large aircraft's rotation point—land well prior to rotation point.

e. Landing behind a departing large aircraft—crossing runway: Note large aircraft's rotation point— if past the intersection—continue the approach—land prior to the intersection. If large aircraft rotates prior to the intersection, avoid flight below the large aircraft's flight path. Abandon the approach unless a landing is assured well before reaching the intersection.

f. Departing behind a large aircraft: Note large aircraft's rotation point—rotate prior to large aircraft's rotation point—continue climb above

and stay upwind of the large aircraft's climb path until turning clear of his wake. Avoid subsequent headings which will cross below and behind a large aircraft. Be alert for any critical takeoff situation which could lead to a vortex encounter.

g. Intersection takeoffs—same runway: Be alert to adjacent large aircraft operations particularly upwind of your runway. If intersection takeoff clearance is received, avoid subsequent heading which will cross below a large aircraft's path.

h. Departing or landing after a large aircraft executing a low missed approach or touch-and-go landing: Because vortices settle and move laterally near the ground, the vortex hazard may exist along the runway and in your flight path after a large aircraft has executed a low missed approach or a touch-and-go landing, particularly in light quartering wind conditions. You should assure that an interval of at least 2 minutes has elapsed before your takeoff or landing.

i. En route VFR—(thousand-foot altitude plus 500 feet). Avoid flight below and behind a large aircraft's path. If a large aircraft is observed above on the same track (meeting or overtaking) adjust your position laterally, preferably upwind.

HELICOPTERS

A hovering helicopter generates a downwash from its main rotor(s) similar to the prop blast of a conventional aircraft. However, in forward flight, this energy is transformed into a pair of trailing vortices similar to wing-tip vortices of fixed wing aircraft. Pilots of small aircraft should avoid the vortices as well as the downwash.

PILOT RESPONSIBILITY

1. Government and industry groups are making concerted efforts to minimize or eliminate the hazards of trailing vortices. However, the flight disciplines necessary to assure vortex avoidance during VFR operations must be exercised by the pilot. Vortex visualization and avoidance procedures should be exercised by the pilot using the same degree of concern as in collision avoidance.

Wake turbulence may be encountered by aircraft in flight as well as when operating on the airport movement area.

2. Pilots are reminded that in operations conducted behind all aircraft, acceptance from ATC of:

a. Traffic information, or

b. Instructions to follow an aircraft, or

c. The acceptance of a visual approach clearance, is an acknowledgment that the pilot will ensure safe takeoff and landing intervals and accepts the responsibility of providing his own wake turbulence separation.

AIR TRAFFIC WAKE TURBULENCE SEPARATIONS

Air traffic controllers are required to apply specific separation intervals for aircraft operating behind a heavy jet because of the possible effects of wake turbulence.

1. The following separation is applied to aircraft operating directly behind a heavy jet at the same altitude or directly behind and less than 1,000 feet below:

a. Heavy jet behind another heavy jet—4 miles.

b. Small/Large aircraft behind a heavy jet—5 miles.

In addition, controllers provide a 6-mile separation for small aircraft landing behind a heavy jet and a 4-mile separation for small aircraft landing behind a large aircraft. This extra mile of separation is required at the time the preceding aircraft is over the landing threshold.

2. Aircraft departing behind heavy jets are provided two minutes or the appropriate 4- or 5-mile radar separation. Controllers may disregard the separation if the pilot of a departing aircraft initiates a request to deviate from the separation requirement and indicates acceptance of responsibility for maneuvering his aircraft so as to avoid the possible wake turbulence hazard. However, occasions will arise when the controller must still hold the aircraft in order to provide separation required for other than wake turbulence purposes.

MEDICAL FACTS FOR PILOTS

FITNESS FOR FLIGHT

Medical Certification

All pilots except those flying gliders and free air balloons must possess valid medical certificates in order to exercise the privileges of their airman certificates. The periodic medical examinations required for medical certification are conducted by designated Aviation Medical Examiners, who are physicians with a special interest in aviation safety and training in aviation medicine.

The standards for medical certification are contained in Part 67 of the Federal Aviation Regulations. Pilots who have a history of certain medical conditions described in these standards are mandatorily disqualified from flying. These medical conditions include a personality disorder manifested by overt acts, a psychosis, alcoholism, drug dependence, epilepsy, an unexplained disturbance of consciousness, myocardial infarction, angina pectoris, and diabetes requiring medication for its control. Other medical conditions may be temporarily disqualifying, such as acute infections, anemia, and peptic ulcer. Pilots who do not meet medical standards may still be qualified under special issuance provisions or the exemption process. This may require that either additional medical information be provided or practical flight tests be conducted.

Student pilots should visit an Aviation Medical Examiner as soon as possible in their flight training in order to avoid unnecessary training expenses should they not meet the medical standards. For the same reason, the student pilot who plans to enter commercial aviation should apply for the highest class of medical certificate that might be necessary in the pilot's career.

CAUTION: The Federal Aviation Regulations prohibit a pilot who possesses a current medical certificate from performing crewmember duties while the pilot has a known medical condition or increase of a known medical condition that would make the pilot unable to meet the standards for the medical certificate.

Illness

Even a minor illness suffered in day-to-day living can seriously degrade performance of many piloting tasks vital to safe flight. Illness can produce fever and distracting symptoms that can impair judgment, memory, alertness, and the ability to make calculations. Although symptoms from an illness may be under adequate control with a medication, the medication itself may decrease pilot performance.

The safest rule is not to fly while suffering from any illness. If this rule is considered too stringent for a particular illness, the pilot should contact an Aviation Medical Examiner for advice.

Medication

Pilot performance can be seriously degraded by both prescribed and over-the-counter medications, as well as by the medical conditions for which they are taken. Many medications, such as tranquilizers, sedatives, strong pain relievers, and cough-suppressant preparations, have primary effects that may impair judgment, memory, alertness, coordination, vision, and the ability to make calculations. Others, such as antihistamines, blood pressure drugs, muscle relaxants, and agents to control diarrhea and motion sickness, have side effects that may impair the same critical functions. Any medication that depresses the nervous system, such as a sedative, tranquilizer, or antihistamine, can make a pilot much more susceptible to hypoxia (see below).

The Federal Aviation Regulations prohibit pilots from performing crewmember duties while using any medication that affects the faculties in any way contrary to safety. The safest rule is not to fly as a crewmember while taking any medication, unless approved to do so by the FAA.

Alcohol

Extensive research has provided a number of facts about the hazards of alcohol consumption and flying. As little as one ounce of liquor, one bottle of beer, or four ounces of wine can impair flying skills, with the alcohol consumed in these drinks being detectable in the breath and blood for at least three hours. Even after the body completely destroys a moderate amount of alcohol, a pilot can still be severely impaired for many hours by hangover. There is simply no way of increasing the destruction of alcohol or alleviating a hangover. Alcohol also renders a pilot much more susceptible to disorientation and hypoxia (see below).

A consistently high alcohol-related fatal aircraft accident rate serves to emphasize that alcohol and flying are a potentially lethal combination. The Federal Aviation Regulations prohibits pilots from performing crewmember duties within eight hours after drinking any alcoholic beverage or while under the influence of alcohol. However, due to the slow destruction of alcohol, a pilot may still be under the influence eight hours after drinking a moderate amount of alcohol. Therefore, an excellent rule is to allow at least 12 to 24 hours between "bottle and throttle," depending on the amount of alcoholic beverage consumed.

Fatigue

Fatigue continues to be one of the most treacherous hazards to flight safety, as it may not be apparent to a pilot until serious errors are made. Fatigue is best described as either acute (short-term) or chronic (long-term).

A normal occurrence of everyday living, acute fatigue is the tiredness felt after long periods of physical and mental strain, including strenuous muscular effort, immobility, heavy mental workload, strong emotional pressure, monotony, and lack of sleep. Consequently, coordination and alertness, so vital to safe pilot performance, can be reduced. Acute fatigue is prevented by adequate rest and sleep, as well as regular exercise and proper nutrition.

Chronic fatigue occurs when there is not enough time for full recovery between episodes of acute fatigue. Performance continues to fall off, and judgment becomes impaired so that unwarranted risks may be taken. Recovery from chronic fatigue requires a prolonged period of rest.

Stress

Stress from the pressures of everyday living can impair pilot performance, often in very subtle ways. Difficulties, particularly at work, can occupy thought processes enough to markedly decrease alertness. Distraction can so interfere with judgment that unwarranted risks are taken, such as flying into deteriorating weather conditions to keep on schedule. Stress and fatigue (see above) can be an extremely hazardous combination.

Most pilots do not leave stress "on the ground." Therefore when more than usual difficulties are being experienced, a pilot should consider delaying flight until these difficulties are satisfactorily resolved.

Emotion

Certain emotionally upsetting events, including a serious argument, death of a family member, separation or divorce, loss of job and financial catastrophe, can render a pilot unable to fly an aircraft safely. The emotions of anger, depression, and anxiety from such events not only decrease alertness but also may lead to taking risks that border on self-destruction. Any pilot who experiences an emotionally upsetting event should not fly until satisfactorily recovered from it.

Personal Checklist

Aircraft accident statistics show that pilots should be conducting preflight checklists on themselves as well as their aircraft, for pilot impairment contributes to many more accidents than failures of aircraft systems. A personal checklist that can be easily committed to memory includes all of the categories of pilot impairment discussed in this section.

EFFECTS OF ALTITUDE

Hypoxia

Hypoxia is a state of oxygen deficiency in the body sufficient to impair functions of the brain and other organs. Hypoxia from exposure to altitude is due only to the reduced barometric pressures encountered at altitude, for the concentration of oxygen in the atmosphere remains about 21 percent from the ground out to space.

Although a deterioration in night vision occurs at a cabin pressure altitude as low as 5,000 feet, other significant effects of altitude hypoxia usually do not occur in the normal healthy pilot below 12,000 feet. From 12,000 to 15,000 feet of altitude, judgment, memory, alertness, coordination and ability to make calculations are impaired, and headache, drowsiness, dizziness and either a sense of well-being (euphoria) or belligerence occur. The effects appear following increasingly shorter periods of exposure to increasing altitude. In fact, pilot performance can seriously deteriorate within 15 minutes at 15,000 feet.

At cabin pressure altitudes above 15,000 feet, the periphery of the visual field grays out to a point where only central vision remains (tunnel vision). A blue coloration (cyanosis) of the fingernails and lips develops. The ability to take corrective and protective action is lost in 20 to 30 minutes at 18,000 feet and 5 to 12 minutes at 20,000 feet, followed soon thereafter by unconsciousness.

The altitude at which significant effects of hypoxia occur can be lowered by a number of factors. Carbon monoxide inhaled in smoking or from exhaust fumes (see below), lowered hemoglobin (anemia), and certain medications can reduce the oxygen-carrying capacity of the blood to the degree that the amount of oxygen provided to body tissues will already be equivalent to the oxygen provided to the tissues when exposed to a cabin pressure altitude of several thousand feet. Small amounts of alcohol and low doses of certain drugs, such as antihistamines, tranquilizers, sedatives and analgesics can, through their depressant actions, render the brain much more susceptible to hypoxia. Extreme heat and cold, fever, and anxiety increase the body's demand for oxygen, and hence its susceptibility to hypoxia.

The effects of hypoxia are usually quite difficult to recognize, especially when they occur gradually. Since symptoms of hypoxia do not vary in an individual, the ability to recognize hypoxia can be greatly improved by experiencing and witnessing the effects of hypoxia during an altitude chamber "flight." The FAA provides this opportunity through aviation physiology training, which is conducted at the FAA Civil Aeromedical Institute and at many military facilities across the United States. Pilots can apply for this training by contacting the Physiological Operations and Training Section, AAC-143, FAA Civil Aeromedical Institute, P.O. Box 25082, Oklahoma City, Oklahoma, 73125.

Hypoxia is prevented by heeding factors that reduce tolerance to altitude, by enriching the inspired air with oxygen from an appropriate oxygen system and by maintaining a comfortable, safe cabin pressure altitude. For optimum protection, pilots are encouraged to use supplemental oxygen above 10,000 feet during the day, and above 5,000 feet at night. The Federal Aviation Regulations require that the minimum flight crew be provided with and use supplemental oxygen after 30 minutes of exposure to cabin pressure altitudes between 12,500 and 14,000 feet, and immediately on exposure to cabin pressure altitudes above 14,000 feet. Every occupant of the aircraft must be provided with supplement oxygen at cabin pressure altitudes above 15,000 feet.

Ear Block

As the aircraft cabin pressure decreases during ascent, the expanding air in the middle ear pushes the eustachian tube open and, by escaping down it to the nasal passages, equalizes in pressure with the cabin pressure. But during descent, the pilot must periodically open the eustachian tube to equalize pressure. This can be accomplished by swallowing, yawning, tensing muscles in the throat or, if these do not work, by the combination of closing the mouth, pinching the nose closed and attempting to blow through the nostrils (Valsalva maneuver).

Either an upper respiratory infection, such as a cold or sore throat, or a nasal allergic condition can produce enough congestion around the eustachian tube to make equalization difficult. Consequently, the difference in pressure between the middle ear and aircraft cabin can build up to a level that will hold the eustachian tube closed, making equalization difficult if not impossible. This problem is commonly referred to as an "ear block."

An ear block produces severe ear pain and loss of hearing that can last from several hours to several days. Rupture of the ear drum can occur in flight or after landing. Fluid can accumulate in the middle ear and become infected.

An ear block is prevented by not flying with an upper respiratory infection or nasal allergic condition. Adequate protection is usually not provided by decongestant sprays or drops to reduce congestion around the eustachian tube. Oral decongestants have side effects that can significantly impair pilot performance.

If an ear block does not clear shortly after landing, a physician should be consulted.

Sinus Block

During ascent and descent, air pressure in the sinuses equalizes with the aircraft cabin pressure through small openings that connect the sinuses to the nasal passages. Either an upper respiratory infection, such as a cold or sinusitis, or a nasal allergic condition can produce enough congestion around an opening to slow equalization and, as the difference in pressure between the sinus and cabin mounts, eventually plug the opening. This "sinus block" occurs most frequently during descent.

A sinus block can occur in the frontal sinuses, located above each eyebrow, or in the maxillary sinuses, located in each upper cheek. It will usually produce excruciating pain over the sinus area. A maxillary sinus block can also make the upper teeth ache. Bloody mucus may discharge from the nasal passages.

A sinus block is prevented by not flying with an upper respiratory infection or nasal allergic condition. Adequate protection is usually not provided by decongestant sprays or drops to reduce congestion around the sinus openings. Oral decongestants have side effects that can impair pilot performance.

If a sinus block does not clear shortly after landing, a physician should be consulted.

Decompression Sickness After Scuba Diving

A pilot or passenger who intends to fly after SCUBA diving should allow the body sufficient time to rid itself of excess nitrogen absorbed during diving. If not, decompression sickness due to evolved gas can occur during exposure to low altitude and create a serious inflight emergency.

The recommended waiting time before flight to cabin pressure altitudes of 8,000 feet or less is at least 2 hours after diving which has not required controlled ascent (non-decompression diving), and at least 24 hours after diving which has required controlled ascent (decompression diving). The waiting time before flight to cabin pressure altitudes above 8,000 feet should be at least 24 hours after any SCUBA diving.

HYPERVENTILATION IN FLIGHT

Hyperventilation, or an abnormal increase in the volume of air breathed in and out of the lungs, can occur subconsciously when a stressful situation is encountered in flight. As hyperventilation "blows off" excessive carbon dioxide from the body, a pilot can experience symptoms of lightheadedness, suffocation, drowsiness, tingling in the extremities, and coolness—and react to them with even greater hyperventilation. Incapacitation can eventually result from incoordination, disorientation, and painful muscle spasms. Finally, unconsciousness can occur.

The symptoms of hyperventilation subside within a few minutes after the rate and depth of breathing are consciously brought back under control. The buildup of carbon dioxide in the body can be hastened by controlled breathing in and out of a paper bag held over the nose and mouth.

Early symptoms of hyperventilation and hypoxia are similar. Moreover, hyperventilation and hypoxia can occur at the same time. Therefore, if a pilot is using an oxygen system when symptoms are experienced, the oxygen regulator should immediately be set to deliver 100 percent oxygen, and then the system checked to assure that it has been functioning effectively before giving attention to rate and depth of breathing.

CARBON MONOXIDE POISONING IN FLIGHT

Carbon monoxide is a colorless, orderless and tasteless gas contained in exhaust fumes. When breathed even in minute quantities over a period of time, it can significantly reduce the ability of the blood to carry oxygen. Consequently, effects of hypoxia occur (see above).

Most heaters in light aircraft work by air flowing over the manifold. Use of these heaters while exhaust fumes are escaping through manifold cracks and seals is responsible every year for several non-fatal and fatal aircraft accidents from carbon monoxide poisoning.

A pilot who detects the odor of exhaust or experiences symptoms of headache, drowsiness, or dizziness while using the heater should suspect carbon monoxide poisoning, and immediately shut off the heater and open air vents. If symptoms are severe, or continue after landing, medical treatment should be sought.

ILLUSIONS IN FLIGHT

Introduction

Many different illusions can be experienced in flight. Some can lead to spatial disorientation. Others can lead to landing errors. Illusions rank among the most common factors cited as contributing to aircraft accidents.

Illusions Leading To Spatial Disorientation

Various complex motions and forces and certain visual scenes encountered in flight can creat illusions of motion and position. Spatial disorientation from these illusions can be prevented only by visual reference to reliable, fixed points on the ground or to flight instruments.

"The leans." An abrupt correction of a banked attitude, which has been entered too slowly to stimulate the balance organs in the inner ear, can create the illusion of banking in the opposite direction. The disoriented pilot will roll the aircraft back into its original dangerous attitude or, if level flight is maintained, will feel compelled to lean in the perceived vertical plane until this illusion subsides.

"Coriolis illusion." An abrupt head movement in a constant-rate turn that has ceased stimulating the balance organs can create the illusion of rotation or movement in an entirely different plane. The disoriented pilot will maneuver the aircraft into a dangerous attitude in an attempt to stop rotation. This most overwhelming of all illusions in flight may be prevented by not making sudden, extreme head movements, particularly while making prolonged constant-rate turns under IFR conditions.

"Graveyard spin." A proper recovery from a spin that has ceased stimulating the balance organs can create the illusion of spinning in the opposite direction. The disoriented pilot will return the aircraft to its original spin.

"Graveyard spiral." An observed loss of altitude during a coordinated constant-rate turn that has ceased stimulating the balance organs can create the illusion of being in a descent with the wings level. The disoriented pilot will pull back on the controls, tightening the spiral and increasing the loss of altitude.

"Somatogravic illusion." A rapid acceleration during takeoff can create the illusion of being in a nose-up attitude. The disoriented pilot will push the aircraft into a nose-low, or dive attitude. A rapid deceleration by a rapid reduction of the throttles can have the opposite effect, with the disoriented pilot pulling the aircraft into a nose-up, or stall attitude.

"Inversion illusion." An abrupt transition from climb to straight and level flight can create the illusion of tumbling backwards. The disoriented pilot will push the aircraft abruptly into a nose-low attitude, possibly intensifying this illusion.

"Elevator illusion." An abrupt upward vertical acceleration, usually by an updraft, can create the illusion of being in a climb. The disoriented pilot will push the aircraft into a nose-low attitude. An abrupt downward vertical acceleration, usually by a downdraft, has the opposite effect, with the disoriented pilot pulling the aircraft into a nose-up attitude.

"False horizon." Sloping cloud formations, an obscured horizon, a dark scene spread with ground lights and stars, and certain geometric patterns of ground lights can create illusions of not being aligned correctly with the actual horizon. The disoriented pilot will place the aircraft in a dangerous attitude.

"Autokinesis." In the dark, a static light will appear to move about when stared at for many seconds. The disoriented pilot will lose control of the aircraft in attempting to align it with the light.

Illusions Leading To Landing Errors

Various surface features and atmospheric conditions encountered in landing can create illusions of incorrect height above and distance from the runway threshold. Landing errors from these illusions can be prevented by anticipating them during approaches, aerial visual inspection of unfamiliar airports before landing, using electronic glideslope or VASI systems when available, and maintaining optimum proficiency in landing procedures.

"Runway width illusion." A narrower-than-usual runway can create the illusion of the aircraft being at a greater height. The pilot who does not recognize this illusion will fly a lower approach, with the risk of striking objects along the approach path or landing short. A wider-than-usual runway can have the opposite effect, with the risk of leveling out high and landing hard or overshooting the runway.

"Runway and terrain slopes illusion." An upsloping runway, upsloping terrain, or both, can create the illusion of greater height. The pilot who does not recognize this illusion will fly a lower approach. A downsloping runway, downsloping approach terrain, or both, can have the opposite effect.

"Featureless terrain illusion." An absence of ground features, as when landing over water, darkened areas and terrain made featureless by snow, can create the illusion of greater height. The pilot who does not recognize this illusion will fly a lower approach.

"Atmospheric illusions." Rain on the windscreen can create the illusion of greater height, and atmospheric haze the illusion of greater distance. The pilot who does not recognize these illusions will fly a lower approach. Penetration of fog can create the illusion of pitching up. The pilot who does not recognize this illusion will steepen the approach, often quite abruptly.

"Ground lighting illusions." Lights along a straight path, such as a road, and even lights on moving trains can be mistaken for runway and approach lights. Bright runway and approach lighting systems, especially where few lights illuminate the surrounding terrain, may create the illusion of lesser distance. The pilot who does not recognize this illusion will fly a high approach.

VISION IN FLIGHT

Introduction

Of the body senses, vision is the most important for safe flight. Major factors that determine how effectively vision can be used are the level of illumination and the technique of scanning the sky for other aircraft.

Vision Under Dim and Bright Illumination

Under conditions of dim illumination, small print and colors on aeronautical charts and aircraft instruments become unreadable unless adequate cockpit lighting is available. Moreover, another aircraft must be much closer to be seen unless its navigation lights are on.

In darkness, vision becomes more sensitive to light, a process called dark adaptation. Although exposure to total darkness for at least 30 minutes is required for complete dark adaptation, the pilot can achieve a moderate degree of dark adaptation within 20 minutes under dim red cockpit lighting. Since red light severely distorts colors, especially on aeronautical charts, and can cause serious difficulty in focusing the eyes on objects inside the aircraft, its use is advisable only where optimum outside night vision capability is necessary. Even so, white cockpit lighting must be available when needed for map and instrument reading, especially under IFR conditions. Dark adaptation is impaired by exposure to cabin pressure altitudes above 5,000 feet, carbon monoxide inhaled in smoking and from exhaust fumes, deficiency of Vitamin A in the diet, and by prolonged exposure to bright sunlight. Since any degree of dark adaptation is lost within a few seconds of viewing a bright light, the pilot should close one eye when using a light to preserve some degree of night vision.

Excessive illumination, especially from light reflected off the canopy, surfaces inside the aircraft, clouds, water, snow, and desert terrain, can produce glare, with uncomfortable squinting, watering of the eyes, and even temporary blindness. Sunglasses for protection from glare should absorb at least 85 percent of visible light (15 percent transmittance) and all colors equally (neutral transmittance), with negligible image distortion from refractive and prismatic errors.

Scanning For Other Aircraft

Scanning the sky for other aircraft is a key factor in collision avoidance. It should be used continuously by the pilot and copilot (or right seat passenger) to cover all areas of the sky visible from the cockpit.

Effective scanning is accomplished with a series of short, regularly spaced eye movements that bring successive areas of the sky into the central visual field. Each movement should not exceed 10 degrees, and each area should be observed for at least one second to enable detection. Although horizontal back-and-forth eye movements seem preferred by most pilots, each pilot should develope a scanning pattern that is most comfortable and then adhere to it to assure optimum scanning.

GOOD OPERATING PRACTICES

ALERTNESS

Be alert at all times, especially when the weather is good. Most pilots pay attention to business when they are operating in full IFR weather conditions, but strangely, air collisions almost invariably have occurred under ideal weather conditions. Unlimited visibility appears to encourage a sense of security which is not at all justified. Considerable information of value may be obtained by listening to advisories being issued in the terminal area, even though controller workload may prevent a pilot from obtaining individual service.

JUDGEMENT IN VFR FLIGHT

1. **General.** Use reasonable restraint in exercising the prerogative of VFR flight, especially in terminal areas. The weather minimums and distances from clouds are minimums. Giving yourself a greater margin in specific instances is just good judgment.

2. **Approach Area.** Conducting a VFR operation in a Control Zone when the official visibility is 3 or 4 miles is not prohibited, but good judgment would dictate that you keep out of the approach area.

3. **Reduced visibility.** It has always been recognized that precipitation reduces forward visibility. Consequently, although again it may be perfectly legal to cancel your IFR flight plan at any time you can proceed VFR, it is good practice, when precipitation is occurring, to continue IFR operation into a terminal area until you are reasonably close to your destination.

4. Simulated Instrument Flights. In conducting simulated instrument flights, be sure that the weather is good enough to compensate for the restricted visibility of the safety pilot and your greater concentration on your flight instruments. Give yourself a little greater margin when your flight plan lies in or near a busy airway or close to an airport.

5. Obstructions to VFR Flight. Extreme caution should be exercised when flying less than 2,100 feet above ground level (AGL) because there are more than 300 skeletal structures (radio and television antenna towers) exceeding 1,000 feet AGL with some extending higher than 2,000 feet AGL. In addition, more than 50 towers which exceed 1,000 feet AGL are either under construction or planned. Similar proposals are planned on a continuing basis. Most skeletal structures are supported by guy wires. The wires are difficult to see in good weather and can be totally obscured during periods of dusk and reduced visibility. These wires can extend about 1,500 feet horizontally from a structure; therefore, all skeletal structures should be avoided by at least 2,000 feet.

USE OF CLEARING PROCEDURES

1. Before Takeoff. Prior to taxiing onto a runway or landing area in preparation for takeoff, pilots should scan the approach areas for possible landing traffic, executing appropriate clearing maneuvers to provide him a clear view of the approach areas.

2. Climbs and Descents. During climbs and descents in flight conditions which permit visual detection of other traffic, pilots should excute gentle banks, left and right at a frequency which permits continuous visual scanning of the airspace about them.

3. Straight and level. Sustained periods of straight and level flight in conditions which permit visual detection of other traffic should be broken at intervals with appropriate clearing procedures to provide effective visual scanning.

4. Traffic pattern. Entries into traffic patterns while descending create specific collision hazards and should be avoided.

5. Traffic at VOR sites. All operators should emphasize the need for sustained vigilance in the vicinity of VORs and airway intersections due to the convergence of traffic.

6. Training operations. Operators of pilot training programs are urged to adopt the following practices:

a. Pilots undergoing flight instruction at all levels should be re-

quested to verbalize clearing procedures (call out, "clear" left, right, above, or below) to instill and sustain the habit of vigilance during maneuvering.

b. High-wing airplane, momentarily raise the wing in the direction of the intended turn and look.

c. Low-wing airplane, momentarily lower the wing in the direction of the intended turn and look.

d. Appropriate clearing procedures should precede the execution of all turns including chandelles, lazy eights, stalls, slow flight, climbs, straight and level spins, and other combination maneuvers.

GIVING WAY

If you think another aircraft is too close to you, give way instead of waiting for the other pilot to respect the right-of-way to which you may be entitled. It is a lot safer to pursue the right-of-way angle after you have completed your flight.

USE OF FEDERAL AIRWAYS

Pilots not operating on an IFR flight plan, and when in level cruising flight, are cautioned to conform with VFR cruising altitudes appropriate to direction of flight. During climb or descent, pilots are encouraged to fly to the right side of the center line of the radial forming the airway in order to avoid IFR and VFR cruising traffic operating along the center line of the airway.

FOLLOW IFR PROCEDURES EVEN WHEN OPERATING VFR

1. To maintain IFR proficiency, pilots are urged to practice IFR procedures whenever possible, even when operating VFR. Some suggested practices include:

a. Obtain a complete preflight and weather briefing. Check the NOTAMS.

b. File a flight plan. This is an excellent low cost insurance policy. The cost is the time it takes to fill it out. The insurance includes the knowledge that someone will be looking for you if you become overdue at destination.

c. Use current charts.

d. Use the navigation aids. Practice maintaining a good course—keep the needle centered.

flight).

e. Maintain a constant altitude (appropriate for the direction of flight).

f. Estimate enroute position times.

g. Make accurate and frequent position reports to the FSS's along your route of flight.

2. Simulated IFR flight is recommended (under the hood); however, pilots are cautioned to review and adhere to the requirements specified in FAR 91.21 before and during such flight.

VFR AT NIGHT

When flying VFR at night, in addition to the altitude appropriate for the direction of flight, pilots should maintain an altitude which is at or above the minimum enroute altitude as shown on charts. This is especially true in mountainous terrain, where there is usually very little ground reference. Don't depend on your being able to see those built-up rocks or TV towers in time to miss them.

FLIGHT OUTSIDE THE UNITED STATES AND U.S. TERRITORIES

When conducting flights, particularly extended flights, outside the U.S. and its territories, full account should be taken of the amount and quality of air navigation services available in the airspace to be traversed. Every effort should be made to secure information on the location and range of navigational aids, availability of communications and meteorological services, the provision of air traffic services, including alerting service, and the existence of search and rescue services.

The filing of a flight plan—always good practice—takes on added significance for extended flights outside U.S. airspace and is, in fact, usually required by the laws of the countries being visited or overflown. It is also particularly important in the case of such flights that pilots leave a complete itinerary and schedule of the flight with someone directly concerned, keep that person advised of the flight's progress and inform him that if serious doubt arises as to the safety of the flight he should first contact the appropriate Flight Service Station.

AVOID FLIGHT BENEATH UNMANNED BALLOONS

The majority of unmanned free balloons currently being operated have, extending below them, either a suspension device to which the payload or instrument package is attached, or a trailing wire antenna, or both. In many instances these balloon subsystems may be invisible to the pilot until his aircraft is close to the balloon, thereby creating a potentially dangerous situation. Therefore, good judgment on the part of the pilot dictates that aircraft should remain well clear of all unmanned free balloons and flight below them should be avoided at all times.

Pilots are urged to report any unmanned free balloons sighted to the nearest FAA ground facility with which communication is established. Such information will assist FAA ATC facilities to identify and flight follow unmanned free balloons operating in the airspace.

FISH AND WILDLIFE SERVICE REGULATION

The Fish and Wildlife Service has the following regulation in effect governing the flight of aircraft on and over wildlife areas:

"The unauthorized operation of aircraft at low altitudes over, or the unauthorized landing of aircraft on a wildlife refuge area is prohibited, except in the event of emergency."

The Fish and Wildlife Service requests that pilots maintain a minimum altitude of 2,000 feet above the terrain of a wildlife refuge area.

PARACHUTE JUMP AIRCRAFT OPERATIONS

Pilots of aircraft engaged in parachute jump operations are reminded that all reported altitudes must be with reference to mean sea level, or flight level as appropriate, to enable ATC to provide meaningful traffic information.

STUDENT PILOTS RADIO IDENTIFICATION

The FAA desires to help the student pilot in acquiring sufficient practical experience in the environment in which he will be required to operate. To receive additional assistance while operating in areas of concentrated air traffic, a student pilot need only identify himself as a student pilot during his initial call to an FAA radio facility. For instance, "Dayton Tower, this is Fleetwing 1234, Student Pilot, over." This special identification will alert FAA air traffic control personnel and enable them to provide the student pilot with such extra assistance and consideration as he may need. This procedure is not mandatory.

3. Don't fly near or above abrupt changes in terrain. Severe turbulence can be expected, especially in high wind conditions.

4. Some canyons run into a dead-end. Don't fly so far up a canyon that you get trapped. ALWAYS BE ABLE TO MAKE A 180 DEGREE TURN!

5. Plan your trip for the early morning hours. As a rule, the air starts to get bad at about 10 a.m., and grows steadily worse until around 4 p.m., then gradually improves until dark. Mountain flying at night in a single engine light aircraft is asking for trouble.

6. When landing at a high altitude field, the same indicated airspeed should be used as at low elevation fields. *Remember:* that due to the less dense air at altitude, this same indicated airspeed actually results in a higher true airspeed, a faster landing speed, and more important, a longer landing distance. During gusty wind conditions which often prevail at high altitude fields, a power approach and power landing is recommended. Additionally, due to the faster groundspeed, your takeoff distance will increase considerably over that required at low altitudes.

7. *Effects of Density Altitude.* Performance figures in the aircraft owner's handbook for length of takeoff run, horsepower, rate of climb, etc., are generally based on standard atmosphere conditions (50° F. pressure 29.92 inches of mercury) at sea level. However, inexperienced pilots as well as experienced pilots may run into trouble when they encounter an altogether different set of conditions. This is particularly true in hot weather and at higher elevations. Aircraft operations at altitudes above sea level and at higher than standard temperatures are commonplace in mountainous areas. Such operations quite often result in a drastic reduction of aircraft performance capabilities because of the changing air density. Density altitude is a measure of air density. It is not to be confused with pressure altitude—true altitude or absolute altitude. It is not to be used as a height reference, but as a determining criteria in the performance capability of an aircraft. Air density decreases with altitude. As air density decreases, density altitude increases. The further effects of high temperature and high humidity are cumulative, resulting in an increasing high density altitude condition. High density altitude reduces all aircraft performance parameters. To the pilot, this means that—the normal horsepower output is reduced, propeller efficiency is reduced and a higher true airspeed is required to sustain the aircraft throught its operating parameters. It means an increase in runway length requirements for takeoff and landings, and a decreased rate of climb. (Note.—A turbo-charged aircraft engine provides some slight advantage in that it provides sea level horsepower up to a specified altitude above sea level.) An average small airplane, for ex-

OPERATION OF AIRCRAFT ROTATING BEACON

1. There have been several incidents in which small aircraft were overturned or damaged by prop/jet blast forces from taxiing large aircraft. A small aircraft taxiing behind any large aircraft with its engines operating could meet with the same results. In the interest of preventing ground upsets and injuries to ground personnel due to prop/jet engine blast forces, the FAA has recommended to air carriers/commercial operators that they establish procedures for the operation of the aircraft rotating beacon any time the engines are in operation.

2. General aviation pilots utilizing aircraft equipped with rotating beacons are also encouraged to participate in this program and operate the beacon any time the aircraft engines are in operation as an alert to other aircraft and ground personnel that prop/jet engine blast forces may be present. Caution must be exercised by all personnel not to rely solely on the rotating beacon as an indication that aircraft engines are in operation, since participation in this program is voluntary.

MOUNTAIN FLYING

Your first experience of flying over mountainous terrain (particularly if most of your flight time has been over the flatlands of the midwest) could be a never-to-be-forgotten nightmare if proper planning is not done and if you are not aware of the potential hazards awaiting. Those familiar section lines are not present in the mountains; those flat, level fields for forced landings are practically non-existent; abrupt changes in wind direction and velocity occur; severe updrafts and downdrafts are common, particularly near or above abrupt changes of terrain such as cliffs or rugged areas; even the clouds look different and can build up with startling rapidity. Mountain flying need not be hazardous if you follow the recommendations below:

1. File a flight plan. Plan your route to avoid topography which would prevent a safe forced landing. The route should be over populated areas and well known mountain passes. Sufficient altitude should be maintained to permit gliding to a safe landing in the event of engine failure.

2. Don't fly a light aircraft when the winds aloft, at your proposed altitude, exceed 35 miles per hour. Expect the winds to be of much greater velocity over mountain passes than reported a few miles from them. Approach mountain passes with as much altitude as possible. Downdrafts of from 1,500 to 2,000 feet per minute are not uncommon on the leeward side.

ample, requiring 1,000 feet for takeoff at sea level under standard atmospheric conditions will require a takeoff run of approximately 2,000 feet at an operational altitude of 5,000 feet.

NOTE: *All flight service stations will compute the current density altitude upon request.*

ASK FOR ASSISTANCE

"I'm not lost, I just don't know for sure where I am, but a familiar landmark will show up soon." "I *think* I have enough fuel to get there." "I think it will be smoother if I go above these clouds, there are bound to be some holes I can get down through when I get near home." "I'd look pretty silly if I asked for help and then found out I didn't really need it." The first time one of these thoughts pop into your mind, *it is time* to ask for assistance. Do not wait until the situation has deteriorated into an emergency before letting ATC know of your predicament. A little embarrassment is better than a big accident!

TO CONTACT AN FSS

Flight Service Stations are allocated frequencies for different functions, for Airport Advisory Service the pilot should contact the FSS on 123.6 MHz for example. Other FSS frequencies are listed in Airport/Facility Directory. If you are in doubt as to what frequency to use to contact an FSS, transmit on 122.1 MHz and advise them of the frequency you are receiving on.

ALTIMETER ERRORS

1. The importance of frequently obtaining current altimeter settings can not be overemphasized. If you do not reset your altimeter when flying *from* an area of high pressure or high temperatures *into* an area of low temperatures or low pressure, *your aircraft will be closer to the surface than the altimeter indicates.* An inch error on the altimeter equals 1,000 feet of altitude. To quote an old saw: "GOING FROM A HIGH TO A LOW, LOOK OUT BELOW."

2. A reverse situation—without resetting the altimeter when going from a low temperature or pressure area into a high temperature or high pressure area, the aircraft will be higher than the altimeter indicates.

3. The possible result of the situation in **1.** above, is obvious, particularly if operating at the minimum altitude. In situation **2.** above, the result may not be as spectacular, but consider an instrument approach: If your altimeter is in error you may still be on instruments when reaching the minimum altitude (as indicated on the altimeter), whereas you might have been in the

clear and able to complete the approach if the altimeter setting was correct. FAR 91.81 defines current altimeter setting practices.

AIRCRAFT CHECKLISTS

Most Owners' Manuals contain recommended checklists for the particular type of aircraft. As checklists vary with different types of aircraft and equipments, it is not practical to recommend a complete set of checklists to cover all aircraft. Additions to the procedures in your Owners' Manual may better fill your personal preference or requirements. However, it is most important that checklists be designed to include all of the items in the Owners' Manual. By using these lists for every flight, the possibility of overlooking important items will be lessened.

VFR IN A CONGESTED AREA?—LISTEN!

When operating VFR in highly congested areas, whether you intend to land at an airport within the area or are just flying through, it is recommended that extra vigilance be maintained and that you monitor an appropriate control frequency. Normally the appropriate frequency is an approach control frequency. By such monitoring action you can "get the picture" of the traffic in your area. When the approach controller has radar, traffic advisories may be given to VFR pilots who request them, subject to the provisions included in RADAR TRAFFIC INFORMATION SERVICE.

OPERATION LIGHTS ON

FAA has initiated a voluntary pilot safety program, "Operation Lights On," to enhance the "see-and-be-seen" concept of averting collisions both in the air and on the ground, and to reduce bird strikes. All pilots are encouraged to turn on their anti-collision lights any time the engine(s) are running day or night. All pilots are further encouraged to turn on their landing lights when operating within 10 miles of any airport (day and night), in conditions of reduced visibility and in areas where flocks of birds may be expected , i.e. coastal areas, lake areas, swamp areas, around refuse dumps, etc.

Although turning on aircraft lights does enhance the "see-and-be-seen" concept, pilots should not become complacent about keeping a sharp lookout for other aircraft. Not all aircraft are equipped with lights and some pilots may not have their lights turned on. The aircraft manufacturers' recommendations for operation of landing lights and electrical systems should be observed.

APPENDIX I

Obtaining FAA Publications

1. *The Checklist.* Advisory Circular 00–2, The Advisory Circular Checklist, contains a list of current FAA Advisory Circulars together with their status as of a given date, their description, and ordering instructions. The checklist is updated tri-annually, and provides detailed instructions on how to obtain Advisory Circulars. It also contains a list of GPO bookstores located throughout the United States which stock many government publications. The checklist may be obtained *free* upon request from the U.S. Department of Transportation, Publications Section, M–443.1, Washington, D.C. 20590.

2. *Federal Aviation Regulations.* The following FAR Parts are those that may be of most interest to the pilot. They pertain primarily to the operation and maintenance of the aircraft and to requirements for obtaining a pilot's certificate or an airframe and powerplant mechanic certificate.

Part 1   Definitions and Abbreviations.

Part 21   Certification Procedures for Products and Parts.

Part 23   Airworthiness Standards: Normal, Utility, and Acrobatic Category Airplanes.

Part 33   Airworthiness Standards: Aircraft Engines.

Part 35   Airworthiness Standards: Propellers.

Part 39   Airworthiness Directives.

Part 43   Maintenance, Preventive Maintenance, Rebuilding, and Alteration.

Part 45   Identification and Registration Marking.

Part 47   Aircraft Registration.

Part 61   Certification: Pilots and Flight Instructors.

Part 65   Certification: Airmen other than Flight Crewmembers.

Part 91   General Operating and Flight Rules.

The FARs may be purchased from the Superintendent of Documents,

U.S. Government Printing Office, Washington, D.C. 20402. A check or money order payable to the Superintendent of Documents should be included with each order. Refer to Advisory Circular 00–44, Status of Federal Aviation Regulations, (latest edition) for the correct pricing and ordering information.

3. *The Advisory Circulars.* Advisory Circulars are issued by the FAA to inform the aviation public, in a systematic way, of nonregulatory material of interest. The contents of Advisory Circulars are not binding on the public unless incorporated into a regulation.

Request *free* Advisory Circulars from: U.S. Department of Transportation, Publications Section, M 443.1, Washington, D.C. 20590. Persons who want to be placed on the FAA's mailing list for future circulars should write to:

U.S. Department of Transportation
Distribution Requirements Section
M-482.2
Washington, D.C.   20590

Be sure to identify the subject matter desired, as separate mailing lists are maintained for each Advisory Circular subject series.

Order "for sale" Advisory Circulars from: Superintendent of Documents, U.S. Government Printing Office, Washington, D.C. 20402; or from any of the GOP bookstores located throughout the United States. Use AC 00–2, Advisory Circular Checklist, (latest edition) to obtain the cost of each circular.

4. *Exam-O-Grams.* Exam-O-Grams are prepared on subjects which prove particularly troublesome to applicants on written tests. They provide information on items which are operationally important but commonly misunderstood.

Exam-O-Grams may be purchased from:

Superintendent of Documents
U.S. Government Printing Office
Washington, D.C. 20402